DYNAMIC GAMES: THEORY AND APPLICATIONS

GERAD 25th Anniversary Series

- **Essays and Surveys in Global Optimization**
  Charles Audet, Pierre Hansen, and Gilles Savard, editors
- **Graph Theory and Combinatorial Optimization**
  David Avis, Alain Hertz, and Odile Marcotte, editors
- **Numerical Methods in Finance**
  Hatem Ben-Ameur and Michèle Breton, editors
- **Analysis, Control and Optimization of Complex Dynamic Systems**
  El-Kébir Boukas and Roland Malhamé, editors
- **Column Generation**
  Guy Desaulniers, Jacques Desrosiers, and Marius M. Solomon, editors
- **Statistical Modeling and Analysis for Complex Data Problems**
  Pierre Duchesne and Bruno Rémillard, editors
- **Performance Evaluation and Planning Methods for the Next Generation Internet**
  André Girard, Brunilde Sansò, and Félisa Vázquez-Abad, editors
- **Dynamic Games: Theory and Applications**
  Alain Haurie and Georges Zaccour, editors
- **Logistics Systems: Design and Optimization**
  André Langevin and Diane Riopel, editors
- **Energy and Environment**
  Richard Loulou, Jean-Philippe Waaub, and Georges Zaccour, editors

DYNAMIC GAMES: THEORY AND APPLICATIONS

Edited by
ALAIN HAURIE
Université de Genève & GERAD, Switzerland

GEORGES ZACCOUR
HEC Montréal & GERAD, Canada

Springer

ISBN 978-1-4419-3756-8 e-ISBN 978-0-387-24602-4

9 8 7 6 5 4 3 2 1

springeronline.com

Foreword

GERAD celebrates this year its 25th anniversary. The Center was created in 1980 by a small group of professors and researchers of HEC Montréal, McGill University and of the École Polytechnique de Montréal. GERAD's activities achieved sufficient scope to justify its conversion in June 1988 into a Joint Research Centre of HEC Montréal, the École Polytechnique de Montréal and McGill University. In 1996, the Université du Québec à Montréal joined these three institutions. GERAD has fifty members (professors), more than twenty research associates and post doctoral students and more than two hundreds master and Ph.D. students.

GERAD is a multi-university center and a vital forum for the development of operations research. Its mission is defined around the following four complementarily objectives:

- The original and expert contribution to all research fields in GERAD's area of expertise;
- The dissemination of research results in the best scientific outlets as well as in the society in general;
- The training of graduate students and post doctoral researchers;
- The contribution to the economic community by solving important problems and providing transferable tools.

GERAD's research thrusts and fields of expertise are as follows:

- Development of mathematical analysis tools and techniques to solve the complex problems that arise in management sciences and engineering;
- Development of algorithms to resolve such problems efficiently;
- Application of these techniques and tools to problems posed in related disciplines, such as statistics, financial engineering, game theory and artificial intelligence;
- Application of advanced tools to optimization and planning of large technical and economic systems, such as energy systems, transportation/communication networks, and production systems;
- Integration of scientific findings into software, expert systems and decision-support systems that can be used by industry.

One of the marking events of the celebrations of the 25th anniversary of GERAD is the publication of ten volumes covering most of the Center's research areas of expertise. The list follows: **Essays and Surveys in Global Optimization**, edited by C. Audet, P. Hansen and G. Savard; **Graph Theory and Combinatorial Optimization**, edited by D. Avis, A. Hertz and O. Marcotte; **Numerical Methods in Finance**, edited by H. Ben-Ameur and M. Breton; **Analysis, Control and Optimization of Complex Dynamic Systems**, edited by E.K. Boukas and R. Malhamé; **Column Generation**, edited by G. Desaulniers, J. Desrosiers and M.M. Solomon; **Statistical Modeling and Analysis for Complex Data Problems**, edited by P. Duchesne and B. Rémillard; **Performance Evaluation and Planning Methods for the Next Generation Internet**, edited by A. Girard, B. Sansò and F. Vázquez-Abad; **Dynamic Games: Theory and Applications**, edited by A. Haurie and G. Zaccour; **Logistics Systems: Design and Optimization**, edited by A. Langevin and D. Riopel; **Energy and Environment**, edited by R. Loulou, J.-P. Waaub and G. Zaccour.

I would like to express my gratitude to the Editors of the ten volumes, to the authors who accepted with great enthusiasm to submit their work and to the reviewers for their benevolent work and timely response. I would also like to thank Mrs. Nicole Paradis, Francine Benoît and Louise Letendre and Mr. André Montpetit for their excellent editing work.

The GERAD group has earned its reputation as a worldwide leader in its field. This is certainly due to the enthusiasm and motivation of GERAD's researchers and students, but also to the funding and the infrastructures available. I would like to seize the opportunity to thank the organizations that, from the beginning, believed in the potential and the value of GERAD and have supported it over the years. These are HEC Montréal, École Polytechnique de Montréal, McGill University, Université du Québec à Montréal and, of course, the Natural Sciences and Engineering Research Council of Canada (NSERC) and the Fonds québécois de la recherche sur la nature et les technologies (FQRNT).

Georges Zaccour
Director of GERAD

Avant-propos

Le Groupe d'études et de recherche en analyse des décisions (GERAD) fête cette année son vingt-cinquième anniversaire. Fondé en 1980 par une poignée de professeurs et chercheurs de HEC Montréal engagés dans des recherches en équipe avec des collègues de l'Université McGill et de l'École Polytechnique de Montréal, le Centre comporte maintenant une cinquantaine de membres, plus d'une vingtaine de professionnels de recherche et stagiaires post-doctoraux et plus de 200 étudiants des cycles supérieurs. Les activités du GERAD ont pris suffisamment d'ampleur pour justifier en juin 1988 sa transformation en un Centre de recherche conjoint de HEC Montréal, de l'École Polytechnique de Montréal et de l'Université McGill. En 1996, l'Université du Québec à Montréal s'est jointe à ces institutions pour parrainer le GERAD.

Le GERAD est un regroupement de chercheurs autour de la discipline de la recherche opérationnelle. Sa mission s'articule autour des objectifs complémentaires suivants :

- la contribution originale et experte dans tous les axes de recherche de ses champs de compétence;
- la diffusion des résultats dans les plus grandes revues du domaine ainsi qu'auprès des différents publics qui forment l'environnement du Centre;
- la formation d'étudiants des cycles supérieurs et de stagiaires post-doctoraux;
- la contribution à la communauté économique à travers la résolution de problèmes et le développement de coffres d'outils transférables.

Les principaux axes de recherche du GERAD, en allant du plus théorique au plus appliqué, sont les suivants :

- le développement d'outils et de techniques d'analyse mathématiques de la recherche opérationnelle pour la résolution de problèmes complexes qui se posent dans les sciences de la gestion et du génie;
- la confection d'algorithmes permettant la résolution efficace de ces problèmes;
- l'application de ces outils à des problèmes posés dans des disciplines connexes à la recherche opérationnelle telles que la statistique, l'ingénierie financière, la théorie des jeux et l'intelligence artificielle;
- l'application de ces outils à l'optimisation et à la planification de grands systèmes technico-économiques comme les systèmes énergé-

tiques, les réseaux de télécommunication et de transport, la logistique et la distributique dans les industries manufacturières et de service;

- l'intégration des résultats scientifiques dans des logiciels, des systèmes experts et dans des systèmes d'aide à la décision transférables à l'industrie.

Le fait marquant des célébrations du 25e du GERAD est la publication de dix volumes couvrant les champs d'expertise du Centre. La liste suit : **Essays and Surveys in Global Optimization**, édité par C. Audet, P. Hansen et G. Savard; **Graph Theory and Combinatorial Optimization**, édité par D. Avis, A. Hertz et O. Marcotte; **Numerical Methods in Finance**, édité par H. Ben-Ameur et M. Breton; **Analysis, Control and Optimization of Complex Dynamic Systems**, édité par E.K. Boukas et R. Malhamé; **Column Generation**, édité par G. Desaulniers, J. Desrosiers et M.M. Solomon; **Statistical Modeling and Analysis for Complex Data Problems**, édité par P. Duchesne et B. Rémillard; **Performance Evaluation and Planning Methods for the Next Generation Internet**, édité par A. Girard, B. Sansò et F. Vázquez-Abad; **Dynamic Games: Theory and Applications**, édité par A. Haurie et G. Zaccour; **Logistics Systems: Design and Optimization**, édité par A. Langevin et D. Riopel; **Energy and Environment**, édité par R. Loulou, J.-P. Waaub et G. Zaccour.

Je voudrais remercier très sincèrement les éditeurs de ces volumes, les nombreux auteurs qui ont très volontiers répondu à l'invitation des éditeurs à soumettre leurs travaux, et les évaluateurs pour leur bénévolat et ponctualité. Je voudrais aussi remercier Mmes Nicole Paradis, Francine Benoît et Louise Letendre ainsi que M. André Montpetit pour leur travail expert d'édition.

La place de premier plan qu'occupe le GERAD sur l'échiquier mondial est certes due à la passion qui anime ses chercheurs et ses étudiants, mais aussi au financement et à l'infrastructure disponibles. Je voudrais profiter de cette occasion pour remercier les organisations qui ont cru dès le départ au potentiel et la valeur du GERAD et nous ont soutenus durant ces années. Il s'agit de HEC Montréal, l'École Polytechnique de Montréal, l'Université McGill, l'Université du Québec à Montréal et, bien sûr, le Conseil de recherche en sciences naturelles et en génie du Canada (CRSNG) et le Fonds québécois de la recherche sur la nature et les technologies (FQRNT).

Georges Zaccour
Directeur du GERAD

Contents

Contributing Authors

EITAN ALTMAN
INRIA, France
altman@sophia.inria.fr

JEAN-PIERRE AUBIN
Réseau de Recherche Viabilité, Jeux, Contrôle, France
J.P.Aubin@wanadoo.fr

MIREILLE BOSSY
INRIA, France
Mireille.Bossy@sophia.inria.fr

MICHÈLE BRETON
HEC Montréal and GERAD, Canada
Michele.Breton@hec.ca

DEAN A. CARLSON
The University of Toledo, USA
dcarlson@math.utoledo.edu

HERBERT DAWID
University of Bielefeld
hdawid@wiwi.uni-bielefeld.de

CHRISTOPHE DEISSENBERG
University of Aix-Marseille II, France
deissenb@univ-aix.fr

RACHID EL AZOUZI
Univesité d'Avignon, France
elazouzi@lia.univ-avignon.fr

SJUR DIDRIK FLÅM
Bergen University, Norway
sjur.flaam@econ.uib.no

ALAIN HAURIE
Université de Genève and GERAD, Switzerland
Alain.Haurie@hec.unige.ch

ALAIN JEAN-MARIE
University of Montpellier 2, France
ajm@lirmm.fr

SALMA KARRAY
University of Ontario Institute of Technology, Canada
salma.karray@uoit.ca

GEORGE LEITMANN
University of California at Berkeley, USA
gleit@uclink4.berkeley.edu

NADIA MAÏZI
École des Mines de Paris, France
Nadia.Maizi@ensmp.fr

GUIOMAR MARTÍN-HERRÁN
Universidad de Valladolid, Spain
guiomar@eco.uva.es

GEERT JAN OLSDER
Delft University of Technology, The Netherlands
G.J.Olsder@ewi.tudelft.nl

LEON A. PETROSJAN
St.Petersburg State University, Russia
spbuoasis7@peterlink.ru

ODILE POURTALLIER
INRIA, France
Odile.Pourtallier@sophia.inria.fr

PAVEL ŠEVČIK
University of Aix-Marseille II, France
paulenfrance@yahoo.fr

SIHEM TABOUBI
HEC Montréal, Canada
sihem.taboubi@hec.ca

ETIENNE TANRÉ
INRIA, France
Etienne.Tanre@sophia.inria.fr

MABEL TIDBALL
INRA-LAMETA, France
tidball@ensam.inra.fr

ABDALLA TURKI
HEC Montréal and GERAD, Canada
Abdalla.Turki@hec.ca

DAVID W.K. YEUNG
Hong Kong Baptist University
wkyeung@hkbu.edu.hk

GEORGES ZACCOUR
HEC Montréal and GERAD, Canada
Georges.Zaccour@gerad.ca

Preface

This volume collects thirteen chapters dealing with a wide range of topics in (mainly) differential games. It is divided in two parts. Part I groups six contributions which deal essentially, but not exclusively, with theoretical or methodological issues arising in different dynamic games. Part II contains seven application-oriented chapters in economics and management science.

Part I

In Chapter 1, Aubin deals with cooperative games defined on networks, which could be of different kinds (socio-economic, neural or genetic networks), and where he allows for coalitions to evolve over time. Aubin provides a class of control systems, coalitions and multilinear connectionist operators under which the architecture of the network remains viable. He next uses the viability/capturability approach to study the problem of characterizing the dynamic core of a dynamic cooperative game defined in characteristic function form.

In Chapter 2, Carlson and Leitmann provide a direct method for open-loop dynamic games with dynamics affine with respect to controls. The direct method was first introduced by Leitmann in 1967 for problems of calculus of variations. It has been the topic of recent contributions with the aim to extend it to differential games setting. In particular, the method has been successfully adapted for differential games where each player has its own state. Carlson and Leitmann investigate here the utility of the direct method in the case where the state dynamics are described by a single equation which is affine in players' strategies.

In Chapter 3, El Azouzi et al. consider the problem of routing in networks in the context where a number of decision makers having theirown utility to maximize. If each decision maker wishes to find a minimal path for each routed object (e.g., a packet), then the solution concept is the Wardrop equilibrium. It is well known that equilibria may exhibit inefficiencies and paradoxical behavior, such as the famous Braess paradox (in which the addition of a link to a network results in worse performance to all users). The authors provide guidelines for the network administrator on how to modify the network so that it indeed results in improved performance.

Flåm considers in Chapter 4 production or market games with transferable utility. These games, which are actually of frequent occurrence and great importance in theory and practice, involve parties concerned with the issue of finding a fair sharing of efficient production costs. Flåm shows that, in many cases, explicit core solutions may be defined by shadow prices, and reached via quite natural dynamics.

Jean-Marie and Tidball discuss in Chapter 5 the relationships between conjectures, conjectural equilibria, consistency and Nash equilibria in the classical theory of discrete-time dynamic games. They propose a theoretical framework in which they define conjectural equilibria with several degrees of consistency. In particular, they introduce feedback-consistency, and prove that the corresponding conjectural equilibria and Nash-feedback equilibria of the game coincide. Finally, they discuss the relationship between these results and previous studies based on differential games and supergames.

In Chapter 6, Petrosjan defines on a game tree a cooperative game in characteristic function form with incomplete information. He next introduces the concept of imputation distribution procedure in connection with the definitions of time-consistency and strongly time-consistency. Petrosjan derives sufficient conditions for the existence of time-consistent solutions. He also develops a regularization procedure and constructs a new characteristic function for games where these conditions cannot be met. The author also defines the regularized core and proves that it is strongly time-consistent. Finally, he investigates the special case of stochastic games.

Part II

Bossy et al. consider in Chapter 7 a deregulated electricity market formed of few competitors. Each supplier announces the maximum quantity he is willing to sell at a certain fixed price. The market then decides the quantities to be delivered by the suppliers which satisfy demand at minimal cost. Bossy et al. characterize Nash equilibrium for the two scenarios where in turn the producers maximize their market shares and profits. A close analysis of the equilibrium results points out towards some difficulties in predicting players' behavior.

Breton and Turki analyze in Chapter 8 a differentiated duopoly where firms engage in research and development (R&D) to reduce their production cost. The authors first derive and compare Bertrand and Cournot equilibria in terms of quantities, prices, investments in R&D, consumer's surplus and total welfare. The results are stated with reference to productivity of R&D and the degree of spillover in the industry. Breton and Turki also assess the robustness of their results and those obtained in the literature. Their conclusion is that the relative efficiencies of Bertrand and Cournot equilibria are sensitive to the specifications that are used, and hence the results are far from being robust.

In Chapter 9, Dawid et al. consider a dynamic model of environmental taxation where the firms are of two types: believers who take the tax announcement by the Regulator at face value and non-believers who perfectly anticipate the Regulator's decisions at a certain cost. The authors assume that the proportion of the two types evolve overtime depending on the relative profits of both groups. Dawid et al. show that the Regulator can use misleading tax announcements to steer the economy to an equilibrium which is Pareto-improving compared with the solutions proposed in the literature.

In Chapter 10, Haurie shows how a multi-timescale hierarchical non-cooperative game paradigm can contribute to the development of integrated assessment models of climate change policies. He exploits the fact that the climate and economic subsystems evolve at very different time scales. Haurie formulates the international negotiation at the level of climate control as a piecewise deterministic stochastic game played in the "slow" time scale, whereas the economic adjustments in the different nations take place in a "faster" time scale. He shows how the negotiations on emissions abatement can be represented in the slow time scale whereas the economic adjustments are represented in the fast time scale as solutions of general economic equilibrium models. He finally provides some indications on the integration of different classes of models that could be made, using an hierarchical game theoretic structure.

In Chapter 11, Karray and Zaccour consider a differential game model for a marketing channel formed by one manufacturer and one retailer. The latter sells the manufacturer's national brand and may also introduce a private label offered at a lower price. The authors first assess the impact of a private label introduction on the players' payoffs. Next, in the event where it is beneficial for the retailer to propose his brand to consumers and detrimental to the manufacturer, they investigate if a cooperative advertising program could help the manufacturer to mitigate the negative impact of the private label.

Martín-Herrán and Taboubi (Chapter 12) aim at determining equilibrium shelf-space allocation in a marketing channel with two competing manufacturers and one retailer. The formers control advertising expenditures in order to build a brand image. They also offer to the retailer an incentive designed to increase their share of the shelf space. The problem is formulated as a Stackelberg infinite-horizon differential game with the manufacturers as leaders. Stationary feedback equilibria are characterized and numerical experiments are conducted to illustrate how the players set their marketing efforts.

In Chapter 13, Yeung considers a duopoly in which the firms agree to form a cartel. In particular, one firm has absolute and marginal cost advantage over the other forcing one of the firms to become a dormant firm. The author derives a subgame consistent solution based on the Nash bargaining axioms. Subgame consistency is a fundamental element in the solution of cooperative stochastic differential games. In particular, it ensures that the extension of the solution policy to a later starting time and any possible state brought about by prior optimal behavior of the players would remain optimal. Hence no players will have incentive to deviate from the initial plan.

Acknowledgements

The Editors would like to express their gratitude to the authors for their contributions and timely responses to our comments and suggestions. We wish also to thank Francine Benoît and Nicole Paradis of GERAD for their expert editing of the volume.

ALAIN HAURIE
GEORGES ZACCOUR

Chapter 1

DYNAMICAL CONNECTIONIST NETWORK AND COOPERATIVE GAMES

Jean-Pierre Aubin

Abstract Socio-economic networks, neural networks and genetic networks describe collective phenomena through constraints relating actions of several players, coalitions of these players and multilinear connectionist operators acting on the set of actions of each coalition. Static and dynamical cooperative games also involve coalitions. Allowing "coalitions to evolve" requires the embedding of the finite set of coalitions in the compact convex subset of "fuzzy coalitions". This survey present results obtained through this strategy.

We provide first a class of control systems governing the evolution of actions, coalitions and multilinear connectionist operators under which the architecture of a network remains viable. The controls are the "viability multipliers" of the "resource space" in which the constraints are defined. They are involved as "tensor products" of the actions of the coalitions and the viability multiplier, allowing us to encapsulate in this dynamical and multilinear framework the concept of Hebbian learning rules in neural networks in the form of "multi-Hebbian" dynamics in the evolution of connectionist operators. They are also involved in the evolution of coalitions through the "cost" of the constraints under the viability multiplier regarded as a price, describing a *"nerd behavior"*.

We use next the viability/capturability approach for studying the problem of characterizing the dynamic core of a dynamic cooperative game defined in a characteristic function form. We define the dynamic core as a set-valued map associating with each fuzzy coalition and each time the set of imputations such that their payoffs at that time to the fuzzy coalition are larger than or equal to the one assigned by the characteristic function of the game and study it.

1. Introduction

Collective phenomena deal with the coordination of actions by a finite number n of players labelled $i = 1, \ldots, n$ using the architecture of a network of players, such as socio-economic networks (see for instance,

Aubin (1997, 1998a), Aubin and Foray (1998), Bonneuil (2000, 2001)), neural networks (see for instance, Aubin (1995, 1996, 1998b), Aubin and Burnod (1998)) and genetic networks (see for instance, Bonneuil (1998b, 2005), Bonneuil and Saint-Pierre (2000)). This coordinated activity requires a network of communications or connections of actions $x_i \in X_i$ ranging over n finite dimensional vector spaces X_i as well as coalitions of players.

The simplest general form of a coordination is the requirement that a relation between actions of the form $g(A(x_1, \ldots, x_n)) \in M$ must be satisfied. Here

1. $A : \prod_{i=1}^{n} X_i \mapsto Y$ is a connectionist operator relating the *individual* actions in a collective way,

2. $M \subset Y$ is the subset of the resource space Y and g is a map, regarded as a propagation map.

We shall study this coordination problem in a dynamic environment, by allowing actions $x(t)$ and connectionist operators $A(t)$ to evolve according to dynamical systems we shall construct later. In this case, the coordination problem takes the form

$$\forall\, t \geq 0, \quad g(A(t)(x_1(t), \ldots, x_n(t))) \in M$$

However, in the fields of motivation under investigation, the number n of variables may be very large. Even though the connectionist operators $A(t)$ defining the "architecture" of the network are allowed to operate *a priori* on all variables $x_i(t)$, they actually operate at each instant t on a coalition $S(t) \subset N := \{1, \ldots, n\}$ of such variables, varying naturally with time according to the nature of the coordination problem.

On the other hand, a recent line of research, *dynamic cooperative game theory* has been opened by Leon Petrosjan (see for instance Petrosjan (1996) and Petrosjan and Zenkevitch (1996)), Alain Haurie (Haurie (1975)), Georges Zaccour, Jerzy Filar and others. We quote the first lines of Filar and Petrosjan (2000): *"Bulk of the literature dealing with cooperative games (in characteristic function form) do not address issues related to the evolution of a solution concept over time. However, most conflict situations are not "one shot" games but continue over some time horizon which may be limited* a priori *by the game rules, or terminate when some specified conditions are attained."* We propose here a concept of dynamic core of a dynamical fuzzy cooperative game as a set-valued map associating with each fuzzy coalition and each time the set of imputations such that their payoffs at that time to the fuzzy coalition are larger than or equal to the one assigned by the characteristic function of the game. We shall characterize this core through the (generalized) derivatives of a valuation function associated with the game, provide its explicit formula, characterize its epigraph as a viable-capture basin of the epigraph of the characteristic function of the fuzzy dynamical cooperative game, use the tangential properties of such basins for proving that the valuation function is a solution to a Hamilton-Jacobi-Isaacs partial differential equation and use this function and its derivatives for characterizing the dynamic core.

In a nutshell, this survey deals with the evolution of fuzzy coalitions for both regulate the viable architecture of a network and the evolutions of imputations in the dynamical core of a dynamical fuzzy cooperative game.

Outline

The survey is organized as follows:

1. We begin by recalling what are fuzzy coalitions in the framework of convexification procedures,
2. we proceed by studying the evolution of networks regulated by viability multipliers, showing how Hebbian rules emerge in this context
3. and by introducing fuzzy coalitions of players in this network and showing how a herd behavior emerge in this framework.
4. We next define dynamical cores of dynamical fuzzy cooperative games (with side-payments)
5. and explain briefly why the viability/capturability approach is relevant to answer the questions we have raised.

2. Fuzzy coalitions

The first definition of a coalition which comes to mind, being that of a subset of players $S \subset N$, is not adequate for tackling dynamical models of evolution of coalitions since the 2^n coalitions range over a finite set, preventing us from using analytical techniques.

One way to overcome this difficulty is to embed the family of subsets of a (discrete) set N of n players to the space $\mathbf{R}^n$:

This canonical embedding is more adapted to the nature of the power set $\mathcal{P}(N)$ than to the universal embedding of a discrete set M of m elements to $\mathbf{R}^m$ by the Dirac measure associating with any $j \in M$ the jth element of the canonical basis of $\mathbf{R}^m$. The convex hull of the image of M by this embedding is the probability simplex of $\mathbf{R}^m$. Hence

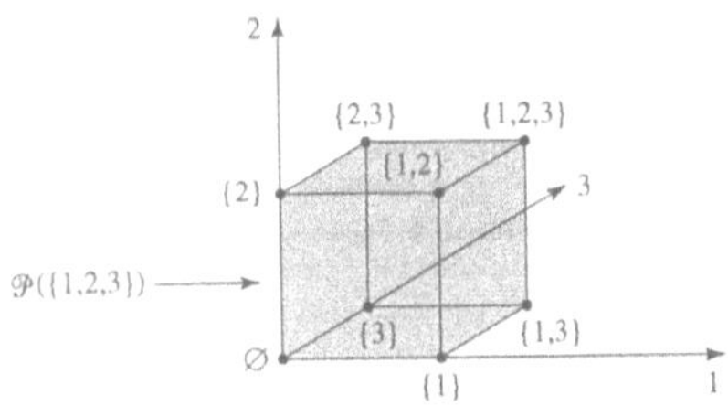

We embed the family of subsets of a (discrete) set N of n players to the space R^n through the map χ associating with any coalition $S \in \mathcal{P}(N)$ its characteristic function $\chi_S \in \{0,1\}^n \subset \mathsf{R}^n$, since R^n can be regarded as the set of functions from N to R.
By definition, the family of fuzzy sets is the convex hull $[0,1]^n$ of the power set $\{0,1\}^n$ in R^n.

fuzzy sets offer a "dedicated convexification" procedure of the discrete power set $M := \mathcal{P}(N)$ instead of the universal convexification procedure of frequencies, probabilities, mixed strategies derived from its embedding in $\mathbf{R}^m = \mathbf{R}^{2^n}$.

By definition, the family of fuzzy sets[1] is the convex hull $[0,1]^n$ of the power set $\{0,1\}^n$ in $\mathbf{R}^n$. Therefore, we can write any fuzzy set in the form

$$\chi = \sum_{S \in \mathcal{P}(N)} m_S \chi_S \text{ where } m_S \geq 0 \,\&\, \sum_{S \in \mathcal{P}(N)} m_S = 1$$

The memberships are then equal to

$$\forall\, i \in N, \quad \chi_i = \sum_{S \ni i} m_S$$

Consequently, if m_S is regarded as the probability for the set S to be formed, the membership of player i to the fuzzy set χ is the sum of the probabilities of the coalitions to which player i belongs. Player i participates fully in χ if $\chi_i = 1$, does not participate at all if $\chi_i = 0$ and participates in a fuzzy way if $\chi_i \in]0,1[$. We associate with a fuzzy coalition χ the set $P(\chi) := \{i \in N | \chi_i \neq 0\} \subset N$ of players i participating in the fuzzy coalition χ.

We also introduce the membership

$$\gamma_S(\chi) := \prod_{j \in S} \chi_j$$

[1]This concept of fuzzy set was introduced in 1965 by L. A. Zadeh. Since then, it has been wildly successful, even in many areas outside mathematics!. We found in *"La lutte finale"*, Michel Lafon (1994), p.69 by A. Bercoff the following quotation of the late François Mitterand, president of the French Republic (1981-1995): *"Aujourd'hui, nous nageons dans la poésie pure des sous ensembles flous"* ... (Today, we swim in the pure poetry of fuzzy subsets)!

of a coalition S in the fuzzy coalition χ as the product of the memberships of players i in the coalition S. It vanishes whenever the membership of one player does and reduces to individual memberships for one player coalitions. When two coalitions are disjoint ($S \cap T = \emptyset$), then $\gamma_{S \cup T}(\chi) = \gamma_S(\chi)\gamma_T(\chi)$. In particular, for any player $i \in S$, $\gamma_S(\chi) = \chi_i \gamma_{S \setminus i}(\chi)$.

Actually, this idea of using fuzzy coalitions has already been used in the framework of static cooperative games with and without side-payments in Aubin (1979, 1981a,b), and Aubin (1998, 1993), Chapter 13. Further developments can be found in Mares (2001) and Mishizaki and Sokawa (2001), Basile (1993, 1994, 1995), Basile, De Simone and Graziano (1996), Florenzano (1990)). Fuzzy coalitions have also been used in dynamical models of cooperative games in Aubin and Cellina (1984), Chapter 4 and of economic theory in Aubin (1997), Chapter 5.

This idea of fuzzy sets can be adapted to more general situations relevant in game theory. We can, for instance, introduce negative memberships when players enter a coalition with aggressive intents. This is mandatory if one wants to be realistic! A positive membership is interpreted as a cooperative participation of the player i in the coalition, while a negative membership is interpreted as a non-cooperative participation of the ith player in the generalized coalition. In what follows, one can replace the cube $[0, 1]^n$ by any product $\prod_{i=1}^{n} [\lambda_i, \mu_i]$ for describing the cooperative or noncooperative behavior of the consumers.

We can still enrich the description of the players by representing each player i by what psychologists call her 'behavior profile' as in Aubin, Louis-Guerin and Zavalloni (1979). We consider q 'behavioral qualities' $k = 1, \ldots, q$, each with a unit of measurement. We also suppose that a behavioral quantity can be measured (evaluated) in terms of a real number (positive or negative) of units. A behavior profile is a vector $a = (a_1, \ldots, a_q) \in \mathbf{R}^q$ which specifies the quantities a_k of the q qualities k attributed to the player. Thus, instead of representing each player by a letter of the alphabet, she is described as an element of the vector space $\mathbf{R}^q$. We then suppose that each player may implement all, none, or only some of her behavioral qualities when she participates in a social coalition. Consider n players represented by their behavior profiles in $\mathbf{R}^q$. Any matrix $\chi = (\chi_i^k)$ describing the levels of participation $\chi_i^k \in [-1, +1]$ of the behavioral qualities k for the n players i is called a **social coalition**. Extension of the following results to social coalitions is straightforward.

Technically, the choice of the scaling $[0, 1]$ inherited from the tradition built on integration and measure theory is not adequate for describing convex sets. When dealing with convex sets, we have to replace the characteristic functions by indicators taking their values in $[0, +\infty]$ and take their convex combinations to provide an alternative allowing us to speak of "fuzzy" convex sets. Therefore, "toll-sets" are nonnegative cost functions assigning to each element its cost of belonging, $+\infty$ if it does not belong to the toll set. The set of elements with finite positive cost do form the "fuzzy boundary" of the toll set, the set of elements with zero cost its "core". This has been done to adapt viability theory to "fuzzy viability theory".

Actually, the Cramer transform

$$C_\mu(p) := \sup_{\chi \in \mathbf{R}^n} \left(\langle p, \chi \rangle - \log \left(\int_{\mathbf{R}^n} e^{\langle x, y \rangle} d\mu(y) \right) \right)$$

maps probability measures to toll sets. In particular, it transforms convolution products of density functions to inf-convolutions of extended functions, Gaussian functions to squares of norms, etc. See Chapter 10 of Aubin (1991) and Aubin and Dordan (1996) for more details and information on this topic.

The components of the state variable $\chi := (\chi_1, \ldots, \chi_n) \in [0,1]^n$ are the rates of participation in the fuzzy coalition χ of player $i = 1, \ldots, n$.

Hence convexification procedures and the need of using functional analysis justifies the introduction of fuzzy sets and its extensions. In the examples presented in this survey, we use only classical fuzzy sets.

3. Regulation of the evolution of a network

3.1 Definition of the architecture of a network

We introduce

1. n finite dimensional vector spaces X_i describing the action spaces of the players
2. a finite dimensional vector space Y regarded as a resource space and a subset $M \subset Y$ of scarce resources[2].

DEFINITION 1.1 *The* architecture of dynamical network *involves the evolution*

1. *of actions* $x(t) := (x_1(t), \ldots, x_n(t)) \in \prod_{i=1}^n X_i$,

[2]For simplicity, the set M of scarce resources is assumed to be constant. But sets $M(t)$ of scarce resources could also evolve through *mutational equations* and the following results can be adapted to this case. Curiously, the overall architecture is not changed when the set of available resources evolves under a mutational equation. See Aubin (1999) for more details on mutational equations.

2. *of connectionist operators* $A_{S(t)}(t) : \prod_{i=1}^{n} X_i \mapsto Y$,
3. *acting on* coalitions $S(t) \subset N := \{1, \ldots, n\}$ *of the n players*

and requires that

$$\forall t \geq 0, \quad g\left(\{A_S(t)(x(t))\}_{S \subset N}\right) \in M$$

where $g : \prod_{S \subset N} Y_S \mapsto Y$.

We associate with any coalition $S \subset N$ the product $X^S := \prod_{i \in S} X_i$ and denote by $A_S \in \mathcal{L}_S(X^S, Y)$ the space of S-linear operators $A_S : X^S \mapsto Y$, i.e., operators that are linear with respect to each variable x_i, $(i \in S)$ when the other ones are fixed. Linear operators $A_i \in \mathcal{L}(X_i, Y)$ are obtained when the coalition $S := \{i\}$ is reduced to a singleton, and we identify $\mathcal{L}_\emptyset(X^\emptyset, Y) := Y$ with the vector space Y.

In order to tackle mathematically this problem, we shall

1. restrict the connectionist operators $A := \sum_{S \subset N} A_S$ to be multi-affine, i.e., the sum over all coalitions of S-linear operators[3] $A_S \in \mathcal{L}_S(X^S, Y)$,
2. allow coalitions S to become fuzzy coalitions so that they can evolve continuously.

So, a network is not only any kind of a relationship between variables, but involves both connectionist operators operating on coalitions of players.

3.2 Constructing the dynamics

The question we raise is the following: Assume that we know the intrinsic laws of evolution of the variables x_i (independently of the constraints), of the connectionist operator $A_S(t)$ and of the coalitions $S(t)$. Is the above architecture viable under these dynamics, in the sense that the collective constraints defining the architecture of the dynamical network are satisfied at each instant.

There is no reason why let on his own, collective constraints defining the above architecture are viable under these dynamics. Then the question arises how to reestablish the viability of the system.

One may

1. either delineate those states (actions, connectionist operators, coalitions) from which begin viable evolutions,

---

[3] Also called (or regarded as) tensors.They are nothing other than matrices when the operators are linear instead of multilinear. Tensors are the matrices of multilinear operators, so to speak, and their "entries" depend upon several indexes instead of the two involved in matrices.

2. or correct the dynamics of the system in order that the architecture of the dynamical network is viable under the altered dynamical system.

The first approach leads to take the viability kernel of the constrained subset of K of states (x_i, A_S, S) satisfying the constraints defining the architecture of the network. We refer to Aubin (1997, 1998a) for this approach. We present in this section a class of methods for correcting the dynamics without touching on the architecture of the network.

One may indeed be able, with a lot of ingeniousness and intimate knowledge of a given problem, and for "simple constraints", to derive dynamics under which the constraints are viable.

However, we can investigate whether there is a kind of mathematical factory providing classes of dynamics "correcting" the initial (intrinsic) ones in such a way that the viability of the constraints is guaranteed. One way to achieve this aim is to use the concept of "viability multipliers" $q(t)$ ranging over the dual Y^* of the resource space Y that can be used as "controls" involved for modifying the initial dynamics. This allows us to provide an explanation of the formation and the evolution of the architecture of the network and of the active coalitions as well as the evolution of the actions themselves.

A few words about viability multipliers are in order here: If a constrained set K is of the form

$$K := \{x \in X \text{ such that } h(x) \in M\}$$

where $h : X \mapsto Z := \mathbf{R}^m$ is the constrained map form the state space X to the resource space Z and $M \subset Z$ is a subset of available resources, we regard elements $u \in Z^\star = Z$ in the dual of the resource space Z (identified with Z) as viability multipliers, since they play a role analogous to Lagrange multipliers in optimization under constraints.

Recall that the minimization of a function $x \mapsto J(x)$ over a constrained set K is equivalent to the minimization without constraints of the function

$$x \mapsto J(x) + \sum_{k=1}^{m} \frac{\partial h_k(x)}{\partial x_j} u_k$$

for an adequate Lagrange multiplier $u \in Z^\star = Z$ in the dual of the resource space Z (identified with Z). See for instance Aubin (1998, 1993), Rockafellar and Wets (1997) among many other references on this topic.

In an analogous way, but with unrelated methods, it has been proved that a closed convex subset K is viable under the control system

$$x_j'(t) = f_j(x(t)) + \sum_{k=1}^{m} \frac{\partial h_k(x(t))}{\partial x_j} u_k(t)$$

obtained by adding to the initial dynamics a term involving regulons that belong to the dual of the same resource space Z. See for instance Aubin and Cellina (1984) and Aubin (1991, 1997) below for more details. Therefore, these viability multipliers used as regulons benefit from the same economic interpretation of virtual prices as the ones provided for Lagrange multipliers in optimization theory.

The viability multipliers $q(t) \in Y^*$ can thus be regarded as regulons, i.e., regulation controls or parameters, or virtual prices in the language of economists. These are chosen at each instant in order that the viability constraints describing the network can be satisfied at each instant. The main theorem guarantees this possibility. Another theorem tells us how to choose at each instant such regulons (the regulation law). Even though viability multipliers do not provide all the dynamics under which a constrained set is viable, they do provide important and noticeable classes of dynamics exhibiting interesting structures that deserve to be investigated and tested in concrete situations.

3.3 An economic interpretation

Although the theory applies to general networks, the problem we face has an economic interpretation that may help the reader in interpreting the main results that we summarize below.

Actors here are economic agents (producers) $i = 1, \ldots, n$ ranging over the set $N := \{1, \ldots, n\}$. Each coalition $S \subset N$ of economic agents is regarded as a production unit (a firm) using resources of their agents to produce (or not produce) commodities. Each agent $i \in N$ provides a resource vector (capital, competencies, etc.) $x_i \in X$ in a resource space $X_i := \mathbb{R}^{m_i}$ used in production processes involving coalitions $S \subset N$ of economic agents (regarded as firms employing economic agents)

We describe the production process of a firm $S \subset N$ by a S-linear operator $A_S : \prod_{i=1}^{n} X_i \mapsto Y$ associating with the resources $x := (x_1, \ldots, x_n)$ provided by the economic agents a commodity $A_S(x)$. The supply constraints are described by a subset $M \subset Y$ of the commodity space, representing the set of commodities that must be produced by the firms: Condition

$$\sum_{S \subset N} A_S(t)(x(t)) \in M$$

express that at each instant, the total production must belong to M.

The connectionist operators among economic agents are the input-output production processes operating on the resources provided by the economic agents to the production units described by coalitions of economic agents. The architecture of the network is then described by the supply constraints requiring that at each instant, agents supply adequate resources to the firms in order that the production objectives are fulfilled.

When fuzzy coalitions χ_i of economic agents[4] are involved, the supply constraints are described by

$$\sum_{S \subset N} \left(\prod_{j \in S} \chi_j(t) \right) A_S(t)(x(t)) \in M \tag{1.1}$$

since the production operators are assumed to be multilinear.

3.4 Linear connectionist operators

We summarize the case in which there is only one player and the operator $A : X \mapsto Y$ is affine studied in Aubin (1997, 1998a,b):

$$\forall\, x \in X, \quad A(x) := Wx + y \ \text{ where } W \in \mathcal{L}(X,Y) \ \&\ y \in Y$$

The coordination problem takes the form:

$$\forall\, t \geq 0, \quad W(t)x(t) + y(t) \in M$$

where both the state x, the resource y and the connectionist operator W evolve. These constraints are not necessarily viable under an arbitrary dynamic system of the form

$$\begin{cases} (i) & x'(t) = c(x(t)) \\ (ii) & y'(t) = d(y(t)) \\ (iii) & W'(t) = \alpha(W(t)) \end{cases} \tag{1.2}$$

We can reestablish viability by involving multipliers $q \in Y^*$ ranging over the dual $Y^* := Y$ of the resource space Y to correct the initial dynamics. We denote by $W^* \in \mathcal{L}(Y^*, X^*)$ the transpose of W:

$$\forall\, q \in Y^*, \quad \forall\, x \in X, \quad \langle W^* q, x \rangle := \langle q, Wx \rangle$$

by $x \otimes q \in \mathcal{L}(X^*, Y^*)$ the tensor product defined by

$$x \otimes q : p \in X^* := X \mapsto (x \otimes q)(p) := \langle p, x \rangle q$$

the matrix of which is made of entries $(x \otimes q)_i^j = x_i q^j$.

---

[4]Whenever the resources involved in production processes are proportional to the intensity of labor, one could interpret in such specific economic models the rate of participation χ_i of economic agent i as (the rate of) labor he uses in the production activity.

The contingent cone $T_M(x)$ to $M \subset Y$ at $y \in M$ is the set of directions $v \in Y$ such that there exist sequences $h_n > 0$ converging to 0, and v_n converging to v satisfying $y + h_n v_n \in M$ for every n. The (regular) normal cone to $M \subset Y$ at $y \in M$ is defined by

$$N_M(y) := \{q \in Y^* | \forall\, v \in T_M(y),\ \langle q, v\rangle \leq 0\}$$

(see Aubin and Frankowska (1990) and Rockafellar and Wets (1997) for more details on these topics).

We proved that the viability of the constraints can be reestablished when the initial system (1.2) is replaced by the control system

$$\begin{cases} (i) & x'(t) = c(x(t)) - W^*(t)q(t) \\ (ii) & y'(t) = d(y(t)) - q(t) \\ (iii) & W'(t) = \alpha(W(t)) - x(t) \otimes q(t) \\ & \quad \text{where } q(t) \in N_M(W(t)x(t) + y(t)) \end{cases}$$

where $N_M(y) \subset Y^*$ denotes the normal cone to M at $y \in M \subset Y$ and $x \otimes q \in \mathcal{L}(X, Y^*)$ denotes the tensor product defined by

$$x \otimes q : p \in X^* := X \mapsto (x \otimes q)(x) := \langle p, x\rangle q$$

the matrix of which is made of entries $(x \otimes q)_i^j = x_i q^j$. In other words, the correction of a dynamical system for reestablishing the viability of constraints of the form $W(t)x(t) + y(t) \in M$ involves the rule proposed by Hebb in his classic book *The organization of behavior* in 1949 as the basic learning process of synaptic weight and called the Hebbian rule: Taking $\alpha(W) = 0$, the evolution of the synaptic matrix $W := (w_i^j)$ obeys the differential equation

$$\frac{d}{dt} w_i^j(t) = -x_i(t) q^j(t)$$

The Hebbian rule states that the velocity of the synaptic weight is the product of pre-synaptic activity and post-synaptic activity. Such a learning rule "pops up" (or, more pedantically, emerges) whenever the synaptic matrices are involved in regulating the system in order to maintain the "homeostatic" constraint $W(t)x(t) + y(t) \in M$. (See Aubin (1996) for more details on the relations between Hebbian rules and tensor products in the framework of neural networks).

Viability multipliers $q(t) \in Y^\star$ regulating viable evolutions satisfy the regulation law

$$\forall\, t \geq 0, \quad q(t) \in R_M(A(t), x(t))$$

where the regulation map R_M is defined by

$$R_M(A, x) = (AA^\star + \|x\|^2 \mathsf{I})^{-1}(Ac(x) + \alpha(A)(x) - T_M(A(x)))$$

One can even require that on top of it, the viability multiplier satisfies

$$q(t) \in N_M(A(t)x(t)) \cap R_M(A(t),\, x(t)))$$

The norm $\|q(t)\|$ of the viability multiplier $q(t)$ measures the intensity of the viability discrepancy of the dynamic since

$$\begin{cases} (i) & \|c(x(t)) - x'(t)\| \leq \|A^*(t)\| \, \|q(t)\| \\ (ii) & \|\alpha(A(t)) - A'(t)\| = \|x(t)\| \, \|q(t)\| \end{cases}$$

When $\alpha(A) \equiv 0$, the viability multipliers with minimal norm in the regulation map provide both the smallest error $\|c(x(t)) - x'(t)\|$ and the smallest velocities of the connection matrix because $\|A'(t)\| = \|x(t)\|$ $\|q(t)\|$. The inertia of the connection matrix, which can be regarded as an index of dynamic connectionist complexity, *is proportional to the norm of the viability multiplier.*

3.5 Hierarchical architecture and complexity

The constraints are of the form

$$A_{\mathbb{H}}^{\mathbb{H}-1} \cdots A_h^{h-1} \ldots A_2^1 x_1 \in M_{\mathbb{H}}$$

This describes for instance a production process associating with the resource x_1 the intermediate outputs $x_2 := A_2^1 x_1$, which itself produces an output $x_3 := A_2^1 x_2$, and so on, until the final output $x_{\mathbb{H}} := A_{\mathbb{H}}^{\mathbb{H}-1} \cdots A_h^{h-1} \ldots A_2^1 x_1$ which must belong to the production set $M_{\mathbb{H}}$.

The evolution without constraints of the commodities and the operators is governed by dynamical systems of the form

$$\begin{cases} (i) & x_h'(t) = c_h(x_h(t)) \\ \\ (ii) & \dfrac{d}{dt} A_{h+1}^h(t) = \alpha_{h+1}^h(A_h(t)) \end{cases}$$

The constraints

$$\forall\, t \geq 0, \quad A_{\mathbb{H}}^{\mathbb{H}-1}(t) \cdots A_h^{h-1}(t) \ldots A_2^1(t) x_1(t) \in M_{\mathbb{H}}$$

are viable under the system

$$\begin{cases} x_1'(t) = c_1(x_1(t)) + A_2^1(t)^\star(t) p^1(t) & (h = 1) \\ \\ x_h'(t) = c_h(x_h(t)) - p^{h-1}(t) + A_{h+1}^h(t)^\star p^h(t) & (h = 1, \ldots, \mathbb{H} - 1) \\ \\ x_{\mathbb{H}}'(t) = c_{\mathbb{H}}(x_{\mathbb{H}}(t)) - p^{\mathbb{H}-1}(t) & (h = \mathbb{H}) \\ \\ \dfrac{d}{dt} A_{h+1}^h(t) = \alpha_{h+1}^h(A_h(t)) + x_h(t) \otimes p^h(t) & (h = 1, \ldots, \mathbb{H} - 1) \end{cases}$$

involving viability multipliers $p^h(t)$ (intermediate "shadow price"). The input-output matrices $A^h_{h+1}(t)$ obey dynamics involving the tensor product of $x_h(t)$ and $p^h(t)$.

The viability multiplier $p^h(t)$ *at level* $h (h = 1, \ldots, \mathbb{H}-1)$ *both regulate the evolution at level* h *and send a message at upper level* $h+1$.

We can tackle actually more complex hierarchical situations with non ordered hierarchies. Assume that $X := \prod_{h=1}^{\mathbb{H}}$, $Y := \prod_{k=1}^{\mathbb{K}}$ and that $A := (A^k_h)$ where $A^k_h \in \mathcal{L}(X_k, Y_h)$. We introduce a set-valued map $J : \{1, \ldots, \mathbb{H}\} \rightsquigarrow \{1, \ldots, \mathbb{K}\}$.

The constraints are defined by

$$\forall\, h = 1, \ldots, \mathbb{H}, \quad \sum_{k \in J(h)} A^k_h(t) x_k(t) \in M_h \subset Y_h$$

We consider a system of differential equations

$$\begin{cases} (i) \quad x'_h(t) = c_h(x_h(t)), \quad h = 1, \ldots, \mathbb{H} \\ \\ (ii) \quad \dfrac{d}{dt} A^k_h(t) = \alpha^k_h(A^k_h(t)) \end{cases}$$

Then the constraints

$$\forall\, h = 1, \ldots, \mathbb{H}, \ldots \sum_{k \in J(h)} A^k_h(t) x_k(t) \in M_h \subset Y_h$$

are viable under the corrected system

$$\begin{cases} (i) \quad x'_h(t) = c_h(x_h(t)) - \sum_{k \in J^{-1}(h)} A^h_k(t)^\star p^k, \\ \qquad\qquad\qquad\qquad\qquad\qquad h = 1, \ldots, \mathbb{H},\ k = 1, \ldots, \mathbb{K} \\ (ii) \quad \dfrac{d}{dt} A^k_h(t) = \alpha^k_h(A^k_h(t)) - x_k(t) \otimes p^h(t), (h,k) \in \mathrm{Graph}(J) \end{cases}$$

3.6 Connectionist tensors

In order to handle more explicit and tractable formulas and results, we shall assume that the connectionist operator $A : X := \prod_{i=1}^n X_i \rightsquigarrow Y$ is multiaffine.

For defining such a multiaffine operator, we associate with any coalition $S \subset N$ its characteristic function $\chi_S : N \mapsto \mathbf{R}$ associating with any $i \in N$

$$\chi_S(i) := \begin{cases} 1 & \text{if} \quad i \in S \\ 0 & \text{if} \quad i \notin S \end{cases}$$

It defines a linear operator $\chi_S \circ \in \mathcal{L}\left(\prod_{i=1}^n X_i, \prod_{i=1}^n X_i\right)$ that associates with any $x = (x_1, \ldots, x_n) \in \prod_{i=1}^n X_i$ the sequence $\chi_S \circ x \in \mathbf{R}^n$ defined by

$$\forall\, i = 1, \ldots, n, \quad (\chi_S \circ x)_i := \begin{cases} x_i & \text{if } i \in S \\ 0 & \text{if } i \notin S \end{cases}$$

We associate with any coalition $S \subset N$ the subspace

$$X^S := x_S \circ \prod_{i=1}^n X_i = \left\{ x \in \prod_{i=1}^n X_i \text{ such that} \forall\, i \notin S, x_i = 0 \right\}$$

since $x_S \circ$ is nothing other that the canonical projector from $\prod_{i=1}^n X_i$ onto X^S. In particular, $X^N := \prod_{i=1}^n X_i$ and $X^\emptyset := \{0\}$.

Let Y be another finite dimensional vector space. We associate with any coalition $S \subset N$ the space $\mathcal{L}_S(X^S, Y)$ of S-linear operators A_S. We extend such a S-linear operator A_S to a n-linear operator (again denoted by) $A_S \in \mathcal{L}_n\left(\prod_{i=1}^n X_i, Y\right)$ defined by:

$$\forall\, x \in \prod_{i=1}^n X_i, \quad A_S(x) = A_S(x_1, \ldots, x_n) := A_S(\chi_S \circ x)$$

A multiaffine operator $A \in \mathcal{A}_n\left(\prod_{i=1}^n X_i, Y\right)$ is a sum of S-linear operators $A_S \in \mathcal{L}_S(X^S, Y)$ when S ranges over the family of coalitions:

$$A(x_1, \ldots, x_n) := \sum_{S \subset N} A_S(\chi_S \circ x) = \sum_{S \subset N} A_S(x)$$

We identify $A_\emptyset$ with a constant $A_\emptyset \in Y$.

Hence the collective constraint linking multiaffine operators and actions can be written in the form

$$\forall\, t \geq 0, \quad \sum_{S \subset N} A_S(t)(x(t)) \in M$$

For any $i \in S$, we shall denote by $(x_{-i}, u_i) \in X^N$ the sequence $y \in X^N$ where $y_j := x_j$ when $j \neq i$ and $y_i = u_i$ when $j = i$. The linear operator $A_S\,(x_{-i}) \in \mathcal{L}(X_i, Y)$ is defined by $u_i \mapsto A_S(x_{-i})u_i := A_S(x_{-i}, u_i)$. We shall use its transpose $A_S\,(x_{-i})^* \in \mathcal{L}(Y^*, X_i^*)$ defined by

$$\forall\, q \in Y^*, \quad \forall\, u_i \in X_i, \langle A_S\,(x_{-i})^*\, q, u_i \rangle = \langle q, A_S\,(x_{-i})\, u_i \rangle$$

We associate with $q \in Y^*$ and elements $x_i \in X_i$ the multilinear operator[5]

$$x_1 \otimes \cdots \otimes x_n \otimes q \in \mathcal{L}_n\Big(\prod_{i=1}^{n} X_i^*, Y^*\Big)$$

associating with any

$$p := (p_1, \ldots, p_n) \in \prod_{i=1}^{n} X_i^*$$

the element $\left(\prod_{i=1}^{n} \langle p_i, x_i\rangle\right) q$:

$$x_1 \otimes \cdots \otimes x_n \otimes q : p := (p_1, \ldots, p_n) \in \prod_{i=1}^{n} X_i^*$$
$$\mapsto (x_1 \otimes \cdots \otimes x_n \otimes q)(p) := \left(\prod_{i=1}^{n} \langle p_i, x_i\rangle\right) q \in Y^*$$

This multilinear operator $x_1 \otimes \cdots \otimes x_n \otimes q$ is called the tensor product of the x_i's and q.

We recall that the duality product on

$$\mathcal{L}_n\Big(\prod_{i=1}^{n} X_i^*, Y^*\Big) \times \mathcal{L}_n\Big(\prod_{i=1}^{n} X_i, Y\Big)$$

for pairs $(x_1 \otimes \cdots \otimes x_n \otimes q, A)$ can be written in the form:

$$\langle x_1 \otimes \cdots \otimes x_n \otimes q, A\rangle := \langle q, A(x_1, \ldots, x_n)\rangle$$

3.7 Multi-Hebbian learning process

Assume that we start with intrinsic dynamics of the actions x_i, the resources y, the connectionist matrices W and the fuzzy coalitions χ:

$$\begin{cases} (i) & x_i'(t) = c_i(x(t)), \quad i = 1, \ldots, n \\ (ii) & A_S'(t) = \alpha_S(A(t)), \quad S \subset N \end{cases}$$

Using viability multipliers, we can modify the above dynamics by introducing regulons that are elements $q \in Y^*$ of the dual Y^* of the space Y:

[5] We recall that the space $\mathcal{L}_n\left(\prod_{i=1}^{n} X_i, Y\right)$ of n-linear operators from $\prod_{i=1}^{n} X_i$ to Y is isometric to the tensor product $\bigotimes_{i=1}^{n} X_i^* \otimes Y$, the dual of which is $\bigotimes_{i=1}^{n} X_i \otimes Y^*$, that is isometric with $\mathcal{L}_n\left(\prod_{i=1}^{n} X_i^*, Y^*\right)$.

THEOREM 1.1 *Assume that the functions c_i, κ_i and α_S are continuous and that $M \subset Y$ are closed. Then the constraints*

$$\forall\, t \geq 0, \quad \sum_{S \subset N} A_S(t)(x(t)) \in M$$

are viable under the control system

$$\begin{cases} (i) \quad x_i'(t) = c_i(x_i(t)) - \sum_{S \ni i} A_S(t)(x_{-i}(t))^* q(t), \quad i = 1, \ldots, n \\ \\ (ii) \quad A_S'(t) = \alpha_S(A(t)) - \left(\bigotimes_{j \in S} x_j(t) \right) \otimes q(t), \quad S \subset N \\ \quad \textit{where } q(t) \in N_M(\sum_{S \subset N} A_S(t)(x(t))) \end{cases}$$

Remark: Multi-Hebbian Rule — When we regard the multilinear operator A_S as a tensor product of components $A^j_{S_{\Pi_{i \in S} i_k}}$, $j = 1, \ldots, p$, $i_k = 1, \ldots, n_i$, $i \in S$, differential equation (ii) can be written in the form: $\forall\, i \in S, j = 1, \ldots, p$, $k = 1, \ldots, n_i$,

$$\frac{d}{dt} A^j_{S_{\Pi_{i \in S} i_k}} = \alpha_{S_{\Pi_{i \in S} i_k}}(A(t)) - \left(\prod_{i \in S} x_{i_k}(t) \right) q^j(t)$$

The correction term of the component $A^j_{S_{\Pi_{i \in S} i_k}}$ of the S-linear operator is the product of the components $x_{i_k}(t)$ actions x_i in the coalition S and of the component q^j of the viability multiplier. *This can be regarded as a multi-Hebbian rule in neural network learning algorithms*, since for linear operators, we find the product of the component x_k of the pre-synaptic action and the component q^j of the post-synaptic action. □

Indeed, when the vector spaces $X_i := \mathbf{R}^{n_i}$ are supplied with basis e^{i_k}, $k = 1, \ldots, n_i$, when we denote by $e^*_{i_k}$ their dual basis, and when $Y := \mathbf{R}^p$ is supplied with a basis f^j, and its dual supplied with the dual basis f^*_j, then the tensor products $\left(\bigotimes_{i \in S} e^{i_k} \right) \otimes f^*_j$ $(j = 1, \ldots, p,\ k = 1, \ldots, n_i)$ form a basis of $\mathcal{L}_S\left(X^{S^*}, Y^*\right)$.

Hence the components of the tensor product $\left(\bigotimes_{i \in S} x_i \right) \otimes q$ in this basis are the products $\left(\prod_{i \in S} x_{i_k} \right) q^j$ of the components q^j of q and x_{i_k} of the x_i's, where $q^j := \langle q, f^j \rangle$ and $x_{i_k} := \langle e^*_{i_k}, x_i \rangle$. Indeed, we can write

$$\left(\bigotimes_{i\in S} x_i\right)\otimes q = \sum_{j=1}^{p}\sum_{i\in S}\sum_{k=1}^{n_i}\left(\langle q, f^j\rangle \prod_{i\in S}\langle e^*_{i_k}, x_i\rangle\right)\left(\bigotimes_{i=1}^{n} e^{i_k}\right)\otimes f^*_j$$

4. Regulation involving fuzzy coalitions

Let $A \in \mathcal{A}_n\left(\prod_{i=1}^n X_i, Y\right)$, a sum of S-linear operators $A_S \in \mathcal{L}_S(X^S, Y)$ when S ranges over the family of coalitions, be a multiaffine operator.

When χ is a fuzzy coalition, we observe that

$$A(\chi \circ x) = \sum_{S\subset P(\chi)} \gamma_S(\chi) A_S(x) = \sum_{S\subset P(\chi)} \left(\prod_{j\in S}\chi_j\right) A_S(x)$$

We wish to encapsulate the idea that at each instant, only a number of fuzzy coalitions χ are active. Hence the collective constraint linking multiaffine operators, fuzzy coalitions and actions can be written in the form

$$\forall\, t \geq 0,\ \sum_{S\subset P(\chi(t))} \gamma_S(\chi(t)) A_S(t)(x(t)) = \sum_{S\subset P(\chi(t))} \left(\prod_{j\in S}\chi_j(t)\right) A_S(t)(x(t)) \in M$$

4.1 Constructing viable dynamics

Assume that we start with intrinsic dynamics of the actions x_i, the resources y, the connectionist matrices W and the fuzzy coalitions χ:

$$\begin{cases} (i) & x_i'(t) = c_i(x(t)), \quad i = 1,\ldots,n \\ (ii) & \chi_i'(t) = \kappa_i(\chi(t)), \quad i = 1,\ldots,n \\ (iii) & A_S'(t) = \alpha_S(A(t)), \quad S \subset N \end{cases}$$

Using viability multipliers, we can modify the above dynamics by introducing regulons that are elements $q \in Y^*$ of the dual Y^* of the space Y:

THEOREM 1.2 *Assume that the functions c_i, κ_i and α_S are continuous and that $M \subset Y$ are closed. Then the constraints*

$$\forall\, t \geq 0,\ \sum_{S\subset P(\chi(t))} A_S(t)(\chi(t)\circ x(t)) = \sum_{S\subset P(\chi(t))} \left(\prod_{j\in S}\chi_j(t)\right) A_S(t)(x(t)) \in M$$

are viable under the control system

$$\begin{cases} (i) \quad x_i'(t) = c_i(x_i(t)) - \sum_{S\ni i}\Big(\prod_{j\in S}\chi_j(t)\Big)A_S(t)(x_{-i}(t))^* q(t), \\ \qquad\qquad\qquad\qquad\qquad\qquad\qquad\qquad i = 1,\ldots,n \\ (ii) \quad \chi_i'(t) = \kappa_i(\chi(t)) - \sum_{S\ni i}\Big(\prod_{j\in S\setminus i}\chi_j(t)\Big)\langle q(t), A_S(t)\,(x(t))\rangle, \\ \qquad\qquad\qquad\qquad\qquad\qquad\qquad\qquad i = 1,\ldots,n \\ (iii) \quad A_S'(t) = \alpha_S(A(t)) - \Big(\prod_{j\in S}\chi_j(t)\Big)\Big(\bigotimes_{j\in S}x_j(t)\Big)\otimes q(t),\ S\subset N \\ \qquad \textit{where } q(t)\in N_M\Big(\sum_{S\subset P(\chi(t))}\big(\prod_{j\in S}\chi_j(t)\big)A_S(t)\big(x(t)\big)\Big) \end{cases}$$

Let us comment on these formulas. First, the viability multipliers $q(t) \in Y^*$ can be regarded as regulons, i.e., regulation controls or parameters, or virtual prices in the language of economists. They are chosen adequately at each instant in order that the viability constraints describing the network can be satisfied at each instant, and the above theorem guarantees this possibility. The next section tells us how to choose at each instant such regulons (the regulation law).

For each player i, the velocities $x_i'(t)$ of the state and the velocities $\chi_i'(t)$ of its membership in the fuzzy coalition $\chi(t)$ are corrected by subtracting

1. the sum over all coalitions S to which he belongs of the $A_S(t)$ $(x_{-i}(t))^*q(t)$ weighted by the membership $\gamma_S(\chi(t))$:

$$x_i'(t) = c_i(x_i(t)) - \sum_{S\ni i}\gamma_S(\chi(t))A_S(t)(x_{-i}(t))^*q(t)$$

2. the sum over all coalitions S to which he belongs of the costs $\langle q(t), A_S(t)\,(x(t))\rangle$ of the constraints associated with connectionist tensor A_S of the coalition S weighted by the membership $\gamma_{S\setminus i}(\chi(t))$:

$$\chi_i'(t) = \kappa_i(\chi(t)) - \sum_{S\ni i}\gamma_{S\setminus i}(\chi(t))\,\langle q(t), A_S(t)\,(x(t))\rangle$$

The (algebraic) increase of player i's membership in the fuzzy coalition aggregates over all coalitions to which he belongs the cost of their constraints weighted by the products of memberships of the other players in the coalition.

It can be interpreted as an incentive for economic agents to increase or decrease his participation in the economy in terms of the cost of constraints and of the membership of other economic agents, encapsulating a mimetic — or "herd", panurgean — behavior (from a famous story by François Rabelais (1483-1553), where Panurge sent overboard the head sheep, followed by the whole herd).

Panurge ... jette en pleine mer son mouton criant et bellant. Tous les aultres moutons, crians et bellans en pareille intonation, commencerent soy jecter et saulter en mer après, à la file ... comme vous sçavez estre du mouton le naturel, tous jours suyvre le premier, quelque part qu'il aille. Aussi li dict Aristoteles, lib. 9, de Histo. animal. estre le plus sot et inepte animant du monde.

As for the correction of the velocities of the connectionist tensors A_S, their correction is a weighted "multi-Hebbian" rule: for each component $A^j_{S_{\Pi_{i\in S} i_k}}$ of A_S, the correction term is the product of the membership $\gamma S(\chi(t))$ of the coalition S, of the components $x_{i_k}(t)$ and of the component $q^j(t)$ of the regulon:

$$\frac{d}{dt}A^j_{S_{\Pi_{i\in S} i_k}} = \alpha_{S_{\Pi_{i\in S} i_k}}(A(t)) - \gamma_S(\chi(t))\left(\prod_{i\in S} x_{i_k}(t)\right)q^j(t)$$

4.2 The regulation map

Actually, the viability multipliers $q(t)$ regulating viable evolutions of the actions $x_i(t)$, the fuzzy coalitions $\chi(t)$ and the multiaffine operators $A(t)$ obey the regulation law (an "adjustment law", in the vocabulary of economists) of the form

$$\forall\, t \geq 0, \quad q(t) \in R_M(x(t), \chi(t), A(t))$$

where $R_M : X^N \times \mathbf{R}^n \times \mathcal{A}_n(X^N, Y) \rightsquigarrow Y^\star$ is the regulation map R_M that we shall compute.

For this purpose, we introduce the operator $h : X^N \times \mathbf{R}^n \times \mathcal{A}_n(X^N, Y)$ defined by

$$h(x, \chi, A) := \sum_{S\subset N} A_S(\chi \circ x)$$

and the linear operator $H(x, \chi, A) : Y^* := Y \mapsto Y$ defined by:

$$\left\{ \begin{array}{ll} H(x, \chi, A) & := \sum_{S \subset N} \left(\prod_{j \in S} \chi_j^2 \|x_j\|^2 \right) \mathbf{I} \\ & \quad + \sum_{R,S \subset N} \sum_{i \in R \cap S} \Big(\gamma_R(\chi)\gamma_S(\chi) A_R(x_{-i}) A_S(x_{-i})^* \\ & \quad + \gamma_{R \backslash i}(\chi)\gamma_{S \backslash i}(\chi) A_R(x) \otimes A_S(x) \Big) \end{array} \right.$$

Then the regulation map is defined by

$$\left\{ \begin{array}{ll} R_M(x, \chi, A) & := H(x, \chi, A)^{-1} \\ & \quad \Big(\sum_{S \subset N} \Big(\alpha_S(A)(x) + \sum_{i \in S} (\gamma_S(\chi) A_S(x_{-i}, c_i(x)) \\ & \quad + \gamma_{S \backslash i}(\chi) \kappa_i(\chi) A_S(x)) \Big) - T_M(h(x, \chi, A)) \Big) \end{array} \right.$$

Indeed, the regulation map R_M associates with any (x, χ, A) the subset $R_M(x, \chi, A)$ of $q \in Y^*$ such that

$$h'(x, \chi, A)((c(x), \kappa(\chi), \alpha(A)) - h'(x, \chi, A)^* q) \in \overline{\text{co}}\,(T_M(h(x)))$$

We next observe that

$$h'(x, \chi, A) h'(x, \chi, A)^* = H(x, \chi, A)$$

and that

$$\left\{ \begin{array}{l} h'(x, \chi, A)(c(x), \kappa(\chi), \alpha(A)) \\ \quad = \sum_{S \subset N} \Big(\alpha_S(A)(x) + \sum_{i \in S} (\gamma_S(\chi) A_S(x_{-i}, c_i(x)) + \gamma_{S \backslash i}(\chi) \kappa_i(\chi) A_S(x)) \Big) \end{array} \right.$$

Remark: Links between viability and Lagrange multipliers — The point made in this paper is to show how the mathematical methods presented in a general way can be useful in designing other models, as the Lagrange multiplier rule does in the static framework. By comparison, we see that if we minimize a collective utility function:

$$\sum_{i=1}^{n} \mathbf{u}_i(x_i) + \sum_{i=1}^{n} \mathbf{v}_i(\chi_i) + \sum_{S \subset N} \mathbf{w}_S(A_S)$$

under constraints (1.1), then first-order optimality conditions at a optimum $((x_i)_i, (\chi_i)_i, (A_S)_{S \subset N})$ imply the existence of Lagrange multipliers p such that:

$$\begin{cases} \nabla \mathbf{u}_i(x_i) = \sum_{S \ni i} \left(\prod_{j \in S} \chi_j \right) A_S(x_{-i}(t))^* p, & i = 1, \ldots, n \\ \\ \nabla \mathbf{v}_i(\chi_i) = \sum_{S \ni i} \left(\prod_{j \in S \setminus i} \chi_j \right) \langle p, A_S(x) \rangle, & i = 1, \ldots, n \\ \\ \nabla \mathbf{w}_S(A_S) = \left(\prod_{j \in S} \chi_j \right) \left(\bigotimes_{j \in S} x_j \right) \otimes p, & S \subset N \end{cases}$$

□

5. Dynamical fuzzy cooperative games under tychastic uncertainty

5.1 Static fuzzy cooperative games

DEFINITION 1.2 *A Fuzzy game with side-payments is defined by a characteristic function* $\mathbf{u} : [0,1]^n \mapsto \mathbf{R}_+$ *of a fuzzy game assumed to be positively homogenous.*

When the characteristic function of the static cooperative game $\mathbf{u}$ is concave, positively homogeneous and continuous on the interior of $\mathbf{R}^n_+$, one checks[6] that the generalized gradient $\partial \mathbf{u}(\chi_N)$ is not empty and coincides with the subset of imputations $p := (p_1, \ldots, p_n) \in \mathbf{R}^n_+$ accepted by all fuzzy coalitions in the sense that

$$\forall \chi \in [0,1]^n, \quad \langle p, \chi \rangle = \sum_{i=1}^{n} p_i \chi_i \geq \mathbf{u}(\chi) \tag{1.3}$$

and that, for the grand coalition $\chi_N := (1, \ldots, 1)$,

$$\langle p, \chi_N \rangle = \sum_{i=1}^{n} p_i = \mathbf{u}(\chi_N)$$

It has been shown that in the framework of static cooperative games with side payments involving fuzzy coalitions, *the concepts of Shapley value and core coincide with the (generalized) gradient* $\partial \mathbf{u}(\chi_N)$ *of the "characteristic function"* $\mathbf{u} : [0,1]^n \mapsto \mathbf{R}_+$ *at the "grand coalition"* $\chi_N := (1, \ldots, 1)$, *the characteristic function of* $N := \{1, 2, \ldots, n\}$. The differences between these concepts for usual games is explained by the different ways one "fuzzyfies" a characteristic function defined on the set of usual coalitions.

[6]See Aubin (1981a,b), Aubin (1979), Chapter 12 and Aubin (1998, 1993), Chapter 13.

5.2 Three examples of game rules

In a dynamical context, (fuzzy) coalitions evolve, so that static conditions (1.3) should be replaced by conditions[7] stating that for any evolution $t \mapsto x(t)$ of fuzzy coalitions, the payoff $y(t) := \langle p(t), \chi(t) \rangle$ should be larger than or equal to $\mathbf{u}(\chi(t))$ according (at least) to one of the three following rules:

1. at a prescribed final time T of the end of the game:

$$y(T) := \sum_{i=1}^{n} p_i(T)\chi_i(T) \geq \mathbf{u}(\chi(T))$$

2. during the whole time span of the game:

$$\forall\, t \in [0,T], \quad y(t) := \sum_{i=1}^{n} p_i(t)\chi_i(t) \geq \mathbf{u}(\chi(t))$$

3. at the first winning time $t^* \in [0,T]$ when

$$y(t^*) := \sum_{i=1}^{n} p_i(t^*)\chi_i(t^*) \geq \mathbf{u}(\chi(t^*))$$

   at which time the game stops.

Summarizing, the above conditions require to find — for each of the above three rules of the game — an evolution of an imputation $p(t) \in \mathbf{R}^n$ such that, for all evolutions of fuzzy coalitions $\chi(t) \in [0,1]^n$ starting at χ, the corresponding rule of the game

$$\left\{ \begin{array}{ll} i) & \sum_{i=1}^{n} p_i(T)\chi_i(T) \geq \mathbf{u}(\chi(T)) \\ ii) & \forall\, t \in [0,T], \quad \sum_{i=1}^{n} p_i(t)\chi_i(t) \geq \mathbf{u}(\chi(t)) \\ iii) & \exists t^* \in [0,T] \text{ such that } \sum_{i=1}^{n} p_i(t^*)\chi_i(t^*) \geq \mathbf{u}(\chi(t^*)) \end{array} \right. \qquad (1.4)$$

must be satisfied.

Therefore, for each one of the above three rules of the game (1.4), a concept of dynamical core should provide a set-valued map $\Gamma : \mathbf{R}_+ \times [0,1]^n \rightsquigarrow \mathbf{R}^n$ associating with each time t and any fuzzy coalition χ a set $\Gamma(t,\chi)$ of imputations $p \in \mathbf{R}^n_+$ such that, taking $p(t) \in \Gamma(T-t, \chi(t))$, and in particular, $p(0) \in \Gamma(T, \chi(0))$, the chosen above condition is satisfied. This is the purpose of this study.

[7]Naturally, the privileged role played by the grand coalition in the static case must be abandoned, since the coalitions evolve, so that the grand coalition eventually loses its capital status.

5.3 A general class of game rules

Actually, in order to treat the three rules of the game (1.4) as particular cases of a more general framework, we introduce two nonnegative extended functions **b** and **c** (characteristic functions of the cooperative games) satisfying

$$\forall\,(t,\chi) \in \mathbf{R}_+ \times \mathbf{R}_+^n \times \mathbf{R}^n, \quad 0 \leq \mathbf{b}(t,\chi) \leq \mathbf{c}(t,\chi) \leq +\infty$$

By associating with the initial characteristic function **u** of the game adequate pairs $(\mathbf{b},\mathbf{c})$ of extended functions, we shall replace the requirements (1.4) by the requirement

$$\begin{cases} i) & \forall\,t \in [0,t^*],\ y(t) \geq \mathbf{b}(T-t,\chi(t))(\text{dynamical constraints}) \\ ii) & y(t^*) \geq \mathbf{c}(T-t^*,\chi(t^*))(\text{objective}) \end{cases} \tag{1.5}$$

We extend the functions **b** and **c** as functions from $\mathbf{R} \times \mathbf{R}^n \times \mathbf{R}^n$ to $\mathbf{R}_+ \cup \{+\infty\}$ by setting

$$\forall\,t < 0, \quad \mathbf{b}(t,\chi) = \mathbf{c}(t,\chi) = +\infty$$

so that nonnegativity constraints on time are automatically taken into account.

For instance, *problems with prescribed final time are obtained with objective functions satisfying the condition*

$$\forall\,t > 0, \quad \mathbf{c}(t,\chi) := +\infty$$

In this case, $t^* = T$ and condition (1.5) boils down to

$$\begin{cases} i) & \forall\,t \in [0,T], \quad y(t) \geq b(T-t,\chi(t)) \\ ii) & y(T) \geq c(0,\chi(T)) \end{cases}$$

Indeed, since $y(t^*)$ is finite and since $\mathbf{c}(T-t^*,\chi(t^*))$ is infinite whenever $T - t^* > 0$, we infer from inequality (1.5)ii) that $T - t^*$ must be equal to 0. □

Allowing the characteristic functions to take infinite values (i.e., to be extended), allows us to acclimate many examples.

For example, *the three rules (1.4) associated with a same characteristic function* $\mathbf{u} : [0,1]^n \mapsto \mathbf{R} \cup \{+\infty\}$ *can be written in the form (1.5) by adequate choices of pairs* $(\mathbf{b},\mathbf{c})$ *of functions associated with* **u**. Indeed, denoting by u_∞ the function defined by

$$\mathbf{u}_\infty(t,\chi) := \begin{cases} \mathbf{u}(\chi) & \text{if} \quad t = 0 \\ +\infty & \text{if} \quad t > 0 \end{cases}$$

and by $\mathbf{0}$ the function defined by

$$\mathbf{0}(t,\chi) = \begin{cases} 0 & \text{if} \quad t \geq 0, \\ +\infty & \text{if not} \end{cases}$$

we can recover the three rules of the game

1. We take $\mathbf{b}(t,\chi) := \mathbf{0}(t,\chi)$ and $\mathbf{c}(t,\chi) = \mathbf{u}_\infty(t,\chi)$, we obtain the prescribed final time rule (1.4)i).
2. We take $\mathbf{b}(t,\chi) := \mathbf{u}(\chi)$ and $\mathbf{c}(t,\chi) := \mathbf{u}_\infty(t,\chi)$, we obtain the span time rule (1.4)ii).
3. We take $\mathbf{b}(t,\chi) := \mathbf{0}(t,\chi)$ and $\mathbf{c}(t,\chi) = \mathbf{u}(\chi)$, we obtain the first winning time rule (1.4)iii).

5.4 Dynamics of fuzzy cooperative games

Naturally, games are played under uncertainty. In games arising social or biological sciences, uncertainty is rarely od a probabilistic and stochastic nature (with statistical regularity), but of a tychastic nature, according to a terminology borrowed to Charles Peirce.

State-dependent uncertainty can also be translated mathematically by parameters on which actors, agents, decision makers, etc. have no controls. These parameters are often perturbations, disturbances (as in "robust control" or "differential games against nature") or more generally, tyches (meaning "chance" in classical Greek, from the Goddess Tyche) ranging over a state-dependent tychastic map. They could be called "random variables" if this vocabulary were not already confiscated by probabilists. This is why we borrow the term of *tychastic evolution* to Charles Peirce who introduced it in a paper published in 1893 under the title *evolutionary love*. One can prove that stochastic viability is a (very) particular case of tychastic viability. The size of the tychastic map captures mathematically the concept of "versatility (tychastic volatility)" — instead of "(stochastic) volatility": The larger the graph of the tychastic map, the more "versatile" the system.

Next, we define the dynamics of the coalitions and of the imputations, assumed to be given.

1. the evolution of coalitions $\chi(t) \in \mathbf{R}^n$ is governed by differential inclusions
$$\chi'(t) := f(\chi(t), v(t))\text{where} v(t) \in Q(\chi(t))$$
where $v(t)$ arc tyches,
2. static constraints
$$\forall \chi \in [0,1]^n, \quad p \in P(\chi) \subset \mathbf{R}^n_+$$
and dynamic constraints on the velocities of the imputations $p(t) \in \mathbf{R}^n_+$ of the form
$$\langle p'(t), \chi(t)\rangle = -\mathbf{m}(\chi(t), p(t), v(t))\langle p(t), \chi(t)\rangle$$
stating that the cost $\langle p', \chi\rangle$ of the instantaneous change of imputation of a coalition is proportional to it by a discount factor $\mathbf{m}(\chi, p)$
3. from which we deduce the velocity $y'(t) = \langle p(t), f(\chi(t), v(t))\rangle - \mathbf{m}(\chi(t), p(t))y(t)$ of the payoff $y(t) := \langle p(t), \chi(t)\rangle$ of the fuzzy coalition $\chi(t)$.

The evolution of the fuzzy coalitions is thus parameterized by imputations and tyches, i.e., is governed by a dynamic game

$$\begin{cases} i) & \chi'(t) = f(\chi(t), v(t)) \\ ii) & y'(t) = \langle p(t), f(\chi(t), v(t))\rangle - \mathbf{m}(\chi(t), p(t))y(t) \\ iii) & \text{where } p(t) \in P(\chi(t)) \;\&\; v(t) \in Q(\chi(t)) \end{cases} \tag{1.6}$$

A feedback $\widetilde{p}$ is a selection of the set-valued map P in the sense that for any $\chi \in [0,1]^n$, $\widetilde{p}(\chi) \in P(\chi)$. We thus associate with any feedback $\widetilde{p}$ the set $\mathcal{C}_{\widetilde{p}}(\chi)$ of triples $(\chi(\cdot), y(\cdot), v(\cdot))$ solutions to

$$\begin{cases} i) & \chi'(t) = f(\chi(t), v(t)) \\ ii) & y'(t) = \langle \widetilde{p}(\chi(t)), f(\chi(t), v(t))\rangle - y(t)\mathbf{m}(\chi(t), \widetilde{p}(\chi(t)), v(t)) \\ & \text{where } v(t) \in Q(\chi(t)) \end{cases} \tag{1.7}$$

5.5 Valuation of the dynamical game

We shall characterize the dynamical core of the fuzzy dynamical cooperative game in terms of the derivatives of a valuation function that we now define.

For each rule of the game (1.5), the set $\mathcal{V}^\sharp$ of initial conditions (T, χ, y) such that there exists a feedback $\chi \mapsto \widetilde{p}(\chi) \in P(\chi)$ such that, for all tyches $t \in [0,T] \mapsto v(t) \in Q(\chi(t))$, for all solutions to system (1.7) of differential equations satisfying $\chi(0) = \chi$, $y(0) = y$, the corresponding condition (1.5) is satisfied, is called the guaranteed valuation set[8].

Knowing it, we deduce the valuation function

$$V^\sharp(T,\chi) := \inf\{y | (T,\chi,y) \in \mathcal{V}^\sharp\}$$

providing the cheapest initial payoff allowing to satisfy the viability/capturability conditions (1.5). It satisfies the initial condition:

$$V^\sharp(0,\chi) := \mathbf{u}(\chi)$$

In each of the three cases, we shall compute explicitly the valuation functions as infsup of underlying criteria we shall uncover. For that purpose, we associate with the characteristic function $\mathbf{u} : [0,1]^n \mapsto \mathbf{R} \cup \{+\infty\}$ of the dynamical cooperative game the functional

$$\begin{cases} J_{\mathbf{u}}(t; (\chi(\cdot), v(\cdot)); \widetilde{p})(\chi) := e^{\int_0^t \mathbf{m}(\chi(s),\widetilde{p}(\chi(s)),v(s))ds}\mathbf{u}(\chi(t)) \\ \quad - \displaystyle\int_0^t e^{\int_0^\tau \mathbf{m}(\chi(s),\widetilde{p}(\chi(s)),v(s))ds} \langle \widetilde{p}(\chi(\tau)), f(\chi(\tau), v(\tau)) \rangle d\tau \end{cases}$$

We shall associate with it and with each of the three rules of the game (1.4) the three corresponding valuation functions:

1. **prescribed end rule**: We obtain

$$V^\sharp_{(\mathbf{0},\mathbf{u}_\infty)}(T,\chi) := \inf_{\widetilde{p}(\chi)\in P(\chi)} \sup_{(\chi(\cdot),v(\cdot))\in \mathcal{C}_{\widetilde{p}}(\chi)} J_{\mathbf{u}}(T; (\chi(\cdot), v(\cdot)); \widetilde{p})(\chi) \tag{1.8}$$

2. **time span rule**: We obtain

$$V^\sharp_{(\mathbf{u},\mathbf{u}_\infty)}(T,\chi) := \inf_{\widetilde{p}(\chi)\in P(\chi)} \sup_{(\chi(\cdot),v(\cdot))\in \mathcal{C}_{\widetilde{p}}(\chi)} \sup_{t\in[0,T]} J_{\mathbf{u}}(t; (\chi(\cdot), v(\cdot)); \widetilde{p})(\chi) \tag{1.9}$$

3. **first winning time rule**: We obtain

$$V^\sharp_{(\mathbf{0},\mathbf{u})}(T,\chi) := \inf_{\widetilde{p}(\chi)\in P(\chi)} \sup_{(\chi(\cdot),v(\cdot))\in \mathcal{C}_{\widetilde{p}}(\chi)} \inf_{t\in[0,T]} J_{\mathbf{u}}(t; (\chi(\cdot), v(\cdot)); \widetilde{p})(\chi) \tag{1.10}$$

A general formula for game rules 1.5 does exist, but is too involved to be reproduced in this survey.

---

[8] One can also define the conditional valuation set $\mathcal{V}^\flat$ of initial conditions (T,χ,y) such that for all tyches v, there exists an evolution of the imputation $p(\cdot)$ such that viability/capturability conditions (1.5) are satisfied. We omit this study for the sake of brevity, since it is parallel to the one of guaranteed valuation sets.

5.6 Hamilton-Jacobi equations and dynamical core

Although these functions are only lower semicontinuous, one can define epiderivatives of lower semicontinuous functions (or generalized gradients) in adequate ways and compute the core Γ: for instance, when the valuation function is differentiable, we shall prove that Γ associates with any $(t,\chi) \in \mathbf{R}_+ \times \mathbf{R}^n$ the subset $\Gamma(t,\chi)$ of imputations $p \in P(\chi)$ satisfying

$$\sup_{v \in Q(\chi)} \left(\sum_{i=1}^{n} \left(\frac{\partial V^\sharp(t,\chi)}{\partial \chi_i} - p_i \right) f_i(\chi, v) + \mathbf{m}(\chi, p, v) V^\sharp(t,\chi) \right) \leq \frac{\partial V^\sharp(t,\chi)}{\partial t}$$

The valuation function $V^\sharp$ is actually a solution to the nonlinear Hamilton-Jacobi-Isaacs partial differential equation

$$-\frac{\partial \mathbf{v}(t,\chi)}{\partial t} + \inf_{p \in P(\chi)} \sup_{v \in Q(\chi)} \left(\sum_{i=1}^{n} \left(\frac{\partial \mathbf{v}(t,\chi)}{\partial \chi_i} - p_i \right) f_i(\chi, v) + \mathbf{m}(\chi, p, v) \mathbf{v}(t,\chi) \right) = 0$$

satisfying the initial condition

$$\mathbf{v}(0,\chi) = \mathbf{u}(\chi)$$

on the subset

$$\Omega_{(\mathbf{b},\mathbf{c})}(\mathbf{v}) := \{(t,\chi) | \mathbf{c}(t,\chi) > \mathbf{v}(t,\chi) \geq \mathbf{v}(t,\chi)\}$$

For each of the game rules (1.4), these subsets are written

1. **prescribed end rule**:

$$\Omega_{(\mathbf{0},\mathbf{u}_\infty)}(\mathbf{v}) := \{(t,\chi) | t > 0 \& \mathbf{v}(t,\chi) \geq 0\}$$

2. **time span rule**

$$\Omega_{(\mathbf{u},\mathbf{u}_\infty)}(\mathbf{v}) := \{(t,\chi) | t > 0 \& \mathbf{v}(t,\chi) \geq \mathbf{u}(\chi)\}$$

3. **first winning time rule**

$$\Omega_{(\mathbf{0},\mathbf{u})}(\mathbf{v}) := \{(t,\chi) | t > 0 \& \mathbf{u}(\chi) > \mathbf{v}(t,\chi) \geq 0\}$$

Actually, the solution of the above partial differential equation is taken in the "contingent sense", where the directional derivatives are the contingent epiderivatives $D_{\uparrow}\mathbf{v}(t,\chi)$ of $\mathbf{v}$ at (t,χ). They are defined by

$$D_{\uparrow}\mathbf{v}(t,\chi)(\lambda,v) := \liminf_{h\to 0+, u\to v} \frac{\mathbf{v}(t+h\lambda, \chi+hu)}{h}$$

(see for instance Aubin and Frankowska (1990) and Rockafellar and Wets (1997)).

DEFINITION 1.3 (DYNAMICAL CORE) *Consider the dynamic fuzzy cooperative game with game rules (1.5). The dynamical core* Γ *of the corresponding fuzzy dynamical cooperative game is equal to*

$$\left\{\begin{array}{l} \Gamma(t,\chi) := \Big\{p\in P(\chi) \text{ such that} \\ \quad \sup_{v\in Q(\chi)}(D_{\uparrow}V^{\sharp}(t,\chi)(-1,f(\chi,v)) - \langle p, f(\chi,v)\rangle \\ \quad +\mathbf{m}(\chi,p,v)V^{\sharp}(t,\chi)) \leq 0\Big\} \end{array}\right.$$

where $V^{\sharp}$ *is the corresponding value function.*

We can prove that for each feedback $\widetilde{p}(t,\chi)\in\Gamma(t,\chi)$ being a selection of the dynamical core Γ, all evolutions $(\chi(\cdot),v(\cdot))$ of the system

$$\left\{\begin{array}{ll} i) & \chi'(t) = f(\chi(t),v(t)) \\ ii) & y'(t) = \langle\widetilde{p}(T-t,\chi(t)),\chi(t)\rangle - \mathbf{m}(\chi(t),\widetilde{p}(T-t,\chi(t))y(t) \\ iii) & v(t)\in Q(\chi(t)) \end{array}\right. \tag{1.11}$$

satisfy the corresponding condition (1.5).

5.7 The static case as infinite versatility

Let us consider the case when $\mathbf{m}(\chi,p,v) = 0$ (self-financing of fuzzy coalitions) and when the evolution of coalitions is governed by $f(\chi,v) = v$ and $Q(\chi) = rB$. Then the dynamical core is the subset $\Gamma(t,\chi)$ of imputations $p\in P(\chi)$ satisfying on $\Omega(V^{\sharp})$ the equation[9]

$$r\left\|\frac{\partial V^{\sharp}(t,\chi)}{\partial\chi} - p\right\| = \frac{\partial V^{\sharp}(t,\chi)}{\partial t}$$

Now, assuming that the data and the solution are smooth we deduce formally that letting the versatility $r\to\infty$, we obtain as a limiting case

[9]when $p = 0$, we find the eikonal equation.

that $p = \frac{\partial V^\sharp(t,\chi)}{\partial \chi}$ and that $\frac{\partial V^\sharp(t,\chi)}{\partial t} = 0$. Since $V^\sharp(0,\chi) = \mathbf{u}(\chi)$, we infer that in this case $\Gamma(t,\chi) = \frac{\partial \mathbf{u}(\chi)}{\partial \chi}$, i.e., the Shapley value of the fuzzy static cooperative game when the characteristic function $\mathbf{u}$ is differentiable and positively homogenous, and the core of the fuzzy static cooperative game when the characteristic function $\mathbf{u}$ is concave, continuous and positively homogenous. □

6. The viability/capturability strategy

6.1 The epigraphical approach

The *epigraph* of an extended function $\mathbf{v} : X \mapsto \mathbf{R} \cup \{+\infty$ is defined by

$$\mathcal{E}p(\mathbf{v}) := \{(\chi, \lambda) \in X \times \mathbf{R} | \mathbf{v}(\chi) \leq \lambda\}$$

We recall that *an extended function* $\mathbf{v}$ *is convex (resp. positively homogeneous) if and only if its epigraph is convex (resp. a cone)* and that the epigraph of $\mathbf{v}$ is closed if and only if $\mathbf{v}$ is lower semicontinuous:

$$\forall\, \chi \in X, \quad \mathbf{v}(\chi) = \liminf_{y \to x} \mathbf{v}(y)$$

With these definitions, we can translate the viability/capturability conditions (1.5) in the following geometric form:

$$\left\{ \begin{array}{ll} i) & \forall\, t \in [0,t^\star], \quad (T-t, \chi(t), y(t)) \in \mathcal{E}p(\mathbf{b}) \\ & \text{(viability constraint)} \\ ii) & (T-t^\star, \chi(t^\star), y(t^\star)) \in \mathcal{E}p(\mathbf{c}) \\ & \text{(capturability of a target)} \end{array} \right. \tag{1.12}$$

This "epigraphical approach" proposed by J.-J. Moreau and R.T. Rockafellar in convex analysis in the early 60's [10], has been used in optimal control by H. Frankowska in a series of papers Frankowska (1989a,b, 1993) and Aubin and Frankowska (1996) for studying the value function of optimal control problems and characterize it as generalized solution (episolutions and/or viscosity solutions) of (first-order) Hamilton-Jacobi-Bellman equations, in Aubin (1981c); Aubin and Cellina (1984) Aubin (1986, 1991) for characterizing and constructing Lyapunov functions, in Cardaliaguet (1994, 1996, 1997, 2000) for characterizing the minimal time function, in Pujal (2000) and Aubin, Pujal and Saint-Pierre (2001) in finance and other authors since. This is this approach that we adopt and adapt here, since the concepts of "capturability of

[10]see for instance Aubin and Frankowska (1990) and Rockafellar and Wets (1997) among many other references.

a target" and of "viability" of a constrained set allows us to study this problem under a new light (see for instance Aubin (1991, 1997) for economic applications) for studying the evolution of the state of a tychastic control system subjected to viability constraints in control theory and in dynamical games against nature or robust control (see Quincampoix (1992), Cardaliaguet (1994, 1996, 1997, 2000), Cardaliaguet, Quincampoix and Saint-Pierre (1999). Numerical algorithms for finding viability kernels have been designed in Saint-Pierre (1994) and adapted to our type of problems in Pujal (2000).

The properties and characterizations of the valuation function are thus derived from the ones of guaranteed viable-capture basins, that are easier to study — and that have been studied — in the framework of plain constrained sets K and targets $C \subset K$ (see Aubin (2001a, 2002) and Aubin and Catté (2002) for recent results on that topic).

6.2 Introducing auxiliary dynamical games

We observe that the evolution of $(T - t, \chi(t), y(t))$ made up of the backward time $\tau(t) := T - t$, of fuzzy coalitions $\chi(t)$ of the players, of imputations and of the payoff $y(t)$ is governed by the dynamical game

$$\begin{cases} i) & \tau'(t) = -1 \\ ii) & \forall\, i = 0, \ldots, n, \quad \chi_i'(t) = f_i(\chi(t), v(t)) \\ iii) & y'(t) = -y(t)\mathbf{m}(\chi(t), p(t), v(t)) + \langle p(t), f(\chi(t), v(t))\rangle \\ & \text{where } p(t) \in P(\chi(t)) \;\&\; v(t) \in Q(\chi(t)) \end{cases} \tag{1.13}$$

starting at (T, χ, y). We summarize it in the form of the dynamical game

$$\begin{cases} i) & z'(t) \in g(z(t), u(t), v(t)) \\ ii) & u(t) \in P(z(t)) \;\&\; v(t) \in Q(z(t)) \end{cases}$$

where $z := (\tau, \chi, y) \in \mathbf{R} \times \mathbf{R}^n \times \mathbf{R}$, where the controls $u := p$ are the imputations, where the map $g : \mathbf{R} \times \mathbf{R}^n \times \mathbf{R} \rightsquigarrow \mathbf{R} \times \mathbf{R}^n \times \mathbf{R}^n \times \mathbf{R}$ is defined by

$$g(z, v) = (-1, f(\chi, v), u, -\mathbf{m}(\chi, u, v)y + \langle u, f(\chi, v)\rangle)$$

where u ranges over $P(z) := P(\chi)$ and v over $Q(z) := Q(\chi)$.

We say that a selection $z \mapsto \widetilde{p}(z) \in P(z)$ is a feedback, regarded as a strategy. One associates with such a feedback chosen by the decision maker or the player the evolutions governed by the tychastic differential equation

$$z'(t) = g(z(t), \widetilde{p}(z(t)), v(t))$$

starting at time 0 at z.

6.3 Introducing guaranteed capture basins

We now define the guaranteed viable-capture basin that are involved in the definition of guaranteed valuation subsets.

DEFINITION 1.4 *Let K and $C \subset K$ be two subsets of Z.*

The guaranteed viable-capture basin *of the target C viable in K is the set of elements $z \in K$ such that there exists a continuous feedback $\widetilde{p}(z) \in P(z)$ such that for every $v(\cdot) \in Q(z(\cdot))$, for every solutions $z(\cdot)$ to $z' = g(z, \widetilde{p}(z), v)$, there exists $t^* \in \mathbf{R}_+$ such that the viability/capturability conditions*

$$\begin{cases} i) & \forall t \in [0, t^*], \quad z(t) \in K \\ ii) & \qquad\qquad\quad\; z(t^*) \in C \end{cases}$$

are satisfied.

6.4 The strategy

We thus observe that

PROPOSITION 1.1 *The guaranteed valuation subset $\mathcal{V}^\sharp$ is the guaranteed viable-capture basin under the dynamical game (1.13) of the epigraph of the function* **c** *viable in the epigraph of the function* **b**.

Since we have related the guaranteed valuation problem to the much simpler — although more abstract — study of guaranteed viable-capture basin of a target and other guaranteed viability/capturability issues for dynamical games,

1. we first "solve" these "viability/capturability problems" for dynamical games at this general level, and in particular, study the tangential conditions enjoyed by the guaranteed viable-capture basins,

2. and use set-valued analysis and nonsmooth analysis for translating the general results of viability theory to the corresponding results of the auxiliary dynamical game, in particular translating tangential conditions to give a meaning to the concept of a generalized solution (Frankowska's episolutions or, by duality, viscosity solutions) to Hamilton-Jacobi-Isaacs variational inequalities.

References

Aubin, J.-P. (2004). Dynamic core of fuzzy dynamical cooperative games. *Annals of Dynamic Games*, Ninth International Symposium on Dynamical Games and Applications, Adelaide, 2000.

Aubin, J.-P. (2002). Boundary-value problems for systems of Hamilton-Jacobi-Bellman inclusions with constraints. *SIAM Journal on Control Optimization*, 41(2):425–456.

Aubin, J.-P. (2001a). Viability kernels and capture basins of sets under differential inclusions. *SIAM Journal on Control Optimization*, 40:853–881.

Aubin, J.-P. (2001b). Regulation of the evolution of the architecture of a network by connectionist tensors operating on coalitions of actors. Preprint.

Aubin, J.-P. (1999). *Mutational and Morphological Analysis: Tools for Shape Regulation and Morphogenesis*. Birkhäuser.

Aubin, J.-P. (1998, 1993). *Optima and equilibria*. Second edition, Springer-Verlag.

Aubin, J.-P. (1998a). Connectionist complexity and its evolution. In: *Equations aux dérivées partielles, Articles dédiés à J.-L. Lions*, pages 50–79, Elsevier.

Aubin, J.-P. (1998b). Minimal complexity and maximal decentralization. In: H.J. Beckmann, B. Johansson, F. Snickars and D. Thord (eds.), *Knowledge and Information in a Dynamic Economy*, pages 83–104, Springer.

Aubin, J.-P. (1997). *Dynamic Economic Theory: A Viability Approach*, Springer-Verlag.

Aubin, J.-P. (1996). *Neural Networks and Qualitative Physics: A Viability Approach*. Cambridge University Press.

Aubin, J.-P. (1995). Learning as adaptive control of synaptic matrices. In: M. Arbib (ed.), *The Handbook of Brain Theory and Neural Networks*, pages 527–530, Bradford Books and MIT Press.

Aubin, J.-P. (1993). Beyond neural networks: Cognitive systems. In: J. Demongeot and Capasso (eds.), *Mathematics Applied to Biology and Medicine*, Wuers, Winnipeg.

Aubin, J.-P. (1991). *Viability Theory*. Birkhäuser, Boston, Basel, Berlin.

Aubin, J.-P. (1986). A viability approach to Lyapunov's second method. In: A. Kurzhanski and K. Sigmund (eds.), *Dynamical Systems, Lectures Notes in Economics and Math. Systems*, 287:31–38, Springer-Verlag.

Aubin, J.-P. (1982). An alternative mathematical description of a player in game theory. *IIASA WP-82-122*.

Aubin, J.-P. (1981a). Cooperative fuzzy games. *Mathematics of Operations Research*, 6:1–13.

Aubin, J.-P. (1981b). Locally Lipchitz cooperative games. *Journal of Mathematical Economics*, 8:241–262.

Aubin, J.-P. (1981c). Contingent derivatives of set-valued maps and existence of solutions to nonlinear inclusions and differential inclusions. In: Nachbin L. (ed.), *Advances in Mathematics, Supplementary Studies*, pages 160–232.

Aubin, J.-P. (1979). Mathematical methods of game and economic theory. *North-Holland, Studies in Mathematics and its applications*, 7: 1–619.

Aubin, J.-P.and Burnod, Y. (1998). Hebbian Learning in Neural Networks with Gates. *Cahiers du Centre de Recherche Viabilité, Jeux, Contrôle*, # 981.

Aubin, J.-P. and Catté, F. (2002). Bilateral fixed-points and algebraic properties of viability kernels and capture basins of sets. *Set-Valued Analysis*, 10(4):379–416.

Aubin, J.-P. and Cellina, A. (1984). *Differential Inclusions.* Springer-Verlag.

Aubin, J.-P. and Dordan, O. (1996). Fuzzy systems, viability theory and toll sets. In: H. Nguyen (ed.), *Handbook of Fuzzy Systems, Modeling and Control*, pages 461–488, Kluwer.

Aubin, J.-P. and Foray, D. (1998). The emergence of network organizations in processes of technological choice: A viability approach. In: P. Cohendet, P. Llerena, H. Stahn and G. Umbhauer (eds.), *The Economics of Networks*, pages 283–290, Springer.

Aubin, J.-P. and Frankowska, H. (1990). *Set-Valued Analysis.* Birkhäuser, Boston, Basel, Berlin.

Aubin, J.-P. and Frankowska, H. (1996). The viability kernel algorithm for computing value functions of infinite horizon optimal control problems. *Journal of Mathematical Analysis and Applications*, 201:555–576.

Aubin, J.-P., Louis-Guerin, C., and Zavalloni, M. (1979). Comptabilité entre conduites sociales réelles dans les groupes et les représentations symboliques de ces groupes : un essai de formalisation mathématique. *Mathmatiques et Sciences Humaines*, 68:27–61.

Aubin, J.-P., Pujal, D., and Saint-Pierre, P. (2001). *Dynamic Management of Portfolios with Transaction Costs under Tychastic Uncertainty.* Preprint.

Basile, A. (1993). Finitely additive nonatomic coalition production economies: Core-Walras equivalence. *International Economic Review*, 34: 993–995.

Basile, A. (1994). Finitely additive correpondences. *Proceedings of the American Mathematical Society*, 121:883–891.

Basile, A. (1995). On the range of certain additive correspondences. *Universita di Napoli*, (to appear).

Basile, A., De Simone, A., and Graziano, M.G. (1996). On the Aubin-like characterization of competitive equilibria in infinite-dimensional economies. *Rivista di Matematica per le Scienze Economiche e Sociali*, 19:187–213.

Bonneuil, N. (2001). History, differential inclusions and Narrative. *History and Theory*, Theme issue 40, *Agency after Postmodernism*, pages 101–115, Wesleyan University.

Bonneuil, N. (1998a). Games, equilibria, and population regulation under viability constraints: An interpretation of the work of the anthropologist Fredrik Barth. *Population: An English selection, special issue of New Methods in Demography*, pages 151–179.

Bonneuil, N. (1998b). Population paths implied by the mean number of pairwise nucleotide differences among mitochondrial sequences. *Annals of Human Genetics*, 62:61–73.

Bonneuil, N. (2005). Possible coalescent populations implied by pairwise comparisons of mitochondrial DNA sequences. *Theoretical Population Biology*, (to appear).

Bonneuil, N. (2000). Viability in dynamic social networks. *Journal of Mathematical Sociology*, 24:175–182.

Bonneuil, N., and Saint-Pierre, P. (2000). Protected polymorphism in the theo-locus haploid model with unpredictable firnesses. *Journal of Mathematical Biology*, 40:251–377.

Bonneuil, N. and Saint-Pierre, P. (1998). Domaine de victoire et stratégies viable dans le cas d'une correspondance non convexe : application à l'anthropologie des pêcheurs selon Fredrik Barth. *Mathématiques et Sciences Humaines*, 132:43–66.

Cardaliaguet, P. (1994). Domaines dicriminants en jeux différentiels. Thèse de l'Université de Paris-Dauphine.

Cardaliaguet, P. (1996). A differential game with two players and one target. *SIAM Journal on Control and Optimization*, 34(4):1441–1460.

Cardaliaguet, P. (1997). On the regularity of semi-permeable surfaces in control theory with application to the optimal exit-time problem (Part II). *SIAM Journal on Control Optimization*, 35(5):1638–1652.

Cardaliaguet, P. (2000). Introduction à la théorie des jeux différentiels. *Lecture Notes*, Université Paris-Dauphine.

Cardaliaguet, P., Quincampoix, M., and Saint-Pierre, P. (1999). Set-valued numerical methods for optimal control and differential games. In: *Stochastic and differential games. Theory and Numerical Methods, Annals of the International Society of Dynamical Games*, pages 177–247, Birkhäuser.

Cornet, B. (1976). On planning procedures defined by multivalued differential equations. *Système dynamiques et modèles économiques (C.N.R.S.)*.

Cornet, B. (1983). An existence theorem of slow solutions for a class of differential inclusions. *Journal of Mathematical Analysis and Applications*, 96:130–147.

Filar, J.A. and Petrosjan, L.A. (2000). Dynamic cooperative games. *International Game Theory Review*, 2:47–65.

Florenzano, M. (1990). Edgeworth equilibria, fuzzy core and equilibria of a production economy without ordered preferences. *Journal of Mathematical Analysis and Applications*, 153:18–36.

Frankowska, H. (1987a). L'équation d'Hamilton-Jacobi contingente. *Comptes-Rendus de l'Académie des Sciences*, PARIS, 1(304):295–298.

Frankowska, H. (1987b). Optimal trajectories associated to a solution of contingent Hamilton-Jacobi equations. *IEEE, 26th, CDC Conference*, Los Angeles.

Frankowska, H. (1989a). Optimal trajectories associated to a solution of contingent Hamilton-Jacobi equations. *Applied Mathematics and Optimization*, 19:291–311.

Frankowska, H. (1989b). Hamilton-Jacobi equation: Viscosity solutions and generalized gradients. *Journal of Mathematical Analysis and Applications*, 141:21–26.

Frankowska, H. (1993). Lower semicontinuous solutions of Hamilton-Jacobi-Bellman equation. *SIAM Journal on Control Optimization*, 31(1):257–272.

Haurie, A. (1975). On some properties of the characteristic function and core of multistage game of coalitions. *IEEE Transactions on Automatic Control*, 20:238–241.

Henry, C. (1972). Differential equations with discontinuous right hand side. *Journal of Economic Theory*, 4:545–551.

Mares, M. (2001). *Fuzzy Cooperative Games. Cooperation with Vague Expectations*. Physica Verlag.

Mishizaki, I. and Sokawa, M. (2001). *Fuzzy and Multiobjective Games for Conflict Resolution*. Physica Verlag.

Petrosjan, L.A. (1996). The time consistency (dynamic stability) in differential games with a discount factor. *Game Theory and Applications*, pages 47-53, Nova Sci. Publ., Commack, NY.

Petrosjan, L.A. and Zenkevitch, N.A. (1996). *Game Theory*. World Scientific, Singapore.

Pujal, D. (2000). Valuation et gestion dynamiques de portefeuilles. Thèse de l'Université de Paris-Dauphine.

Quincampoix, M. (1992). Differential inclusions and target problems. *SIAM Journal on Control Optimization*, 30:324–335.

Rockafellar, R.T. and Wets, R. (1997). *Variational Analysis*. Springer-Verlag

Saint-Pierre, P. (1994). Approximation of the viability kernel. *Applied Mathematics & Optimisation*, 29:187–209.

Chapter 2

A DIRECT METHOD FOR OPEN-LOOP DYNAMIC GAMES FOR AFFINE CONTROL SYSTEMS

Dean A. Carlson
George Leitmann

Abstract Recently in Carlson and Leitmann (2004) some improvements on Leitmann's direct method, first presented for problems in the calculus of variations in Leitmann (1967), for open-loop dynamic games in Dockner and Leitmann (2001) were given. In these papers each player has its own state which it controls with its own control inputs. That is, there is a state equation for each player. However, many applications involve the players competing for a single resource (e.g., two countries competing for a single species of fish). In this note we investigate the utility of the direct method for a class of games whose dynamics are described by a single equation for which the state dynamics are affine in the players strategies. An illustrative example is also presented

1. The direct method

In Carlson and Leitmann (2004) a direct method for finding open-loop Nash equilibria for a class of differential N-player games is presented. A particular case included in this study concerns the situation in which the j-th player's dynamics at any time $t \in [t_0, t_f]$ is a vector-valued function $t \to x_j(t) \in \mathbb{R}^{n_j}$ that is described by an ordinary control system of the form

$$\dot{x}_j(t) = f_j(t, \mathbf{x}(t)) \quad + \quad g_j(t, \mathbf{x}(t)) u_j(t) \text{ a.e. } t_0 \leq t \leq t_f \tag{2.1}$$

$$x_j(t_0) = x_{jt_0} \quad \text{and} \quad x_j(t_f) = x_{jt_f} \tag{2.2}$$

with control constraints

$$u_j(t) \in U_j(t) \subset \mathbb{R}^{m_j} \quad \text{a.e.} \quad t \in [t_0, t_f], \tag{2.3}$$

and state constraints

$$\mathbf{x}(t) \in \mathbf{X}(t) \subset \mathbb{R}^{\mathbf{n}} \quad \text{for} \quad t \in [t_0, t_f], \tag{2.4}$$

in which for each $j = 1, 2, \ldots N$ the function $f_j(\cdot, \cdot) : [t_0, t_f] \times \mathbb{R}^{\mathbf{n}} \to \mathbb{R}^{n_j}$ is continuous, $g_j(\cdot, \cdot) : [t_0, t_f] \times \mathbb{R}^{m_j \times n_j}$ is a continuous $m_j \times n_j$ matrix-valued function having a left inverse, and $U_j(\cdot)$ is set-valued mapping, and $\mathbf{X}(t)$ is a given set in $\mathbb{R}^{\mathbf{n}}$ for $t \in [t_0, t_f]$. Here we use the notation $\mathbf{x} = (x_1, x_2, \ldots, x_N) \in \mathbb{R}^{n_1} \times \mathbb{R}^{n_2} \times \mathbb{R}^{n_N} = \mathbb{R}^{\mathbf{n}}$, where $\mathbf{n} = n_1 + n_2 + \ldots + n_N$; similarly $\mathbf{u} = (u_1, u_2 \ldots, u_N) \in \mathbb{R}^{\mathbf{m}}$, $\mathbf{m} = m_1 + m_2 \ldots + m_N$. Additionally we assume that the sets, $M_j = \{(t, \mathbf{x}, u_j) \in [t_0, t_f] \times \mathbb{R}^{\mathbf{n}} \times \mathbb{R}^{m_j} : \quad u_j \in U_j(t)\}$ are closed and nonempty. The objective of each player is to minimize an objective function of the form,

$$J_j(\mathbf{x}(\cdot), u_j(\cdot)) = \int_{t_0}^{t_f} f_j^0(t, \mathbf{x}(t), u_j(t))\, dt, \tag{2.5}$$

where we assume that for each $j = 1, 2, \ldots, N$ the function $f_j^0(\cdot, \cdot, \cdot) : M_j \times \mathbb{R}^{\mathbf{n}} \times \mathbb{R}^{m_j}$ is continuous.

With the above model description we now define the feasible set of admissible trajectory-strategy pairs.

Definition 2.1 *We say a pair of functions* $\{\mathbf{x}(\cdot), \mathbf{u}(\cdot)\} : [t_0, t_f] \to \mathbb{R}^{\mathbf{n}} \times \mathbb{R}^{\mathbf{m}}$ *is an admissible trajectory-strategy pair iff* $t \to \mathbf{x}(t)$ *is absolutely continuous on* $[t_0, t_f]$, $t \to \mathbf{u}(t)$ *is Lebesgue measurable on* $[t_0, t_f]$, *for each* $j = 1, 2, \ldots, N$, *the relations (2.1)–(2.3) are satisfied, and for each* $j = 1, 2, \ldots, N$, *the functionals (2.5) are finite Lebesgue integrals.*

Remark 2.1 For brevity we will refer to an admissible trajectory-strategy pair as an admissible pair. Also, for a given admissible pair, $\{\mathbf{x}(\cdot), \mathbf{u}(\cdot)\}$, we will follow the traditional convention and refer to $\mathbf{x}(\cdot)$ as an admissible trajectory and $\mathbf{u}(\cdot)$ as an admissible strategy.

For a fixed $j = 1, 2, \ldots, N$, $\mathbf{x} \in \mathbb{R}^{\mathbf{n}}$, and $y_j \in \mathbb{R}^{n_j}$ we use the notation $[\mathbf{x}^j, y_j]$ to denote a new vector in $\mathbb{R}^{\mathbf{n}}$ in which $x_j \in \mathbb{R}^{n_j}$ is replaced by $y_j \in \mathbb{R}^{n_j}$. That is,

$$[\mathbf{x}^j, y_j] \doteq (x_1, x_2, \ldots, x_{j-1}, y_j, x_{j+1}, \ldots, x_N).$$

Analogously $[\mathbf{u}^j, v_j] \doteq (u_1, u_2, \ldots, u_{j-1}, v_j, u_{j+1}, \ldots, u_N)$ for all $\mathbf{u} \in \mathbb{R}^{\mathbf{m}}$, $v_j \in \mathbb{R}^{m_j}$, and $j = 1, 2, \ldots, N$. With this notation we now have the following two definitions.

Definition 2.2 *Let* $j = 1, 2, \ldots, N$ *be fixed and let* $\{\mathbf{x}(\cdot), \mathbf{u}(\cdot)\}$ *be an admissible pair. We say that the pair of functions* $\{y_j(\cdot), v_j(\cdot)\} : [t_0, t_f]$

$\to \mathbb{R}^{n_j} \times \mathbb{R}^{m_j}$ is an admissible trajectory-strategy pair for player j relative to $\{\mathbf{x}(\cdot), \mathbf{u}(\cdot)\}$ iff the pair

$$\{[\mathbf{x}(\cdot)^j, y_j(\cdot)], [\mathbf{u}(\cdot)^j, v_j(\cdot)]\}$$

is an admissible pair.

DEFINITION 2.3 *An admissible pair $\{\mathbf{x}^*(\cdot), \mathbf{u}^*(\cdot)\}$ is a Nash equilibrium iff for each $j = 1, 2, \ldots, N$ and each pair $\{y_j(\cdot), v_j(\cdot)\}$ that is admissible for player j relative to $\{\mathbf{x}^*(\cdot), \mathbf{u}^*(\cdot)\}$, it is the case that*

$$\begin{aligned} J_j(\mathbf{x}^*(\cdot), u_j^*(\cdot)) &= \int_{t_0}^{t_f} f_j^0(t, \mathbf{x}^*(t), u_j^*(t))\, dt \\ &\leq \int_{t_0}^{t_f} f_j^0(t, [\mathbf{x}^*(t)^j, y_j(t)], v_j(t))\, dt \\ &= J_j([\mathbf{x}^*(\cdot)^j, y_j(\cdot)], v_j(\cdot)). \end{aligned}$$

Our goal in this paper is to provide a "direct method" which in some cases will enable us to determine a Nash equilibrium. We point out that relative to a fixed Nash equilibrium $\{\mathbf{x}^*(\cdot), \mathbf{u}^*(\cdot)\}$ each of the players in the above game solves an optimization problem taking the form of a standard problem of optimal control. Thus, under suitable additional assumptions, it is relatively easy to derive a set of necessary conditions (in the form of a Pontryagin-type maximum principle) that must be satisfied by all Nash equilibria. Unfortunately these conditions are only necessary and not sufficient. Further, it is well known that non-uniqueness is always a source of difficulty in dynamic games so that in general the necessary conditions are not uniquely solvable (as is often the case in optimal control theory, when sufficient convexity is imposed). Therefore it is important to be able to find usable sufficient conditions for Nash equilibria.

The associated variational game

We observe that, under our assumptions, the algebraic equations,

$$z_j = f_j(t, \mathbf{x}) + g_j(t, \mathbf{x})u_j \quad j = 1, 2, \ldots N, \tag{2.6}$$

can be solved for u_j in terms of t, z_j, and $\mathbf{x}$ to obtain

$$u_j = g_j(t, \mathbf{x})^{-1} (z_j - f_j(t, \mathbf{x})), \quad j = 1, 2, \ldots N, \tag{2.7}$$

where $g_j(t, \mathbf{x})^{-1}$ denotes the inverse of the matrix $g_j(t, \mathbf{x})$. As a consequence we can define the extended real-valued functions $L_j(\cdot, \cdot, \cdot)$: $[t_0, t_f] \times \mathbb{R}^{\mathbf{n}} \times \mathbb{R}^{n_j} \to \mathbb{R} \cup +\infty$ as

$$L_j(t, \mathbf{x}, z_j) = f_j^0(t, \mathbf{x}, g_j(t, \mathbf{x})^{-1}(z_j - f_j(t, \mathbf{x}))) \tag{2.8}$$

if $g_j(t,\mathbf{x})^{-1}(z_j - f_j(t,\mathbf{x})) \in U_j(t)$ with $L_j(t,\mathbf{x},z_j) = +\infty$ otherwise.

With these functions we can consider the N-player variational game in which the objective functional for the jth player is defined by,

$$I_j(\mathbf{x}(\cdot)) = \int_{t_0}^{t_f} L_j(t,\mathbf{x}(t),\dot{x}_j(t))\,dt. \tag{2.9}$$

With this notation we have the following additional definitions.

DEFINITION 2.4 *An absolutely continuous function $\mathbf{x}(\cdot) : [t_0, t_f] \to \mathbb{R}^{\mathbf{n}}$ is said to be admissible for the variational game iff it satisfies the boundary conditions given in equation (2.2) and such that the map $t \to L_j(t,\mathbf{x}(t),\dot{x}_j(t))$ is finitely Lebesgue integrable on $[t_0, t_f]$ for each $j = 1, 2, \ldots, N$.*

DEFINITION 2.5 *Let $\mathbf{x}(\cdot) : [t_0, t_f] \to \mathbb{R}^{\mathbf{n}}$ be admissible for the variational game and let $j \in \{1, 2, \ldots, N\}$ be fixed. We say that $y_j(\cdot) : [t_0, t_f] \to \mathbb{R}^{n_j}$ is admissible for player j relative to $\mathbf{x}(\cdot)$ iff $[\mathbf{x}^j(\cdot), y_j(\cdot)]$ is admissible for the variational game.*

DEFINITION 2.6 *We say that $\mathbf{x}^*(\cdot) : [t_0, t_f] \to \mathbb{R}^{\mathbf{n}}$ is a Nash equilibrium for the variational game iff for each $j = 1, 2, \ldots, N$,*

$$I_j(\mathbf{x}^*(\cdot)) \le I_j([\mathbf{x}^{*j}(\cdot), y_j(\cdot)])$$

for all functions $y_j(\cdot) : [t_0, t_f] \to \mathbb{R}^{n_j}$ that are admissible for player j relative to $\mathbf{x}^*(\cdot)$.

Clearly the variational game and our original game are related. In particular we have the following theorem given in Carlson and Leitmann (2004).

THEOREM 2.1 *Let $\mathbf{x}^*(\cdot)$ be a Nash equilibrium for the variational game defined above. Then there exists a measurable function $\mathbf{u}^*(\cdot) : [t_0, t_f] \to \mathbb{R}^{\mathbf{m}}$ such that the pair $\{\mathbf{x}^*(\cdot), \mathbf{u}^*(\cdot)\}$ is an admissible trajectory-strategy pair for the original dynamic game. Moreover, it is a Nash equilibrium for the original game as well.*

Proof. See Carlson and Leitmann (2004), Theorem 7.1. □

REMARK 2.2 The above result holds in a much more general setting than indicated above. We chose the restricted setting since it is sufficient for our needs in the analysis of the model we will consider in the next section.

With the above result we now focus our attention on the variational game. In 1967, for the case of one player variational games (i.e., the calculus of variations), Leitmann (1967) presented a technique (the "direct method") for determining solutions of these games by comparing their solutions to that of an equivalent problem whose solution is more easily determined than that of the original game. This equivalence was obtained through a coordinate transformation. Since then this method has been used successfully to solve a variety of problems. Recently, Carlson (2002) presented an extension of this method that expands the utility of the approach as well as made some useful comparisons with a technique originally presented by Carathéodory in the early twentieth century (see Carathéodory (1982)). Also, Dockner and Leitmann (2001) extended the original direct method to include the case of open-loop dynamic games. Finally, the extension of Carlson to the method was also modified in Leitmann (2004) to the include the case of open-loop differential games in Carlson and Leitmann (2004).

We begin by stating the following lemma found in Carlson and Leitmann (2004).

LEMMA 2.1 *Let $x_j = z_j(t, \tilde{x}_j)$ be a transformation of class C^1 having a unique inverse $\tilde{x}_j = \tilde{z}_j(t, x_j)$ for all $t \in [t_0, t_f]$ such that there is a one-to-one correspondence $\mathbf{x}(t) \Leftrightarrow \tilde{\mathbf{x}}(t)$, for all admissible trajectories $\mathbf{x}(\cdot)$ satisfying the boundary conditions (2.2) and for all $\tilde{\mathbf{x}}(\cdot)$ satisfying*

$$\tilde{x}_j(t_0) = \tilde{z}_j(t_0, x_{0j}) \quad \text{and} \quad \tilde{x}_j(t_f) = \tilde{z}_j(t_f, x_{t_f j})$$

for all $j=1,2,\ldots,N$. Furthermore, for each $j=1,2,\ldots,N$ let $\tilde{L}_j(\cdot,\cdot,\cdot) : [t_0, t_f] \times \mathbb{R}^{\mathbf{n}} \times \mathbb{R}^{n_j} \to \mathbb{R}$ be a given integrand. For a given admissible $\mathbf{x}^*(\cdot) : [t_0, t_f] \to \mathbb{R}^{\mathbf{n}}$ suppose the transformations $x_j = z_j(t, \tilde{x}_j)$ are such that there exists a C^1 function $H_j(\cdot,\cdot) : [t_0, t_f] \times \mathbb{R}^{n_j} \to \mathbb{R}$ so that the functional identity

$$\begin{aligned} L_j(t, [\mathbf{x}^{*j}(t), x_j(t)], \dot{x}_j(t)) \quad &- \quad \tilde{L}_j(t, [\mathbf{x}^{*j}(t), \tilde{x}_j(t)], \dot{\tilde{x}}_j(t)) \\ &= \quad \frac{d}{dt} H_j(t, \tilde{x}_j(t)) \end{aligned} \tag{2.10}$$

holds on $[t_0, t_f]$. If $\tilde{x}_j^*(\cdot)$ yields an extremum of $\tilde{I}_j([\mathbf{x}^{*j}(\cdot), \cdot])$ with $\tilde{x}_j^*(\cdot)$ satisfying the transformed boundary conditions, then $x_j^*(\cdot)$ with $x_j^*(t) = z_j(t, \tilde{x}^*(t))$ yields an extremum for $I_j([\mathbf{x}^{*j}(\cdot), \cdot])$ with the boundary conditions (2.2).

Moreover, the function $\mathbf{x}^*(\cdot)$ is an open-loop Nash equilibrium for the variational game.

Proof. See Carlson and Leitmann (2004), Lemma 5.1. □

This lemma has three useful corollaries which we state below.

COROLLARY 2.1 *The existence of $H_j(\cdot,\cdot)$ in (2.9) implies that the following identities hold for $(t,\tilde{x}_j) \in (t_0,t_f) \times \mathbb{R}^{n_j}$ and for $j = 1,2,\ldots,N$:*

$$\begin{aligned} L_j(t,[\mathbf{x}^{*j}(t), z_j(t,\tilde{x}_j)], \frac{\partial z_j(t,\tilde{x}_j))}{\partial t} &+ \langle \nabla_{\tilde{x}_j} z_j(t,\tilde{x}_j), \tilde{p}_j\rangle) \\ -\tilde{L}_j(t,[\mathbf{x}^{*j}(t),\tilde{x}_j],\tilde{p}_j) &\equiv \frac{\partial H_j(t,\tilde{x}_j)}{\partial t} + \langle \nabla_{\tilde{x}_j} H_j(t,\tilde{x}_j), \tilde{p}_j\rangle, \end{aligned} \tag{2.11}$$

in which $\nabla_{\tilde{x}_j} H_j(\cdot,\cdot)$ denotes the gradient of $H_j(\cdot,\cdot)$ with respect to the variables $\tilde{x}_j$ and $\langle\cdot,\cdot\rangle$ denotes the usual scalar or inner product in $\mathbb{R}^{n_j}$.

COROLLARY 2.2 *For each $j = 1,2,\ldots,N$ the left-hand side of the identity, (2.11) is linear in $\tilde{p}_j$, that is, it is of the form,*

$$\theta_j(t,\tilde{x}_j) + \langle \psi_j(t,\tilde{x}_j), \tilde{p}_j\rangle$$

and,

$$\frac{\partial H_j(t,\tilde{x}_j)}{\partial t} = \theta_j(t,\tilde{x}_j) \quad \text{and} \quad \nabla_{\tilde{x}_j} H_j(t,\tilde{x}_j) = \psi(t,\tilde{x}_j)$$

on $[t_0,t_f] \times \mathbb{R}^{n_j}$.

COROLLARY 2.3 *For integrands $L_j(\cdot,\cdot,\cdot)$ of the form,*

$$\begin{aligned} L_j(t,[\mathbf{x}^{*j}(t),x_j(t)],\dot{x}_j(t)) &= \dot{x}_j'(t)a_j(t,[\mathbf{x}^{*j}(t),x_j(t)])\dot{x}_j(t) \\ &\quad + b_j(t,[\mathbf{x}^{*j}(t),x_j(t)])'\dot{x}_j(t) \\ &\quad + c_j(t,[\mathbf{x}^{*j}(t),x_j(t)]), \end{aligned}$$

and

$$\begin{aligned} \tilde{L}_j(t,[\mathbf{x}^{*j}(t),x_j(t)],\dot{x}_j(t)) &= \dot{x}_j'(t)\alpha_j(t,[\mathbf{x}^{*j}(t),x_j(t)])\dot{x}_j(t) \\ &\quad + \beta_j(t,[\mathbf{x}^{*j}(t),x_j(t)])'\dot{x}_j(t) \\ &\quad + \gamma_j(t,[\mathbf{x}^{*j}(t),x_j(t)]), \end{aligned}$$

with $a_j(t,[\mathbf{x}^{*j}(t),x_j(t)]) \neq 0$ and $\alpha_j(t,[\mathbf{x}^{*j}(t),x_j(t)]) \neq 0$, the class of transformations that permit us to obtain (2.11) must satisfy,

$$\left[\frac{\partial z_j(t,\tilde{x}_j)}{\partial \tilde{x}_j}\right]' a_j(t,[\mathbf{x}^*(t)^j, z_j(t,\tilde{x}_j)]) \left[\frac{\partial z_j(t,\tilde{x}_j)}{\partial \tilde{x}_j}\right] = \alpha_j(t,[\mathbf{x}^*(t)^j,\tilde{x}_j])$$

for $(t,x_j) \in [t_0,t_1] \times \mathbb{R}^{n_j}$.

A class of dynamic games to which the above method has not been applied is that in which there is a single state equation which is controlled by all of the players. A simple example of such a problem is the competitive harvesting of a renewable resource (e.g., a single species fishery model). In the next section we show how the direct method described above can be applied to a class of these types of models.

2. The model

Consider an N-player game where a single state $x(t) \in \mathbb{R}^n$ satisfies an ordinary control system of the form

$$\dot{x}(t) = F(t, x(t)) + \sum_{i=1}^{N} G_i(t, x(t))u_i(t) \quad \text{a.e.} \quad t_0 \leq t \leq t_f, \tag{2.12}$$

with initial and terminal conditions

$$x(t_0) = x_{t_0} \quad \text{and} \quad x(t_f) = x_{t_f}, \tag{2.13}$$

a fixed state constraint,

$$x(t) \in X(t) \subset \mathbb{R}^n \quad \text{for} \quad t_0 \leq t \leq t_f, \tag{2.14}$$

with $X(t)$ a convex set for each $t_0 \leq t \leq t_f$, and control constraints,

$$u_i(t) \in U_i(t) \subset \mathbb{R}^{m_i} \quad \text{a.e.} \quad t_0 \leq t \leq t_f \quad i = 1, 2, \ldots N. \tag{2.15}$$

In this system each player has a strategy, $u_i(\cdot)$, which influences the state variable $x(\cdot)$ over time.

DEFINITION 2.7 *A set of functions*

$$\{x(\cdot), \mathbf{u}(\cdot)\} \doteq \{x(\cdot), u_1(\cdot), u_2(\cdot), \ldots, u_N(\cdot)\}$$

defined for $t_0 \leq t \leq t_f$ is called an admissible trajectory-strategy pair iff $x(\cdot)$ is absolutely continuous on its domain, $\mathbf{u}(\cdot)$ is Lebesgue measurable on its domain, and the equations (2.12)– (2.15) are satisfied.

We assume that $F(\cdot, \cdot) : [t_0, +\infty) \times \mathbb{R}^n \to \mathbb{R}^n$ and $G_i(\cdot, \cdot, \cdot) : [t_0, +\infty) \times \mathbb{R}^n \times \mathbb{R}^{m_j} \to \mathbb{R}^n$ is sufficiently smooth so that for each selection of strategies $\mathbf{u}(\cdot)$ (i.e., measurable functions) the initial value problem given by (2.12)–(2.13) has a unique solution $x_{\mathbf{u}}(\cdot)$. These conditions can be made more explicit for particular models and are not unduly restrictive. For brevity we do not to indicate these explicitly.

Each of the players in the dynamic game wishes to minimize a performance criterion given of the form,

$$J_j(x(\cdot), u_j(\cdot)) = \int_{t_0}^{t_f} f_j(t, x(t), u_j(t))\, dt, \quad j = 1, 2, \ldots, N, \tag{2.16}$$

in which we assume that $f_j(\cdot, \cdot, \cdot) : [t_0, t_f] \times \mathbb{R}^n \times \mathbb{R}^{m_j} \to \mathbb{R}$ is continuous.

To place the above dynamic game into a form amenable to the direct method consider a set of strictly positive weights, say $\alpha_i > 0$,

$i = 1, 2, \ldots N$, which satisfy $\sum_{i=1}^{N} \alpha_i = 1$ and consider the related ordinary control system

$$\dot{x}_i(t) = F\left(t, \sum_{i=1}^{N} \alpha_i x_i(t)\right) + \frac{1}{\alpha_i} G\left(t, \sum_{i=1}^{N} \alpha_i x_i(t)\right) u_i(t) \quad \text{a.e.} \quad t \geq t_0, \tag{2.17}$$

$i = 1, 2, \ldots, N$, with boundary conditions,

$$x_i(t_0) = x_{t_0} \quad \text{and} \quad x_i(t_f) = x_{t_f}, \quad i = 1, 2, \ldots N, \tag{2.18}$$

and control constraints and state constraints,

$$u_i(t) \in U_i(t) \subset \mathbb{R}^{m_i} \quad \text{a.e.} \quad t_0 \leq t \leq t_f \quad i = 1, 2, \ldots N, \tag{2.19}$$

$$x_i(t) \in X_i(t) \doteq X(t) \subset \mathbb{R}^n \quad \text{for} \quad t_0 \leq t \leq t_f \quad i = 1, 2, \ldots N. \tag{2.20}$$

DEFINITION 2.8 *A set of functions*

$$\{\mathbf{x}(\cdot), \mathbf{u}(\cdot)\} \doteq \{x_1(\cdot), x_2(\cdot), \ldots x_N(\cdot), u_1(\cdot), u_2(\cdot), \ldots, u_N(\cdot)\}$$

defined for $t_0 \leq t \leq t_f$ *is called an admissible trajectory-strategy pair for the related system iff* $\mathbf{x}(\cdot) : [t_0, +\infty) \to \mathbb{R}^{\mathbf{n}}$, *where* $\mathbf{n} = nN$, *is absolutely continuous on its domain,* $\mathbf{u}(\cdot) : [t_0, +\infty) \to \mathbb{R}^{\mathbf{m}}$, *where* $\mathbf{m} = m_1 + m_2 + \ldots + m_N$, *is Lebesgue measurable on its domain, and the equations (2.17)–(2.19) are satisfied.*

For this related system it is easy to see that the conditions guaranteeing uniqueness for the original system would also insure the existence of the solution $\mathbf{x}(\cdot)$ for a fixed set of strategies $u_i(\cdot)$.

PROPOSITION 2.1 *Let* $\{\mathbf{x}(\cdot), \mathbf{u}(\cdot)\}$ *be an admissible trajectory-strategy pair for the related control system. Then the pair,* $\{x(\cdot), \mathbf{u}(\cdot)\}$, *with* $x(t) \doteq \sum_{i=1}^{N} \alpha_i x_i(t)$ *is an admissible trajectory-strategy pair for the original control system. Conversely, if* $\{x(\cdot), \mathbf{u}(\cdot)\}$ *is an admissible trajectory-strategy pair for the original control system, then there exists a function* $\mathbf{x}(\cdot) = (x_1(\cdot), \ldots, x_N(\cdot))$ *so that* $x(t) \doteq \sum_{i=1}^{N} \alpha_i x_i(t)$ *for* $i = 1, 2, \ldots N$ *and* $\{\mathbf{x}(\cdot), \mathbf{u}(\cdot)\}$ *is an admissible trajectory-strategy pair for the related control system.*

Proof. We begin by first letting $\{\mathbf{x}(\cdot), \mathbf{u}(\cdot)\}$ be an admissible trajectory-strategy pair for the related control system. Then defining $x(t) = \sum_{i=1}^{N} \alpha_i x_i(t)$ for $t_0 \leq t \leq t_f$ we observe that

$$\dot{x}(t) = \sum_{i=1}^{N} \alpha_i \dot{x}_i(t)$$

$$
\begin{aligned}
&= \sum_{i=1}^{N} \alpha_i \left(F(t, x(t)) + \frac{1}{\alpha_i} G_i(t, x(t)) u_i(t) \right) \\
&= \sum_{i=1}^{N} \alpha_i F(t, x(t)) + \sum_{i=1}^{N} G_i(t, x(t)) u_i(t) \\
&= F(t, x(t)) + \sum_{i=1}^{N} G_i(t, x(t)) u_i(t),
\end{aligned}
$$

since $\sum_{i=1}^{N} \alpha_i = 1$. Further we also have that

$$
\begin{aligned}
x(t_0) &= \sum_{i=1}^{N} \alpha_i x_{t_0} = x_{t_0}, \\
u_j(t) &\in U_j(t) \quad \text{for almost all} \quad t_0 \leq t \leq t_f \quad \text{and} \quad j = 1, 2, \ldots N, \\
x_j(t) &\in X(t) \quad \text{for} \quad t_0 \leq t \leq t_f \quad \text{and} \quad j = 1, 2, \ldots N,
\end{aligned}
$$

implying that $\{x(\cdot), \mathbf{u}(\cdot)\}$ is an admissible trajectory-strategy pair.

Now assume that $\{x(\cdot), \mathbf{u}(\cdot)\}$ is an admissible trajectory-strategy pair for the original dynamical system (2.12–2.15) and consider the system of differential equations given by (2.17) with the initial conditions (2.18). By our hypotheses this system has a unique solution $\mathbf{x}(\cdot) : [t_0, +\infty) \to \mathbb{R}^N$. Furthermore, from the above computation we know that the function, $y(\cdot) \doteq \sum_{i=1}^{N} \alpha_i x_i(\cdot)$, along with the strategies, $\mathbf{u}(\cdot)$ satisfy the differential equation (2.12) as well as the initial condition (2.13). However, this initial value problem has a unique solution, namely $x(\cdot)$, so that we must have $y(t) \equiv x(t)$ for all $t_0 \leq t \leq t_f$. Further, we also have the constraints, (2.19) and (2.20), holding as well. Hence we have, $\{\mathbf{x}(\cdot), \mathbf{u}(\cdot)\}$ is an admissible trajectory-strategy pair for the related system as desired. □

In light of the above theorem it is clear that to use the direct method to solve the dynamic game described by (2.12)–(2.16) we consider the game described by the dynamic equations (2.17)–(2.19) where now the objective for player j, $j = 1, 2, \ldots N$, is given as

$$
\mathcal{J}_j(\mathbf{x}(\cdot), u_j(\cdot)) = \int_{t_0}^{t_f} f_j^0 \left(t, \sum_{i=1}^{N} \alpha_i x(t), u_j(t) \right) dt. \tag{2.21}
$$

In the next section we demonstrate this process with an example from mathematical economics.

REMARK 2.3 In solving constrained optimization or dynamic games problems one of the biggest difficulties is finding reasonable candidates for the solution that meet the constraints. Perhaps the most often used method is to solve the unconstrained problem and hope that it satisfies the constraints. To understand why this technique works we observe that either in a game or in an optimization problem the set of admissible trajectory-strategy pairs that satisfy the constraints is a subset of the set of all admissible for pairs for the problem without constraints. Consequently, if you can find an admissible trajectory-strategy pair which is an optimal (or Nash equilibrium) solution for the problem without constraints (say via the direct method for the unconstrained problem) and if additionally it actually satisfies the constraints you indeed have a solution for the original problem with constraints. It is this technique that is used in the next section to obtain the Nash equilibrium.

3. Example

We consider two firms which produce an identical product. The production cost for each firm is given by the total cost function,

$$C(u_j) = \frac{1}{2}u_j^2, \quad j = 1, 2,$$

in which u_j refers to a jth firm's production level. Each firm supplies all that it produces to the market at all times. The amount supplied at each time effects the price, $P(t)$, and the total inventory of the market determines the price according to the ordinary control system,

$$\dot{P}(t) = s[a - u_1(t) - u_2(t) - P(t)] \quad \text{a.e.} \quad t \in [t_0, t_f]. \tag{2.22}$$

Here $s > 0$ refers to the speed at which the price adjusts to the price corresponding to the total quantity (i.e., $u_1(t) + u_2(t)$). The model assumes a linear demand rate given by $\Pi = a - X$ where X denotes total supply related to a price P. Thus the dynamics above says that the rate of change of price at time t is proportional to the difference between the actual price $P(t)$ and the idealized price $\Pi(t) = a - u_1(t) - u_2(t)$. We assume that (through negotiation perhaps) the firms have agreed to move from the price P_0 at time t_0 to a price P_f at time t_f. This leads to the boundary conditions,

$$P(t_0) = P_0 \quad \text{and} \quad P(t_f) = P_f. \tag{2.23}$$

Additionally we also impose the constraints

$$u_j(t) \geq 0 \quad \text{for almost all} \quad t \in [t_0, t_f]. \tag{2.24}$$

and

$$P(t) \geq 0 \quad \text{for} \quad t \in [t_0, t_f]. \tag{2.25}$$

The goal of each firm is to maximize its accumulated profit, assuming that it sells all that it produces, over the interval, $[t_0, t_f]$ given by the integral functional,

$$J_j(P(\cdot), u_j(\cdot)) = \int_{t_0}^{t_f} \left[P(t)u_j(t) - \frac{1}{2}u_j^2(t) \right] dt. \tag{2.26}$$

To put the above dynamic game into the framework to use the direct method let $\alpha, \beta > 0$ satisfy $\alpha + \beta = 1$ and consider the ordinary 2-dimensional control system,

$$\begin{aligned} \dot{x}(t) &= -s(\alpha x(t) + \beta y(t) - a) - \frac{s}{\alpha}u_1(t), \text{ a.e. } t_0 \leq t \leq t_f \quad (2.27) \\ \dot{y}(t) &= -s(\alpha x(t) + \beta y(t) - a) - \frac{s}{\beta}u_2(t), \text{ a.e. } t_0 \leq t \leq t_f \quad (2.28) \end{aligned}$$

with the boundary conditions,

$$\begin{aligned} x(t_0) &= y(t_0) = P_0 \quad (2.29) \\ x(t_f) &= y(t_f) = P_f, \quad (2.30) \end{aligned}$$

and of course the control constraints given by (2.24) and state constraints (2.25). The payoffs for each of the player now become,

$$J_j(x(\cdot), y(\cdot), u_j(\cdot)) = \int_{t_0}^{t_f} \left[(\alpha x(t) + \beta y(t))u_j(t) - \frac{1}{2}u_j(t)^2 \right] dt \tag{2.31}$$

for $j = 1, 2$. This gives a dynamic game for which the direct method can be applied.

We now put the above game in the equivalent variational form by solving the dynamic equations (2.27) and (2.28) for the individual strategies. That is we have,

$$\begin{aligned} u_1 &= \alpha(a - (\alpha x + \beta y) - \frac{1}{s}p) \quad (2.32) \\ u_2 &= \beta(a - (\alpha x + \beta y) - \frac{1}{s}q) \quad (2.33) \end{aligned}$$

which gives (after a number of elementary steps of algebra) the new objectives (with negative sign to pose the variational problems as minimization problems) to get

$$\begin{aligned} \mathcal{J}_1(x(\cdot), y(\cdot), \dot{x}(\cdot)) = \int_{t_0}^{t_f} \Bigg\{ & \frac{\alpha^2}{2s^2}\dot{x}(t)^2 + \frac{\alpha^2 a^2}{2} \\ & + \left(\frac{\alpha^2}{2} + \alpha \right) (\alpha x(t) + \beta y(t))^2 \end{aligned}$$

$$+ \left[\frac{\alpha}{s}(\alpha x(t) + \beta y(t)) - \frac{\alpha^2}{s}(a - (\alpha x(t) + \beta y(t)))\right] \dot{x}(t)$$
$$- a(\alpha^2 + \alpha)(\alpha x(t) + \beta y(t))\Big\} dt \tag{2.34}$$

and

$$\mathcal{J}_2(x(\cdot), y(\cdot), \dot{y}(t)) = \int_{t_0}^{t_f} \left\{\frac{\beta^2}{2s^2}\dot{y}(t)^2 + \frac{\beta^2 a^2}{2}\right.$$
$$+ \left(\frac{\beta^2}{2} + \beta\right)(\alpha x(t) + \beta y(t))^2$$
$$+ \left[\frac{\beta}{s}(\alpha x(t) + \beta y(t)) - \frac{\beta^2}{s}(a - (\alpha x(t) + \beta y(t)))\right] \dot{y}(t)$$
$$- a(\beta^2 + \alpha)(\alpha x(t) + \beta y(t))\Big\} dt. \tag{2.35}$$

For the remainder of our discussion we focus on the first player as the computation of the second player is the same. We begin by observing that the integrand for player 1 is

$$L_1(x, y, p) = \left\{\frac{\alpha^2}{2s^2}p^2 + \frac{\alpha^2 a^2}{2} + \left(\frac{\alpha^2}{2} + \alpha\right)(\alpha x + \beta y)^2\right.$$
$$+ \left[\frac{\alpha}{s}(\alpha x + \beta y) - \frac{\alpha^2}{s}(a - (\alpha x + \beta y))\right] p$$
$$-a(\alpha^2 + \alpha)(\alpha x + \beta y)\Big\}. \tag{2.36}$$

Inspecting this integrand we choose $\tilde{L}(\cdot, \cdot, \cdot)$ to be,

$$\tilde{L}(\tilde{x}, \tilde{y}, \tilde{p}) = \frac{\alpha^2}{2s^2}\tilde{p}^2 + \frac{\alpha^2 a^2}{2}$$

from which we immediately deduce, applying Corollary 2.3, that the appropriate transformation, $z_1(\cdot, \cdot)$, must satisfy the partial differential equation,

$$\left(\frac{\partial z_1}{\partial \tilde{x}}\right)^2 = 1$$

giving us that $z_1(t, \tilde{x}) = f(t) \pm \tilde{x}$ and that

$$\frac{\partial z_1}{\partial t} + \frac{\partial z_1}{\partial \tilde{x}}\tilde{p} = \dot{f}(t) \pm \tilde{p}.$$

From this we now compute,

$$
\begin{aligned}
\Delta L_1 &= L_1(f(t) \pm \tilde{x}, y^*(t), \dot{f}(t) \pm \tilde{p}) - \tilde{L}_1(\tilde{x}, y^*(t), \tilde{p}) \\
&= \left\{ \frac{\alpha^2}{2s^2}(\dot{f}(t) \pm \tilde{p})^2 + \frac{\alpha^2 a^2}{2} + \left(\frac{\alpha^2}{2} + \alpha\right)(\alpha(f(t) \pm \tilde{x}) + \beta y^*(t))^2 \right. \\
&\quad + \left[\frac{\alpha}{s}(\alpha(f(t) \pm \tilde{x}) + \beta y^*(t)) \right. \\
&\quad \left. - \frac{\alpha^2}{s}(a - (\alpha(f(t) \pm \tilde{x}) + \beta y^*(t))) \right] (\dot{f}(t) \pm \tilde{p}) - a(\alpha^2 + \alpha) \times \\
&\qquad \left. (\alpha(f(t) \pm \tilde{x}) + \beta y^*(t)) \right\} \\
&\quad - \left\{ \frac{\alpha^2}{2s^2}\tilde{p}^2 + \frac{\alpha^2 a^2}{2} \right\} \\
&= \left\{ \frac{\alpha^2}{2s^2}\dot{f}(t)^2 + \left(\frac{\alpha^2}{2} + \alpha\right)[\alpha(f(t) \pm \tilde{x}) + \beta y^*(t)]^2 - (\alpha^2 + \alpha) \times \right. \\
&\qquad [\alpha(f(t) \pm \tilde{x}) + \beta y^*(t)] \\
&\quad \left. + \left[\left(\frac{\alpha^2}{s} + \frac{\alpha}{s}\right)[\alpha(f(t) \pm \tilde{x}) \quad + \beta y^*(t)] - \frac{\alpha^2 a}{s} \right] \dot{f}(t) \right\} \\
&\quad \pm \left\{ \frac{\alpha^2}{s^2}\dot{f}(t) + \left(\frac{\alpha^2}{s} + \frac{\alpha}{s}\right) \times [\alpha(f(t) \pm \tilde{x}) + \beta y^*(t)] - \frac{\alpha^2 a}{s} \right\} \tilde{p} \\
&\doteq \frac{\partial H_1(t, \tilde{x})}{\partial t} + \frac{\partial H_1(t, \tilde{x})}{\partial \tilde{x}} \tilde{p}.
\end{aligned}
$$

From this we compute the mixed partial derivatives to obtain,

$$
\begin{aligned}
\frac{\partial^2 H_1}{\partial \tilde{x} \partial t}(t, \tilde{x}) &= \pm 2 \left(\frac{\alpha^2}{2} + \alpha\right)[\alpha(f(t) \pm \tilde{x}) + \beta y^*(t)]\, \alpha \\
&\qquad \mp a\alpha(\alpha^2 + \alpha) \pm \alpha \left(\frac{\alpha^2}{s} + \frac{\alpha}{s}\right) \dot{f}(t) \\
&= \pm \left\{ \alpha^3(\alpha + 2)(f(t) \pm \tilde{x}) + \alpha^2 \beta(\alpha + 2) y^*(t) \right. \\
&\qquad \left. - \alpha^2(\alpha + 1)a + \frac{\alpha^2}{s}(\alpha + 1)\dot{f}(t) \right\}
\end{aligned}
$$

and

$$
\begin{aligned}
\frac{\partial^2 H_1}{\partial t \partial \tilde{x}}(t, \tilde{x}) &= \pm \left\{ \frac{\alpha^2}{s^2}\ddot{f}(t) + \left(\frac{\alpha^2}{s} + \frac{\alpha}{s}\right)\left[\alpha \dot{f}(t) + \beta \dot{y}^*(t)\right] \right\} \\
&= \pm \left\{ \frac{\alpha^2}{s^2}\ddot{f}(t) + \frac{\alpha^2}{s}(\alpha + 1)\dot{f}(t) + \frac{\alpha\beta}{s}(\alpha + 1)\dot{y}^*(t) \right\}.
\end{aligned}
$$

Assuming sufficient smoothness and equating the mixed partial derivatives we obtain the following equation:

$$\begin{aligned}\ddot{f}(t) - \alpha s^2(\alpha+2)f(t) &= \beta s^2(\alpha+2)y^*(t) - \frac{\beta s}{\alpha}(\alpha+1)\dot{y}^*(t)\\ &\quad \pm \alpha s^2(\alpha+2)\tilde{x} - as^2(\alpha+1).\end{aligned}$$

A similar analysis for player 2 yields:

$$\begin{aligned}L_2(x,y,q) &= \left\{\frac{\beta^2}{2s^2}q^2 + \frac{\beta^2 a^2}{2} + \left(\frac{\beta^2}{2}+\beta\right)(\alpha x+\beta y)^2\right.\\ &\quad \left[\frac{\beta}{s}(\alpha x+\beta y) - \frac{\beta^2}{s}(a-(\alpha x+\beta y)\right] q \qquad (2.37)\\ &\quad \left. - a(\beta^2+\beta)(\alpha x+\beta y)\right\},\end{aligned}$$

and so choosing

$$\tilde{L}_2(\ddot{x},\ddot{y},\ddot{q}) = \left\{\frac{\beta^2}{2s^2}q^2 + \frac{\beta^2 a^2}{2}\right\}$$

gives us that the transformation $z_2(\cdot,\cdot)$ is obtained by solving the partial differential equation

$$\left(\frac{\partial z_2}{\partial \tilde{y}}\right)^2 = 1,$$

which of course gives us, $z_2(t,\tilde{y}) = g(t) \pm \tilde{y}$. Proceeding as above we arrive at the following differential equation for $g(\cdot)$,

$$\begin{aligned}\ddot{g}(t) - \beta s^2(\beta+2)g(t) &= \alpha s^2(\beta+2)x^*(t) - \frac{\alpha s}{\beta}(1+\beta)\dot{x}^*(t)\\ &\quad \pm \beta s^2(\beta+2)\tilde{y} - as^2(\beta+1).\end{aligned}$$

Now the auxiliary variational problem we must solve consists of minimizing the two functionals,

$$\int_{t_0}^{t_f}\left(\frac{\alpha^2}{2s^2}\dot{\tilde{x}}^2(t) + \frac{\alpha a^2}{2}\right)dt \quad \text{and} \quad \int_{t_0}^{t_f}\left(\frac{\beta^2}{2s^2}\dot{\tilde{y}}^2(t) + \frac{\beta a^2}{2}\right)dt$$

over some appropriately chosen boundary conditions. We observe that these two minimization problems are easily solved if these conditions take the form,

$$\tilde{x}(t_0) = \tilde{x}(t_f) = c_1 \quad \text{and} \quad \tilde{y}(t_0) = \tilde{y}(t_f) = c_2$$

for arbitrary but fixed constants c_1 and c_2. The solutions are in fact,

$$\tilde{x}^*(t) \equiv c_1 \quad \text{and} \quad \tilde{y}^*(t) \equiv c_2$$

According to our theory we then have that the solution to our variational game is,

$$x^*(t) = f(t) \pm c_1 \quad \text{and} \quad y^*(t) = g(t) \pm c_2.$$

In particular, using this information in the equations for $f(\cdot)$ and $g(\cdot)$ with $\tilde{x} = c_1$ and with $\tilde{y} = c_2$ we obtain the following equations for $x^*(\cdot)$ and $y^*(\cdot)$,

$$\begin{aligned} \ddot{x}^*(t) - \alpha s^2(\alpha+2)x^*(t) &= \beta s^2(\alpha+2)y^*(t) \\ &\quad -\frac{\beta s}{\alpha}(\alpha+1)\dot{y}^*(t) - as^2(\alpha+1) \\ \ddot{y}^*(t) - \beta s^2(\beta+2)y^*(t) &= \alpha s^2(\beta+2)x^*(t) \\ &\quad -\frac{\alpha s}{\beta}(1+\beta)\dot{x}^*(t) - as^2(\beta+1), \end{aligned}$$

with the end conditions,

$$x^*(t_0) = y^*(t_0) = P_0 \quad \text{and} \quad x^*(t_f) = y^*(t_f) = P_f.$$

These equations coincide exactly with the Euler-Lagrange equations, as derived by the Maximum Principle for the open-loop variational game without constraints. Additionally we note that as these equations are derived here via the direct method we see that they become sufficient conditions for a Nash equilibrium of the unconstrained system, and hence for the constrained system for solutions which satisfy the constraints (see the comments in Remark 2.3). Moreover, we also observe that we can recover the functions $H_j(\cdot,\cdot)$, for $j = 1, 2$, since we can recover both $f(\cdot)$ and $g(\cdot)$ by the formulas

$$f(t) = x^*(t) \mp c_1 \quad \text{and} \quad g(t) = y^*(t) \mp c_2.$$

The required functions are now recovered by integrating the partial derivatives of $H_1(\cdot,\cdot)$ and $H_2(\cdot,\cdot)$ which can be computed. Consequently, we see that in this instance the solution to our variational game is given by the solutions of the above Euler-Lagrange system, provided the resulting strategies and the price satisfy the requisite constraints. Finally, we can obtain the solution to the original problem by taking,

$$P^*(t) = \alpha x^*(t) + \beta y^*(t),$$

$$u_1^*(t) = \alpha\left(a - P^*(t) - \frac{1}{s}\dot{x}^*(t)\right),$$

and

$$u_2^*(t) = \beta\left(a - P^*(t) - \frac{1}{s}\dot{y}^*(t)\right).$$

Of course, we still must check that these functions meet whatever constraints are required (i.e., $u_i(t) \geq 0$ and $P(t) \geq 0$).

There is one special case of the above analysis in which the solution can be obtained easily. This is the case when $\alpha = \beta = \frac{1}{2}$. In this case the above Euler-Lagrange system becomes,

$$\begin{aligned} \ddot{x}^*(t) - \frac{5}{4}s^2 x^*(t) &= \frac{5}{4}s^2 y^*(t) - \frac{3}{2}s\dot{y}^*(t) - \frac{3}{2}as^2 \\ \ddot{y}^*(t) - \frac{5}{4}s^2 y^*(t) &= \frac{5}{4}s^2 x^*(t) - \frac{3}{2}s\dot{x}^*(t) - \frac{3}{2}as^2. \end{aligned}$$

Using the fact that $P^*(t) = \frac{1}{2}(x^*(t) + y^*(t))$ for all $t \in [t_0, t_f]$ we can multiply each of these equations by $\frac{1}{2}$ and add them together to obtain the following equation for $P^*(\cdot)$,

$$\ddot{P}^*(t) + \frac{3}{2}s\dot{P}^*(t) - \frac{5}{2}s^2 P^*(t) = -\frac{3}{2}as^2,$$

for $t_0 \leq t \leq t_f$. This equation is an elementary non-homogeneous second order linear equation with constant coefficients whose general solution is given by

$$P^*(t) = Ae^{r_+(t-t_0)} + Be^{r_-(t-t_0)} + \frac{3}{5}a$$

in which $r_\pm$ are the characteristics roots of the equation and A and B are arbitrary constants. More specifically, the characteristic roots are roots of the polynomial

$$r^2 + \frac{3}{2}sr - \frac{5}{2}s^2 = 0$$

and are given by

$$r_+ = s \quad \text{and} \quad r_- = -\frac{5}{2}s.$$

Thus, to solve the dynamic game in this case we select A and B so that $P^*(\cdot)$ satisfies the fixed boundary conditions. Further we note that we can also take

$$x^*(t) = y^*(t) = \frac{1}{2}P^*(t)$$

and so obtain the optimal strategies as

$$u_1^*(t) = u_2^*(t) = \frac{1}{2}\left(a - P^*(t) - \frac{1}{s}\dot{P}^*(t)\right).$$

It remains to verify that there exists some choice of parameters for which the optimal price, $P^*(\cdot)$, and the optimal strategies, $u_1^*(\cdot), u_2^*(\cdot)$

remain nonnegative. To this end we observe that we impose the fixed boundary conditions to obtain the following linear system of equations for the unknowns, A and B:

$$\begin{pmatrix} 1 & 1 \\ e^{s(t_f-t_0)} & e^{-\frac{5}{2}s(t_f-t_0)} \end{pmatrix} \begin{pmatrix} A \\ B \end{pmatrix} = \begin{pmatrix} P_0 - \frac{3}{5}a \\ P_f - \frac{3}{5}a \end{pmatrix}.$$

Using Cramer's rule we obtain the following formulas for A and B,

$$\begin{aligned} A &= \frac{1}{D}\left[\left(P_0 - \frac{3}{5}a\right)e^{-\frac{5}{2}s(t_f-t_0)} - \left(P_f - \frac{3}{5}a\right)\right] \\ B &= \frac{1}{D}\left[\left(P_f - \frac{3}{5}a\right) - \left(P_0 - \frac{3}{5}a\right)e^{s(t_f-t_0)}\right] \end{aligned}$$

in which D is the determinant of the coefficient matrix and is given by

$$D = e^{-\frac{5}{2}s(t_f-t_0)} - e^{s(t_f-t_0)} = e^{s(t_f-t_0)}\left(e^{-\frac{7}{2}s(t_f-t_0)} - 1\right).$$

We observe that D is clearly negative since $t_f > t_0$. Also, to insure that $P^*(t)$ is nonnegative for $t \in [t_0, t_f]$ it is sufficient to insure that A and B are both positive. This means we must have,

$$\begin{aligned} 0 &> \left[\left(P_0 - \frac{3}{5}a\right)e^{-\frac{5}{2}s(t_f-t_0)} - \left(P_f - \frac{3}{5}a\right)\right] \\ 0 &> \left[\left(P_f - \frac{3}{5}a\right) - \left(P_0 - \frac{3}{5}a\right)e^{s(t_f-t_0)}\right] \end{aligned}$$

which can be equivalently expressed as,

$$\left(P_0 - \frac{3}{5}a\right)e^{-\frac{5}{2}s(t_f-t_0)} < P_f - \frac{3}{5}a < \left(P_0 - \frac{3}{5}a\right)e^{s(t_f-t_0)}. \qquad (2.38)$$

Observe that as long as P_0 and P_f are chosen to be larger than $\frac{3}{5}a$ this last inequality can be satisfied if we choose $t_f - t_0$ sufficiently large. In this case we have explicitly given the optimal price, $P^*(\cdot)$ in terms of the model parameters P_0, P_f, t_0, t_f, a, and s (all strictly positive). It remains to check that the strategies are nonnegative. To this end we notice that,

$$\dot{P}^*(t) = Ase^{s(t-t_0)} - \frac{5}{2}Bse^{-\frac{5}{2}s(t-t_0)}$$

so that we have, the admissible strategies given by, for $j = 1, 2$,

$$u_j^*(t) = \frac{1}{2}\left[a - P^*(t) - \frac{1}{2s}\dot{P}^*(t)\right]$$

$$
\begin{aligned}
&= \frac{1}{2}\left[a - \left(Ae^{s(t-t_0)} + Be^{-\frac{5}{2}s(t-t_0)}\right)\right. \\
&- \left.\frac{1}{2s}\left(Ase^{s(t-t_0)} - \frac{5}{2}Bse^{-\frac{5}{2}s(t-t_0)}\right)\right] \\
&= \frac{1}{2}\left[a - \frac{3}{2}Ae^{s(t-t_0)} + \frac{1}{4}Be^{-\frac{5}{2}(t-t_0)}\right].
\end{aligned}
$$

Taking the time derivative of $u_j^*(\cdot)$ we obtain

$$
\dot{u}_j^*(t) = \frac{1}{2}\left[-\frac{3}{2}Ase^{s(t-t_0)} - \frac{5}{8}Bse^{-\frac{5}{2}(t-t_0)}\right] < 0,
$$

since A and B are positive. This implies that $u_j^*(t) \geq u_j^*(t_f)$ for all $t \in [t_0, t_f]$. Thus to insure that $u_j^*(\cdot)$ is nonnegative it is sufficient to insure $u_j^*(t_f) \geq 0$ which holds if we have

$$
a - \frac{3}{2}Ae^{s(t_f-t_0)} + \frac{1}{4}Be^{-\frac{5}{2}(t_f-t_0)} \geq 0.
$$

To investigate this inequality we first observe that we have, from the solution $P^*(\cdot)$, that

$$
P_f = Ae^{s(t_f-t_0)} + Be^{\frac{-5}{2}s(t_f-t_0)} + \frac{3}{5}a.
$$

This allows us to rewrite the last inequality in the form,

$$
a - \frac{7}{4}Ae^{-s(t_f-t_0)} + \frac{1}{4}\left(P_f - \frac{3}{5}a\right) \geq 0
$$

or equivalently (using the explicit expression for A),

$$
P_f - \frac{3}{5}a \geq 7\frac{1}{e^{-\frac{7}{2}s(t_f-t_0)} - 1}\left[\left(P_0 - \frac{3}{5}a\right)e^{-\frac{5}{2}s(t_f-t_0)} - \left(P_f - \frac{3}{5}a\right)\right] - 4a.
$$

Solving this inequality for $P_f - \frac{3}{5}a$ we obtain the inequality,

$$
P_f - \frac{3}{5}a \leq \frac{7e^{-\frac{5}{2}s(t_f-t_0)}}{6 + e^{-\frac{7}{2}s(t_f-t_0)}}\left(P_0 - \frac{3}{5}a\right) + 4a\frac{1 - e^{-\frac{7}{2}s(t_f-t_0)}}{6 + e^{-\frac{7}{2}s(t_f-t_0)}}. \quad (2.39)
$$

Thus to insure that the state and control constraints, $P^*(t) \geq 0$ and $u_i(t) \geq 0$ for $t \in [t_0, t_f]$, hold, we must check that the parameters of the system satisfy inequalities (2.38) and (2.39). We have already observed that for $P_0, P_f \geq \frac{3}{5}a$ we can choose $t_f - t_0$ sufficiently large to insure that (2.38) holds. Further, we observe that as $t_f - t_0 \to +\infty$ the right side of (2.39) tends $\frac{2}{3}a$ so that we can always find $t_f - t_0$ sufficiently large so that we have

$$\frac{7e^{-\frac{5}{2}s(t_f-t_0)}}{6+e^{-\frac{7}{2}s(t_f-t_0)}}\left(P_0-\frac{3}{5}a\right)+4a\frac{1-e^{-\frac{7}{2}s(t_f-t_0)}}{6+e^{-\frac{7}{2}s(t_f-t_0)}}\leq\left(P_0-\frac{3}{5}a\right)e^{s(t_f-t_0)}$$

Moreover, it is easy to see that

$$\frac{7e^{-\frac{5}{2}s(t_f-t_0)}}{6+e^{-\frac{7}{2}s(t_f-t_0)}}\left(P_0-\frac{3}{5}a\right)+4a\frac{1-e^{-\frac{7}{2}s(t_f-t_0)}}{6+e^{-\frac{7}{2}s(t_f-t_0)}}\geq\left(P_0-\frac{3}{5}a\right)e^{-\frac{5}{2}s(t_f-t_0)}$$

holds whenever $t_f - t_0$ is sufficiently large. Combining these observations allows us to conclude that for $t_f - t_0$ sufficiently large $\{P^*(\cdot), u_1^*(\cdot), u_2^*(\cdot)\}$ is a Nash equilibrium for the original dynamic game.

4. Conclusion

In this paper we have presented means to utilize the direct method to obtain open-loop Nash equilibria for differential games for which there is a single state whose time evolution is determined by the competitive strategies of several players appearing linearly in the equation. That is a so called affine control system with "many inputs and one output."

References

Carathéodory, C. (1982). *Calculus of Variations and Partial Differential Equations.* Chelsea, New York, New York.

Carlson, D.A. (2002). An observation on two methods of obtaining solutions to variational problems. *Journal of Optimization Theory and Applications*, 114:345–362.

Carlson, D.A. and Leitmann, G. (2004). An extension of the coordinate transformation method for open-loop Nash equilibria. *Journal of Optimization Theory and Applications*, To appear.

Dockner, E.J. and Leitmann, G. (2001) Coordinate transformation and derivation of open-loop Nash equilibrium. *Journal of Optimization Theory and Applications*, 110(1):1–16.

Leitmann, G. (1967). A note on absolute extrema of certain integrals. *International Journal of Non-Linear Mechanics*, 2:55–59.

Leitmann, G. (2004). A direct method of optimization and its application to a class of differential games. *Dynamics of Continuous, Discrete and Impulsive Systems – Series A: Mathematical Analysis*, 11:191–204.

Chapter 3

BRAESS PARADOX AND PROPERTIES OF WARDROP EQUILIBRIUM IN SOME MULTISERVICE NETWORKS

Rachid El Azouzi
Eitan Altman
Odile Pourtallier

Abstract In recent years there has been a growing interest in mathematical models for routing in networks in which the decisions are taken in a non-cooperative way. Instead of a single decision maker (that may represent the network) that chooses the paths so as to maximize a global utility, one considers a number of decision makers having each its own utility to maximize by routing its own flow. This gives rise to the use of non-cooperative game theory and the Nash equilibrium concept for optimality. In the special case in which each decision maker wishes to find a minimal path for each routed object (e.g. a packet) then the solution concept is the Wardrop equilibrium. It is well known that equilibria may exhibit inefficiencies and paradoxical behavior, such as the famous Braess paradox (in which the addition of a link to a network results in worse performance to all users). This raises the challenge for the network administrator of how to upgrade the network so that it indeed results in improved performance. We present in this paper some guidelines for that.

1. Introduction

In this paper, we consider the problem of routing, in which the performance measure to be minimized is some general cost (which could represent the expected delay). We assume that some objects, are routed over shortest paths computed in terms of that cost. An object could correspond to a whole session in case all packets of a connection are assumed to follow the same path. It could correspond to a single packet if each packet could have its own path. A routing approach in which each packet follows a shortest delay path has been advocated in Ad-hoc networks (Gupta and Kumar (1998)), in which, the large amount of mobility of both users as well as of the routers requires to update the routes frequently; it has further been argued that by minimizing the delay of each packet, we minimize re-sequencing delays, that may be harmful in real time applications, but also in data communications (indeed, the throughput of TCP/IP connections may quite deteriorate when packets arrive out of sequence, since the latter is frequently interpreted wrongly as a signal of a loss or of a congestion).

When the above type of routing approach is used then the expected load at different links in the network can be predicted as an equilibrium which can be computed in a way similar to equilibria that arise in road traffic. The latter is known as a Wardrop equilibrium (Wardrop (1952)) (it is known to exist and to be unique under general assumptions on the topology and on the cost; see, Patriksson (1994), p. 74–75). We study in this paper some properties of the equilibrium. In particular, we are interested in the impact of the demand of link capacities and of the topology on the performance measures at equilibrium. This has a particular significance for the network administrator or designer when it comes to upgrading the network. A frequently used heuristic approach for upgrading a network is through *Bottleneck Analysis.* A system bottleneck is defined as "a resource or service facility whose capacity seriously limits the performance of the entire system" (Kobayashi (1978), p. 13). Bottleneck analysis consists of adding capacity to identified bottlenecks until they cease to be bottlenecks. In a non-cooperative framework, however, this approach may have devastating effects; adding capacity to a link (and in particular, to a bottleneck link) may cause delays of all users to increase; in an economic context in which users pay the service provider, this may further cause a decrease in the revenues of the provider. The first problem has already been identified in the road-traffic context by Braess (1968) (see also Dafermos and Nagurney (1984a); Smith (1979)), and have further been studied in the context of queuing networks (Beans, Kelly and Taylor (1997); Calvert, Solomon and Ziedins (1997); Cohen and Jeffries (1997); Cohen and Kelly (1990)). In the latter references both queuing delay as well as rejection probabilities have been considered as performance measure. The focus of Braess paradox on the bottleneck link in a queuing context, as well as the paradoxical impact on the service provider have been studied in Massuda (1999). In all the above references, the paradoxical behavior occurs in models in which the number of users is infinitely large, and the equilibrium concept is that of Wardrop equilibrium (Wardrop (1952)). Yet the problem may occur also in models involving finite number of players (e.g. service providers) for which the Nash equilibrium is the optimality concept. This has been illustrated in Korilis, Lazar and Orda (1995); Korilis, Lazar and Orda (1999). The Braess paradox has further been identified and studied in the context of distributed computing (Kameda, Altman and Kozawa (1999); Kameda et al. (2000)) where arrivals of jobs may be routed and performed on different processors. Several papers that are scheduled to appear in JACM identify the Braess paradox that occurs in the context of non cooperative communication networks and of load balancing, see Kameda and Pourtallier (2002) and Roughgarden and Tardos (2002). Both papers illustrate how harmful the Braess paradox can be. These papers reflect the interest in understanding the degradation that is due to non-cooperative nature of networks, in which added capacity can result in degraded performance. In view of the interest in this identified problem, it seems important to come up with engineering design tools that can predict how to upgrade a network (by adding links or capacity) so as to avoid the harmful paradox. This is what our paper proposes.

The Braess paradox illustrates that the network designer or service providers, or more generally, whoever is responsible to the network topology and link capacities, have to take into consideration the reaction of non-cooperative users to his decisions. Some upgrading guidelines have been proposed in Altman, El Azouzi and Pourtallier (2001); Kameda and Pourtallier (2002); Korilis, Lazar and Orda (1999) so as to avoid the Braess paradox or so as to obtain a better performance. They considered not only the framework of the Wardrop equilibrium, but also the Nash-equilibrium concept in which a finite number of service providers try to minimize the average delays (or cost) for all the flow generated by its subscribers. The results obtained for the Wardrop equilibrium were restricted to a particular cost representing the delay of a M/M/1 queue at each link. In this paper we extend the above results to general costs. We further consider a more general routing structure (between paths and not just between links) and allow for several classes of users (so that the cost of a path or of a link may depend on the class in some way). Some other guidelines for avoiding Braess paradox in the setting of Nash equilibrium have been obtained in Altman, El Azouzi and Pourtallier (2001), yet in that setting the guidelines turn out to be much more restrictive than those we obtain for the setting of Wardrop equilibrium.

The main objective of this present paper is to pursue that direction and to provide new guidelines for avoiding the Braess paradox when upgrading the network. The Braess paradox implies that there is no monotonicity of performance measures with respect to link capacities. Another objective of this paper is to check under what conditions are delays as well as the marginal costs at equilibrium increasing in the demands. The answer to this question turns out to be useful for the analysis of the Braess paradox. Some results on the monotonicity in the demand are already available in Dafermos and Nagurney (1984b).

The paper is organized as follows: In the next section (Section 2), we present the network model, we define the concept of Wardrop equilibrium, and formulate the problem. In Section 3 we then present a framework of that equilibrium that allows for different costs for different classes of users (which may reflect, for example, that packets of different users may have different priorities and thus have different delays due to appropriate buffer management schemes). In Section 4 we then present a sufficient condition for the monotonicity of performance measures when the demands increase. This allows us then to study in Section 5 methods for capacity addition. In Section 6, we demonstrate the efficiency of the proposed capacity addition by means of a numerical example in **BCMP** queuing network.

2. Problem formulation and notation

We consider an open network model that consists of a set $\mathbb{M}$ containing M nodes, and a set $\mathbb{L}$ containing L links. We call the unit that has to be routed a "job". It may stand for a packet (in a packet switched network) or to a whole session (if indeed all packets of a session follow the same path). The network is crossed through by infinitely many jobs that have to choose their routing.

Jobs are classified into K different classes (we will denote $\mathbb{K}$ the set of classes). For example, in the context of road traffic a class may represent the set of a given type of vehicles, such as busses, trucks, cars or bicycles. In the context of telecommunications a class may represent the jobs sent by all the users of a given service provider. We assume that jobs do not change their class while passing through the network. We suppose that the jobs of a given class k may arrive in the system at some different possible points, and leave the system at some different possible points. Nevertheless the origin and destination points of a given job are determined when the job arrives in the network, and cannot change while in the system. We call a pair of one origin and one destination points an O-D pair.

A job with a given O-D pair (od) arrives in the system at node o and leaves it at node d after visiting a series of nodes and links, which we refer to as a *path*, then it leaves the system.

In many previous papers (Orda, Rom and Shimkin (1993); Korilis, Lazar and Orda (1995)), routing could be done at each node. In this paper we follow the approach in which a job of class k with O-D pair

(od) has to choose one of a given finite set of paths (see also Kameda and Zhang (1995); Patriksson (1994)).

In this paper we suppose that the routing decision scheme is completely decentralized: each single job has to decide among a set of possible paths that connect the *O-D* pair of that job. This choice will be made in order to minimize the cost of that job. The solution concept we are thus going to use is the Wardrop equilibrium (Wardrop (1952)).

Let l denote a link of the network connecting a pair of nodes and let p denote a path, consisting of a sequence of links connecting an *O-D* pair of nodes. Let W^k denote the set of *O-D* pairs for the jobs of class k. Denote also W the union $W = \bigcup_k W^k$. The set of paths connecting the *O-D* pair $w \in W^k$ is denoted by $\mathcal{P}_w^k$ and the entire set of paths in the network for the jobs of class k by $\mathcal{P}^k$. There are n_p^k paths in the networks for jobs of class k and n_p paths of all jobs.

Let y_l^k denote the flow of class k on link l and let x_p^k denote the nonnegative flow of class k on path p. The relationship between the link loads by class and the path flows is:

$$y_l^k = \sum_{p \in \mathcal{P}^k} x_p^k \delta_{lp}$$

where $\delta_{lr} = 1$, if link l is contained in path p, and 0, otherwise. Let μ_l^k the service rate of class k at link l. Hence, the utilization of link l for class k is given by $\rho_l^k = y_l^k / \mu_l^k$ and the total utilization on link l is:

$$\rho_l = \sum_{k \in I\!K} \rho_l^k.$$

Let r_w^k denote the demand of class k for *O-D* pair w, where the following conditions are satisfied:

$$r_w^k = \sum_{p \in \mathcal{P}_w^k} x_p^k, \quad \forall\, k, \quad \forall\, w,$$

In addition, let x_p denote the total flow on path p, where

$$x_p = \sum_{k \in I\!K} x_p^k, \quad \forall\, p \in \mathcal{P}.$$

We group the class path flows into the n_p^k-dimensional column vector $\mathbf{X}$ with components: $[x_{r_1}^1, \ldots, x_{r_{n_p^k}}^K]^T$; We refer to such a vector as a *flow configuration.* We also group total path flow into a n_p-dimensional column vector $\mathbf{x}$ with components: $[x_{r_1}, \ldots, x_{r_{n_p}}]^T$. We call this vector the *total path flow vector*. A flow configuration **X** is said feasible, if it satisfies for each *O-D* pair $w \in W^k$,

$$\sum_{p \in \mathcal{P}_w^k} x_p^k = r_w^k. \tag{3.1}$$

We are now ready to describe the cost functions associated with the paths. We consider a feasible flow configuration **X**. Let $T_p^k(\mathbf{X})$ denote the travel cost incurred by a job of class k for using the path p if the flow configuration resulting from the routing of each job is **X**.

3. Wardrop equilibrium for a multi-class network

Each individual job of class k with *O-D* pair w, chooses its routing through the system, by means of the choice of a path $p \in \mathcal{P}_w^k$. A flow configuration **X** follows from the choices of each of the infinitely many jobs.

A flow configuration **X** will be said to be a *Wardrop equilibrium* or *individually optimal*, if none of the jobs has any incentive to change unilaterally its decision. This equilibrium concept was first introduced by Wardrop (1952) in the field of transportation and can be defined through the two principles:

- **Wardrop's first principle:** the cost for crossing the used paths between a source and a destination are equal, the cost for any unused path with same *O-D* pair is larger or equal that that of used ones.

- **Wardrop's second principle:** the cost is minimum for each job.

Formally, in the context of multi-class this can be defined as,

DEFINITION 3.1 *A feasible flow configuration (i.e., satisfying equation (3.1))* $\boldsymbol{X}$, *is a Wardrop equilibrium for the multi-class problem if for any class* k, *any* $w \in W^k$ *and any path* $p \in \mathcal{P}_w^k$ *we have*

$$\begin{cases} T_p^k(\boldsymbol{X}) \geq \lambda_w^k \ \textit{if}\ x_p^k = 0, \\ T_p^k(\boldsymbol{X}) = \lambda_w^k \ \textit{if}\ x_p^k \geq 0, \end{cases} \tag{3.2}$$

where $\lambda_w^k = \min_{p \in \mathcal{P}_w^k} T_p^k(\boldsymbol{X})$. *The minimal cost* λ_w^k *will be referred to as "the travel cost" associated to class* k *and* O-D *pair* w.

We need one of the following assumptions on the cost function:

Assumption A

1. There exists a function T_p that depends only upon the total flow vector $\mathbf{x}$ (and not on the flow sent by each class), such that the average cost per flow unit, for jobs of class k, can be written as $T_p^k(\mathbf{X}) = c^k T_p(\mathbf{x})$, $\forall\, p \in \mathcal{P}^k$, where c^k is some class dependent positive constant.

2. T_p is positive, continuous and strictly increasing. We will denote $\mathbf{T}' = (T_{p_1}, T_{p_2}, \dots)'$ the vector of functions T_p.

Assumption B

1. The average cost per flow unit for jobs of class k that passes through path $p \in \mathcal{P}^k$ is:

$$T_p^k(\mathbf{X}) = \sum_{l \in \mathbb{L}} \frac{\delta_{lp}}{\mu_l^k} T_l(\rho_l),$$

where $T_l(\rho_l)$ is the weighted cost per unit flow in link l (the function T_l does not depend on the class k).

2. $T_l(.)$ is positive, continuous and strictly increasing.

3. μ_l^k can be represented as μ_l / c^k where c^k is some class dependent positive constant, and $0 < \mu_l$ is finite.

We denote $v_w = \sum_{k \in \mathbb{K}} c^k r_w^k$ the weighted total demand for *O-D* pair w.

We make the following observation.

Lemma 3.1 *Consider a cost vector* $\boldsymbol{T}$ *satisfying Assumption* **A** *or* **B**. *Then the Wardrop equilibrium conditions (3.1) and (3.2) become: For all* k, *all* $w \in W^k$ *and all* $p \in \mathcal{P}_w^k$,

$$\begin{aligned} T_p^k(\boldsymbol{X}) &\geq \lambda_w^k, \ \ \textit{if } x_p = 0, \\ T_p^k(\boldsymbol{X}) &= \lambda_w^k, \ \ \textit{if } x_p > 0. \end{aligned} \tag{3.3}$$

Moreover, the ratio λ_w^k / c^k *is independent of class* k, *so that we can define* λ_w *by* $\lambda_w := \lambda_w^k / c^k$.

Proof. Consider first the case of a cost vector $\mathbf{T}$ that satisfies Assumption **A**. Let $w \in W$ and $p \in \mathcal{P}^k$. If $x_p = 0$ then $x_p^k = 0\ \forall\, k \in \mathbb{K}$. The first part of (3.3) follows from the first part of (3.2). Suppose that $x_p > 0$ and, by contradiction, that there exists $\bar{k} \in \mathbb{K}$ such that

$$T_p^{\bar{k}}(\mathbf{X}) = T_p(\mathbf{x}) c^{\bar{k}} > \lambda_w^{\bar{k}}. \tag{3.4}$$

Since $x_p > 0$, there exit $k_0 \in I\!\!K$ such that $x_p^{k_0} > 0$. From the second part of (3.2), we have

$$T_p^{k_0}(\mathbf{X}) = T_p(\mathbf{x})c^{k_0} = \lambda_w^{k_0}. \tag{3.5}$$

Because $r_w^{\bar{k}} > 0$, there exists $p' \in \mathcal{P}_w^{\bar{k}}$ such that $x_{p'}^{\bar{k}} > 0$. Then, from (3.2) we get

$$T_{p'}^{\bar{k}}(\mathbf{X}) = T_{p'}(\mathbf{x})c^{\bar{k}} = \lambda_w^{\bar{k}}. \tag{3.6}$$

It follows from (3.4) and (3.6) that

$$T_p(\mathbf{x}) > T_{p'}(\mathbf{x}). \tag{3.7}$$

Since $\lambda_w^{k_0} \leq T_{p'}(\mathbf{x})c^{k_0}$, from (3.5), we obtain $T_p(\mathbf{x}) \leq T_{p'}(\mathbf{x})$, which contradicts (3.7). This establishes (3.3).

For any $w \in W$, let $p \in \bigcup_k \mathcal{P}_w^k$ be a path such that $x_p > 0$. From (3.3), it comes that for any class k such that $p \in \mathcal{P}_w^k$, $T_p(\mathbf{x}) = \frac{T_p^k(\mathbf{X})}{c^k} = \frac{\lambda_w^k}{c^k}$. The second part of Lemma 3.1 follows, since the terms in the above equation do not depend on k. The proof for a cost function vector satisfying Assumption **B** follows along similar lines. □

4. Impact of throughput variation on the equilibrium

In this section, we study the impact of a variation of the demands r_w^k of some class k on the cost vector $\mathbf{T}(\mathbf{X})$ at the (Wardrop) equilibrium $\mathbf{X}$. The results of this section extend those of Dafermos and Nagurney (1984b), where a simpler cost structure was considered considered. Namely, for any class k, the cost for using a path was the sum of link costs along that path, and the link costs did not depend on k.

The following theorem, states that under Assumption **A** or **B**, an increase in the demands associated to a particular O-D pair w always leads to an increase of the cost associated to w for all classes k.

THEOREM 3.1 *Consider two throughput demand profiles $(\tilde{r}_w^k)_{(w,k)}$ and $(\hat{r}_w^k)_{(w,k)}$. Let $\tilde{\mathbf{X}}$ and $\hat{\mathbf{X}}$ be the Wardrop equilibria associated to these throughput demands, and let $\tilde{\lambda}_w^k$ and $\hat{\lambda}_w^k$ be the class k's travel cost associated to these two equilibria.*

1. *For a cost vector $\boldsymbol{T}$ satisfying Assumption $\boldsymbol{A}$, if $\hat{r}_{\overline{w}} < \tilde{r}_{\overline{w}}$, for some $\overline{w} \in W$ and $\hat{r}_w = \tilde{r}_w$ for all $w \neq \overline{w}$, then $\hat{\lambda}_{\overline{w}}^k < \tilde{\lambda}_{\overline{w}}^k \ \forall k \in I\!\!K$.*

2. *For a cost vector $\boldsymbol{T}$ satisfying Assumption $\boldsymbol{B}$, if $\hat{v}_{\overline{w}} < \tilde{v}_{\overline{w}}$, for some $\overline{w} \in W$ and $\hat{v}_w = \tilde{v}_w$ for all $w \neq \overline{w}$, then $\hat{\lambda}_{\overline{w}}^k < \tilde{\lambda}_{\overline{w}}^k \ \forall k \in I\!\!K$.*

Proof. Consider first the case of a cost vector that satisfies Assumption **A**.

1. From (3.3) and Assumption **A** we have

$$\hat{\lambda}_w = T_p(\hat{x}) \text{ if } \hat{x}_p > 0,$$
$$\hat{\lambda}_w \leq T_p(\hat{x}), \text{ if } \hat{x}_p = 0,$$

and

$$\tilde{\lambda}_w = T_p(\tilde{x}), \text{ if } \tilde{x}_p > 0,$$
$$\tilde{\lambda}_w \leq T_p(\tilde{x}), \text{ if } \tilde{x}_p = 0.$$

Thus

$$\begin{array}{lcl} \hat{x}_p\hat{\lambda}_w = T_p(\hat{\mathbf{x}})\hat{x}_p, & \text{and} & \tilde{x}_p\tilde{\lambda}_w = T_p(\tilde{\mathbf{x}})\tilde{x}_p, \\ \tilde{x}_p\hat{\lambda}_w \leq T_p(\hat{\mathbf{x}})\tilde{x}_p, & & \hat{x}_p\tilde{\lambda}_w \leq T_p(\tilde{\mathbf{x}})\hat{x}_p. \end{array}$$

Now by summing up over $p \in \mathcal{P}_w$, we obtain

$$\hat{r}_w\hat{\lambda}_w = \sum_{p\in\mathcal{P}_w} T_p(\hat{\mathbf{x}})\hat{x}_p,$$
$$\tilde{r}_w\hat{\lambda}_w \leq \sum_{p\in\mathcal{P}_w} T_p(\hat{\mathbf{x}})\tilde{x}_p,$$

and

$$\tilde{r}_w\tilde{\lambda}_w = \sum_{p\in\mathcal{P}_w} T_p(\tilde{\mathbf{x}})\tilde{x}_p,$$
$$\hat{r}_w\tilde{\lambda}_w \leq \sum_{p\in\mathcal{P}_w} T_p(\tilde{\mathbf{x}})\hat{x}_p.$$

By summing up over $w \in W$, it comes

$$\sum_{w\in W} (\tilde{r}_w - \hat{r}_w)(\tilde{\lambda}_w - \hat{\lambda}_w) \geq (\mathbf{T}'(\hat{\mathbf{x}}) - \mathbf{T}'(\tilde{\mathbf{x}}))(\hat{\mathbf{x}} - \tilde{\mathbf{x}}) > 0. \quad (3.8)$$

The last inequality follows from assumption **A**. Since $\hat{r}_w = \tilde{r}_w$ for $w \neq \overline{w}$, inequality (3.8) yields $(\tilde{r}_{\overline{w}} - \hat{r}_{\overline{w}})(\tilde{\lambda}_{\overline{w}} - \hat{\lambda}_{\overline{w}}) > 0$, which implies, since $\tilde{r}_{\overline{w}} > \hat{r}_{\overline{w}}$, that $\tilde{\lambda}_{\overline{w}} > \hat{\lambda}_{\overline{w}}$. It follows that $\tilde{\lambda}^k_{\overline{w}} > \hat{\lambda}^k_{\overline{w}}$ for all $k \in I\!K$.

Consider now the case of a cost vector that satisfies Assumption **B**.

2. From (3.3) and Assumption **B** we have

$$\hat{\lambda}_w = \sum_{l\in I\!L} \frac{\delta_{lp}}{\mu_l} T_l(\hat{\rho}_l) \text{ if } \hat{x}_p > 0,$$
$$\hat{\lambda}_w \leq \sum_{l\in I\!L} \frac{\delta_{lp}}{\mu_l} T_l(\hat{\rho}_l), \text{ if } \hat{x}_p = 0,$$

and

$$\tilde{\lambda}_w = \sum_{l \in \mathbb{L}} \frac{\delta_{lp}}{\mu_l} T_l(\tilde{\rho}_l), \text{ if } \tilde{x}_p > 0,$$
$$\tilde{\lambda}_w \leq \sum_{l \in \mathbb{L}} \frac{\delta_{lp}}{\mu_l} T_l(\tilde{\rho}_l), \text{ if } \tilde{x}_p = 0.$$

Let $z_p = \sum_{k \in \mathbb{K}_p} c^k x_p^k$. The above equations become

$$\hat{z}_p \hat{\lambda}_w = \sum_{l \in \mathbb{L}} \frac{\delta_{lp}}{\mu_l} T_l(\hat{\rho}_l) \hat{z}_p,$$
$$\tilde{z}_p \hat{\lambda}_w \leq \sum_{l \in \mathbb{L}} \frac{\delta_{lp}}{\mu_l} T_l(\hat{\rho}_l) \tilde{z}_p,$$

and

$$\tilde{z}_p \tilde{\lambda}_w = \sum_{l \in \mathbb{L}} \frac{\delta_{lp}}{\mu_l} T_l(\tilde{\rho}_l) \tilde{z}_p,$$
$$\hat{z}_p \tilde{\lambda}_w \leq \sum_{l \in \mathbb{L}} \frac{\delta_{lp}}{\mu_l} T_l(\tilde{\rho}_l) \hat{z}_p.$$

By summing up over $p \in \mathcal{P}_w$, and $w \in W$, we obtain

$$\sum_{w \in W} \hat{v}_w \hat{\lambda}_w = \sum_{l \in \mathbb{L}} \hat{\rho}_l T_l(\hat{\rho}_l),$$
$$\sum_{w \in W} \tilde{v}_w \hat{\lambda}_w \leq \sum_{l \in \mathbb{L}} \hat{\rho}_l T_l(\hat{\rho}_l),$$

and

$$\sum_{w \in W} \tilde{v}_w \tilde{\lambda}_w = \sum_{l \in \mathbb{L}} \tilde{\rho}_l T_l(\tilde{\rho}_l),$$
$$\sum_{w \in W} \hat{v}_w \tilde{\lambda}_w \leq \sum_{l \in \mathbb{L}} \tilde{\rho}_l T_l(\tilde{\rho}_l).$$

Indeed, we have from (3.1), $r_w^k = \sum_{p \in \mathcal{P}_w^k} x_p^k$, multiplying by c^k and summing up over $k \in \mathbb{K}$, we obtain

$$\begin{aligned} v_w &= \sum_{k \in \mathbb{K}} \sum_{p \in \mathcal{P}_w^k} c^k x_p^k = \sum_{p \in \mathcal{P}_w} \sum_{k \in \mathbb{K}_p} c^k x_p^k \\ &= \sum_{p \in \mathcal{P}_w} z_p. \end{aligned}$$

It comes that

$$\sum_{w \in W} (\tilde{v}_w - \hat{v}_w)(\tilde{\lambda}_w - \hat{\lambda}_w) \geq \sum_{l \in \mathbb{L}} (\tilde{\rho}_l - \hat{\rho}_l)(T_l(\tilde{\rho}) - T_l(\hat{\rho}_l)) > 0. \quad (3.9)$$

The last inequality follows from assumption **B**. Proceeding as in the first part of the proof, we obtain $\tilde{\lambda}^k_{\overline{w}} > \hat{\lambda}^k_{\overline{w}}$ for all $k \in I\!\!K$. □

REMARK 3.1 In the case where all classes ship flow from a common source s to a common destination d i.e., $\mathcal{P}^k = \{(sd)\}, \forall\, k$, Theorem 3.1 establishes the monotonicity of performance (given by travel cost $\lambda^k_{(sd)}$) at Wardrop equilibrium for all $k \in I\!\!K$, when the demands of classes increases.

5. Avoiding Braess paradox

The purpose of this section is to provide some methods for adding resources to a general network, with one source s and one destination d, that guarantee improvement in performance. This would guarantee in particular that the well known Braess paradox (in which adding a link results in deterioration of the performance for all users) does not occur.

For some given network with one source and one destination, the designer problem is to distribute some additional capacity among the links of the network so as to improve the performances at the (Wardrop) equilibrium.

Adding capacity in the network can be done by several ways. Among them,

(1) by adding a new direct path from the source s to the destination d,

(2) by improving an existing direct path,

(3) by improving all the paths connecting s to d.

We first consider (1), i.e. the addition of a direct path from s to d that can be used by the jobs of all classes. That direct path could be in fact a whole new network, provided that it is disjoint with the previous network; it may also have new sources and destinations in addition to s and d and new traffic from new classes that use these new sources and destinations. The next theorem shows that this may lead to a decrease of the costs of all paths used at equilibrium.

THEOREM 3.2 *Consider a cost vector* $\mathbf{T}$ *that satisfies Assumption* $\boldsymbol{A}$ *or* $\boldsymbol{B}$. *Let* $\hat{\mathbf{X}}$ *and* $\tilde{\mathbf{X}}$ *be the Wardrop equilibria after and before the addition of a direct path* $\hat{p}$ *from* s *to* d. *Consider* $\tilde{\lambda}^k_{(sd)}$ *and* $\hat{\lambda}^k_{(sd)}$ *the travel costs for class* k *respectively at* $\tilde{\mathbf{X}}$ *and* $\hat{\mathbf{X}}$. *Then,* $\hat{\lambda}^k_{(sd)} \le \tilde{\lambda}^k_{(sd)}$, $\forall\, k \in I\!\!K$. *Moreover the last inequality is strict if* $\hat{x}_{\hat{p}} > 0$.

Proof. Consider the same network $(I\!\!M, I\!\!L)$ with the initial service rate configuration $\tilde{\mu}$ and throughput demand $(\bar{r}^k_{(sd)})_{k\in I\!\!K}$ where $\bar{r}^k_{(sd)} =$

$r^k_{(sd)} - \hat{x}^k_{\hat{p}}$ for all class $k \in I\!K$. Let $\bar{\mathbf{X}}$ represent the Wardrop equilibrium associated to this new throughput demand and $\bar{\lambda}^k_{(sd)}$ the travel cost for class k at Wardrop equilibrium $\bar{\mathbf{X}}$. From Conditions (3.1) and (3.2) we have $\bar{\lambda}^k_{(sd)} = \hat{\lambda}^k_{(sd)}$, $\forall\, k \in I\!K$. If $x_{\hat{p}} = 0$, then $\bar{\lambda}^k_{(sd)} = \tilde{\lambda}^k_{(sd)}$, which implies that $\hat{\lambda}^k_{(sd)} = \tilde{\lambda}^k_{(sd)}$.

Assume, then, that $\hat{x}_{\hat{p}} > 0$. We have $\bar{r}_{(sd)} < r_{(sd)}$ (which will be used for Assumption **A**) and $\bar{v}_{(sd)} < v_{(sd)}$ (which will be used for Assumption **B**), following Theorem 3.1, we conclude that $\hat{\lambda}^k_{(sd)} = \bar{\lambda}^k_{(sd)} < \tilde{\lambda}^k_{(sd)}$, for all $k \in I\!K$ and this completes the proof. □

We now examine the second way of adding capacity to the network, namely, the improvement of an existing direct path. We consider a network $(I\!M, I\!L)$ that contains a direct path, $\hat{p}$, from s to d that can be used by the jobs of all classes. We derive sufficient conditions that guarantee an improvement in the performance when we increase the capacity of this direct path.

THEOREM 3.3 *Let $\boldsymbol{T}$ a cost vector satisfying Assumptions $\boldsymbol{A}$. We consider an improvement of the path $\hat{p}$ so that the cost associated to this path is smaller for all classes, i.e., $\hat{T}_{\hat{p}}(\boldsymbol{x}) < \tilde{T}_{\hat{p}}(\boldsymbol{x})$. Let $\hat{\boldsymbol{X}}$ and $\tilde{\boldsymbol{X}}$ the Wardrop equilibria respectively after and before this improvement. Consider $\tilde{\lambda}^k_{(sd)}$ and $\hat{\lambda}^k_{(sd)}$ the travel cost of class k at equilibria. Then $\hat{\lambda}^k_{(sd)} \le \tilde{\lambda}^k_{(sd)}$, $\forall\, k \in I\!K$. Moreover the inequality is strict if $\hat{x}_{\hat{p}} > 0$ or $\tilde{x}_{\hat{p}} > 0$.*

Proof. From Lemma 3.1 we have

$$\begin{aligned} \lambda_{(sd)} &= T_p(\mathbf{x}), \quad x_p > 0, \\ \lambda_{(sd)} &\le T_p(\mathbf{x}), \quad x_p = 0, \\ \sum_{p \in \cup_k \mathcal{P}^k_{(sd)}} x_p &= r_{(sd)}, \quad x_p \ge 0, \; p \in \mathcal{P}^k. \end{aligned} \tag{3.10}$$

We know from Theorems 3.2 and 3.14 in Patriksson (1994) that $\hat{\mathbf{x}}$ and $\tilde{\mathbf{x}}$ must satisfy the variational inequalities

$$\hat{\mathbf{T}}'(\hat{\mathbf{x}})^T(\mathbf{x} - \hat{\mathbf{x}}) \;\ge\; 0, \quad \forall\, \mathbf{x} \text{ that satisfies (3.10)}, \tag{3.11}$$

$$\tilde{\mathbf{T}}'(\hat{\mathbf{x}})^T(\mathbf{x} - \tilde{\mathbf{x}}) \;\ge\; 0, \quad \forall\, \mathbf{x} \text{ that satisfies (3.10)}. \tag{3.12}$$

By adding (3.11) with $\mathbf{x} = \tilde{\mathbf{x}}$ and (3.12) with $\mathbf{x} = \hat{\mathbf{x}}$, we obtain $[\hat{\mathbf{T}}'(\hat{\mathbf{x}}) - \tilde{\mathbf{T}}'(\tilde{\mathbf{x}})][\hat{\mathbf{x}} - \tilde{\mathbf{x}}] \le 0$, thus

$$[\hat{\mathbf{T}}'(\hat{\mathbf{x}}) - \tilde{\mathbf{T}}'(\hat{\mathbf{x}}) + \tilde{\mathbf{T}}'(\hat{\mathbf{x}}) - \tilde{\mathbf{T}}'(\tilde{\mathbf{x}})][\hat{\mathbf{x}} - \tilde{\mathbf{x}}] \le 0,$$

and

$$[\hat{\mathbf{T}}'(\hat{\mathbf{x}}) - \tilde{\mathbf{T}}'(\hat{\mathbf{x}})][\hat{\mathbf{x}} - \tilde{\mathbf{x}}] \leq [\tilde{\mathbf{T}}'(\hat{\mathbf{x}}) - \tilde{\mathbf{T}}'(\tilde{\mathbf{x}})][\hat{\mathbf{x}} - \tilde{\mathbf{x}}] < 0. \tag{3.13}$$

Since the costs of other paths are unchanged, i.e., $\hat{T}_p = \tilde{T}_p$ for all $p \neq \hat{p}$, Equation (3.13) becomes $(\hat{T}_{\hat{p}}(\hat{\mathbf{x}}) - \tilde{T}_{\hat{p}}(\hat{\mathbf{x}})(\hat{x}_{\hat{p}} - \tilde{x}_{\hat{p}}) < 0$ if $\hat{\mathbf{x}} \neq \tilde{\mathbf{x}}$. Since $\hat{T}_{\hat{p}}(\hat{\mathbf{x}}) < \tilde{T}_{\hat{p}}(\hat{\mathbf{x}})$, we have

$$\hat{x}_{\hat{p}} > \tilde{x}_{\hat{p}} \text{ if } \hat{\mathbf{x}} \neq \tilde{\mathbf{x}}. \tag{3.14}$$

Now we have two cases:

- If $\hat{\mathbf{x}} = \tilde{\mathbf{x}}$ and since $\hat{T}_{\hat{p}}(\mathbf{x}) \neq \tilde{T}_{\hat{p}}(\mathbf{x})\ \forall\, \mathbf{x}$, it follows that $\hat{x}_{\hat{p}} = \tilde{x}_{\hat{p}} = 0$, which implies that $\hat{\lambda}^k_{(sd)} = \tilde{\lambda}^k_{(sd)}$.
- If $\hat{\mathbf{x}} \neq \tilde{\mathbf{x}}$, then from (3.14) we have $\hat{x}_{\hat{p}} > \tilde{x}_{\hat{p}}$. Consider now two networks that differ only by the presence or absence of the direct path $\hat{p}$ from s to d. In both networks we have the same initial capacity configuration and the same set $I\!K$ of classes, with respectively demands $\check{r}^k_{(sd)} = r^k_{(sd)} - \hat{x}^k_{\hat{p}}$ and $\bar{r}^k_{(sd)} = r^k_{(sd)} - \tilde{x}^k_{\hat{p}}$. Let $\check{\lambda}^k_{(sd)}$ and $\bar{\lambda}^k_{(sd)}$ the travel cost of class k associated to these throughput demands. Since $\hat{x}_{\hat{p}} > \tilde{x}_{\hat{p}}$ then $\check{r}_{(sd)} < \bar{r}_{(sd)}$, and from Theorem 3.1 we have
$$\forall\, k \in I\!K, \quad \check{\lambda}^k_{(sd)} < \bar{\lambda}^k_{(sd)}. \tag{3.15}$$
On the other hand, for the network with demands $(\check{r}^k_{(sd)})_{k \in K}$, it is easy to see that the equilibria conditions (3.1) and (3.2) are satisfied by the system flow configuration $\check{\mathbf{X}}$, with $\check{\lambda}^k_{(sd)} = \hat{\lambda}^k_{(sd)}$. Similarly we conclude that the network with demands $(\check{r}^k_{(sd)})_{k \in I\!K}$ has the system flow configuration $\bar{\mathbf{X}}$, with $\bar{\lambda}^k_{(sd)} = \tilde{\lambda}^k_{(sd)}$. Hence from (3.15) we obtain $\hat{\lambda}^k_{(sd)} < \tilde{\lambda}^k_{(sd)}$. □

THEOREM 3.4 *Consider a cost function vector that satisfies Assumptions **B**. Let $\hat{\mu}^k_l$ and $\tilde{\mu}^k_l$, respectively, be the service rate configurations after and before adding the capacity to the path $\hat{p}$, i.e $\hat{\mu}_l > \tilde{\mu}_l$ for $l \in \hat{p}$ and $\hat{\mu}_l = \tilde{\mu}_l$ for $l \notin \hat{p}$. Let $\hat{\boldsymbol{X}}$ and $\tilde{\boldsymbol{X}}$, respectively, the Wardrop equilibria after and before this improvement. Consider $\hat{\lambda}^k_{(sd)}$ and $\tilde{\lambda}^k_{(sd)}$ the travel cost of class k at the equilibria. Then $\hat{\lambda}^k_{(sd)} \leq \tilde{\lambda}^k_{(sd)}$, $\forall\, k \in I\!K$. Moreover the inequality is strict if $\hat{x}_{\hat{p}} > 0$ or $\tilde{x}_{\hat{p}} > 0$.*

Proof. Note that if there exists a link l_1 that belongs to the path $\hat{p}$, such that $\hat{\rho}_{l_1} < \tilde{\rho}_{l_1}$, then $\hat{\rho}_l < \tilde{\rho}_l$ for each link that belongs to the path $\hat{p}$.

Assume, then, that $\hat{\rho}_l \leq \tilde{\rho}_l$ for $l \in \hat{p}$. We have two possibilities. First, if $\tilde{x}_{\hat{p}} = 0$, then $\hat{\lambda}^k_{(sd)} = \tilde{\lambda}^k_{(sd)}$ for all $k \in I\!\!K$. Second, if $\tilde{x}_{\hat{p}} > 0$, then we have

$$\tilde{\lambda}_{(sd)} = \sum_{l \in \hat{p}} \frac{\tilde{T}_l(\tilde{\rho}_l)}{\tilde{\mu}_l} \text{ and } \hat{\lambda}_{(sd)} \leq \sum_{l \in \hat{p}} \frac{\hat{T}_l(\hat{\rho}_l)}{\hat{\mu}_l}.$$

Since $T_l(.)$ is strictly increasing and $\tilde{\mu}_l < \hat{\mu}_l$ for all $l \in \hat{p}$, we have $\hat{\lambda}_{(sd)} \leq \sum_{l \in \hat{p}} \frac{\hat{T}_l(\hat{\rho}_l)}{\hat{\mu}_l} < \sum_{l \in \hat{p}} \frac{\tilde{T}_l(\tilde{\rho}_l)}{\tilde{\mu}_l} = \tilde{\lambda}_{(sd)}$. It follows that $\hat{\lambda}^k_{(sd)} < \tilde{\lambda}^k_{(sd)}$ for all $k \in I\!\!K$.

Now assume that $\hat{\rho}_l > \tilde{\rho}_l$ for $l \in \hat{p}$. Let us consider the two networks that differ only by the presence or absence of the direct path $\hat{p}$ from s to d. In both networks we have the same initial capacity configuration and the same set $I\!\!K$ of classes, with respective demands $\check{r}^k_{(sd)} = r^k_{(sd)} - \hat{x}^k_{\hat{p}}$ and $\bar{r}^k_{(sd)} = r^k_{(sd)} - \tilde{x}^k_{\hat{p}}$. Let $\check{\lambda}^k_{(sd)}$ and $\bar{\lambda}^k_{(sd)}$ the travel costs of class k associated to these throughput demands. Since $\hat{\rho}_l > \tilde{\rho}_l$ and $\hat{\mu}_l > \tilde{\mu}_l$ for $l \in \hat{p}$, we have

$$\begin{aligned}
\bar{v}_{(sd)} - \check{v}_{(sd)} &> \sum_{k \in I\!\!K} c^k \bar{r}^k_{(sd)} - c^k \check{r}^k_{(sd)} \\
&= \sum_{k \in I\!\!K} c^k (r^k_{(sd)} - \tilde{x}^k_{\hat{p}}) - c^k (r^k_{(sd)} - \hat{x}^k_{\hat{p}}) \\
&= \sum_{k \in I\!\!K} c^k (\hat{x}^k_{\hat{p}}) - c^k (\tilde{x}^k_{\hat{p}}) \\
&= \sum_{l \in \hat{p}} (\hat{\mu}_l \hat{\rho}_{\hat{l}} - \tilde{\mu}_l \tilde{\rho}_l) > 0.
\end{aligned}$$

From Theorem 3.1, we conclude that $\check{\lambda}^k_{(sd)} < \bar{\lambda}^k_{(sd)}$ for all $k \in I\!\!K$. Proceeding as in the proof of Theorem 3.3, we obtain $\hat{\lambda}^k_{(sd)} < \tilde{\lambda}^k_{(sd)}$ for all $k \in I\!\!K$. □

We examine the last way of adding capacity to the network, i.e. the addition of capacity on all the paths connecting s to d. Consider a network $(I\!\!M, I\!\!L)$ and a cost vector $\mathbf{T}$ that satisfies Assumption **A** or **B**. We consider the improvement of the capacity of all path so that the following holds:

$$\hat{T}^k_p(\mathbf{X}) = \frac{1}{\alpha} \tilde{T}^k_p(\frac{\mathbf{X}}{\alpha}), \text{ with } \alpha > 1. \tag{3.16}$$

We observe that for any $\alpha > 1$, $\hat{T}^k_p(\mathbf{X}) = \frac{1}{\alpha} \tilde{T}^k_p(\frac{\mathbf{X}}{\alpha}) < \tilde{T}^k_p(\frac{\mathbf{X}}{\alpha}) < \tilde{T}^k_p(\mathbf{X})$.

THEOREM 3.5 *Consider a cost vector $\boldsymbol{T}$ that satisfies Assumption $\boldsymbol{A}$ or $\boldsymbol{B}$. Let $\tilde{\boldsymbol{X}}$ and $\hat{\boldsymbol{X}}$ be the Wardrop equilibria associated respectively to cost* *functions $\tilde{T}_p^k$ and $\hat{T}_p^k$. Consider $\tilde{\lambda}_{(sd)}^k$ and $\hat{\lambda}_{(sd)}^k$ the travel cost of class k at the respective Wardrop equilibria $\tilde{\mathbf{X}}$ and $\hat{\mathbf{X}}$. Then $\hat{\lambda}_{(sd)}^k < \tilde{\lambda}_{(sd)}^k$, $\forall\, k \in I\!K$.*

Proof. We consider now the network $(I\!M, I\!L)$, with travel costs $\tilde{T}_p^k$ and throughput demands $\bar{r}_{(sd)}^k = r_{(sd)}^k/\alpha$, $k \in I\!K$. Let $\bar{\lambda}_{(sd)}^k$ the travel cost of class k associated to these throughput demands. At equilibrium $\hat{\mathbf{X}}$, by redefining the cost and path flows as $\alpha\hat{\lambda}_{(sd)}^k$ and $(1/\alpha)\hat{x}_p^k$, respectively, it is straightforward to show that changing the demands from $r_{(sd)}^k$ to $\bar{r}_{(sd)}^k$ using the cost functions $\hat{T}_p^k(\mathbf{X})$ is equivalent to changing the cost functions from $\hat{T}_p^k(\mathbf{X})$ to $\tilde{T}_p^k(\mathbf{X}/\alpha)$ using the demands $r_{(sd)}^k$. Hence the corresponding travel costs are $\bar{\lambda}_{(sd)}^k = \alpha\hat{\lambda}_{(sd)}^k$. On the other hand, we have $\bar{r}_{(sd)} = r_{(sd)}/\alpha < r_{(sd)}$ and $\bar{v}_{(sd)} = v_{(sd)}/\alpha < v_{(sd)}$ hence from Theorem 3.1, $\bar{\lambda}_{(sd)}^k < \tilde{\lambda}_{(sd)}^k$, $\hat{\lambda}_{(sd)}^k = \bar{\lambda}_{(sd)}^k/\alpha < \bar{\lambda}_{(sd)}^k < \tilde{\lambda}_{(sd)}^k$, which concludes the proof. □

6. An open BCMP queuing network

In this section we study an example of such Braess paradox in networks consisting entirely of **BCMP** (Baskett et al. (1975)) queuing networks (BCMP stand for the initial of the authors) (see also Kelly (1979)).

6.1 BCMP queuing network

We consider an open **BCMP** queuing network model that consists of L service links. Each service center contains either a single-server queue with the processor-sharing (**PS**). We assume that the service rate of each single server is state independent. Jobs are classified into K different classes. The arrival process of jobs of each class forms a Poisson process and is independent of the state of the system.

Let us denote the state of the network by $\mathbf{n} = (n_1, n_2, \ldots, n_L)$ where $\mathbf{n}_l = (\mathbf{n}_l^1, \mathbf{n}_l^2, \ldots, \mathbf{n}_l^K)$ and $n_l = \sum_{l \in L} n_l^k$ where n_l^k denotes the total number of jobs of class k visiting link l. For an open queuing network (Kelly (1979); Baskett et al. (1975)), the equilibrium probability of the network state $\mathbf{n}$ is obtained as follows:

$$p(\mathbf{n}) = \prod_{l \in L} \frac{p_l(n_l)}{G_l},$$

where $p_l(n_l) = n_l! \prod_{l \in L} (\rho_l^k)/n_l^k$ and $G_l = 1/(1-\rho_l)$. Let $E[n_l^k]$ be the average number of class k jobs at link l. We have $E[n_l^k] = \rho_l^k/(1-\rho_l)$.

By using Little's formula, we have

$$T_l^k = \frac{E[n_l^k]}{x_l^k} = \frac{1/\mu_l^k}{(1-\rho_l)},$$

from which the average delay of a class k job that flows through path-class $p \in \mathcal{P}^k$ is given by

$$T_p^k = \sum_{l \in L} \delta_{lp} T_l^k = \sum_{l \in L} \delta_{lp} \frac{1/\mu_l^k}{(1-\rho_l)}.$$

We assume that μ_l^k can be represented by μ_l/c^k, hence the average delays satisfy assumption **B**.

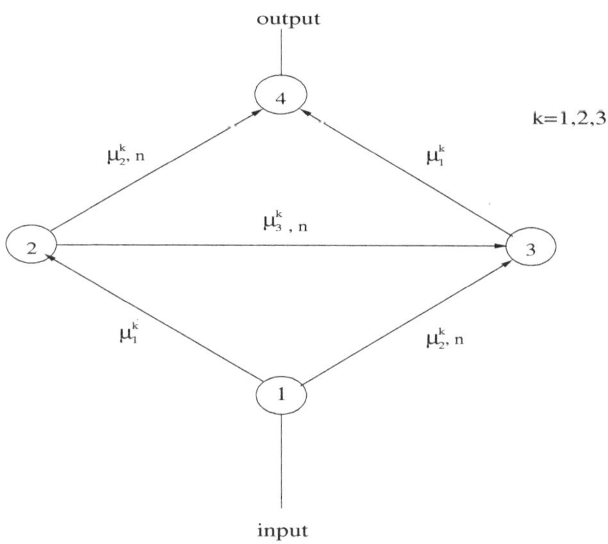

Figure 3.1. Network

6.2 Braess paradox

Consider the networks shown in Figure 3.1. Packets are classified into *three* different classes. Links (1,2) and (3,4) have each the following service rates: $\mu_1^k = \mu_1$, $\mu_1^2 = 2\mu_1$ and $\mu_1^3 = 3\mu_1$ where $\mu_1 = 2.7$. Link (1,3) represents a path of n tandem links, each with the service rates: $\mu_2^1 = \mu_2$, $\mu_2^2 = 2\mu_2$ and $\mu_2^3 = 3\mu_2$ with $\mu_2 = 27$. Similarly link (2,4) is a path made of n consecutive links, each with service rates: $\mu_2^1 = 27$, $\mu_2^2 = 54$ and $\mu_2^3 = 81$. Link (2,3) is path of n consecutive links each with service rate of each class $\mu_3^1 = \mu$, $\mu_3^2 = 2\mu$ and $\mu_3^2 = 3\mu$ where μ varies from 0 (absence of the link) to infinity. We denote $x^k_{p_1}$ the left flow of class k using links (1,2) and (2,4), $x^k_{p_2}$ the right flow of class k using links (1,3) and (3,4), and x_{p_3} the zigzag flow of class k using links (1,2), (2,3) and (3,4). The total cost for each class is given by

$$\mathbf{T}^k = x^k_{p_1} T^k_{p_1} + x^k_{p_2} T^k_{p_2} + x^k_{p_3} T^k_{p_3},$$

where $x^k_{p_1} + x^k_{p_1} + x^k_{p_1} = r^k$.

We first consider the scenario where additional capacity μ is added to path (2,3), for $n = 54$, $r^1 = 0.6$, $r^2 = 1.6$ and $r^3 = 1.8$. In Figure 3.2 we observe that no traffic uses the zigzag path for $0 \leq \mu \leq 36.28$. For $36.28 \leq \mu \leq 96.49$, all three paths are used. For $\mu > 96.49$, all traffic uses the zigzag path. For μ between 36.28 and 96.49, the delay is, paradoxically, worse than it would be without the zigzag path. The delay of class 1 (*resp.* 2,3) decreases to 2.85 (*resp.* 1.42, 0.95) as μ goes to infinity.

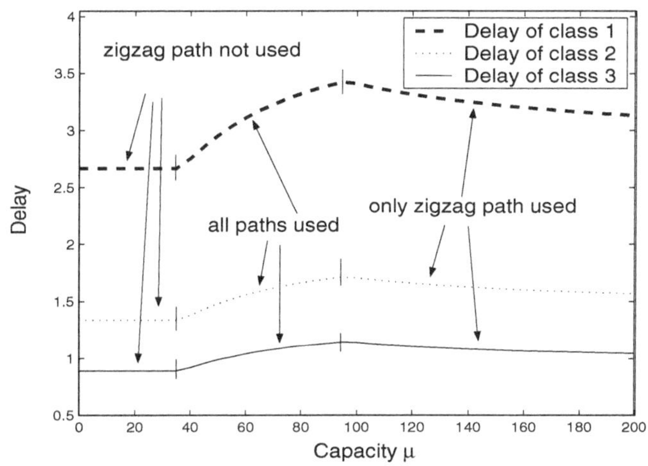

Figure 3.2. Delay of each class as a function of the added capacity in path (2,3)

6.3 Adding a direct path between source and destination

Now we use the method proposed in Theorems 3.2 and 3.3, i.e., the upgrade achieved by adding a direct path connecting source 1 and destination 4.

The results in Theorems 3.2 and 3.3 suggest that yet another good design practice is to focus the upgrades on direct connections between source and destination; and Figure 3.4 illustrates that indeed this approach decreases the delay of each class.

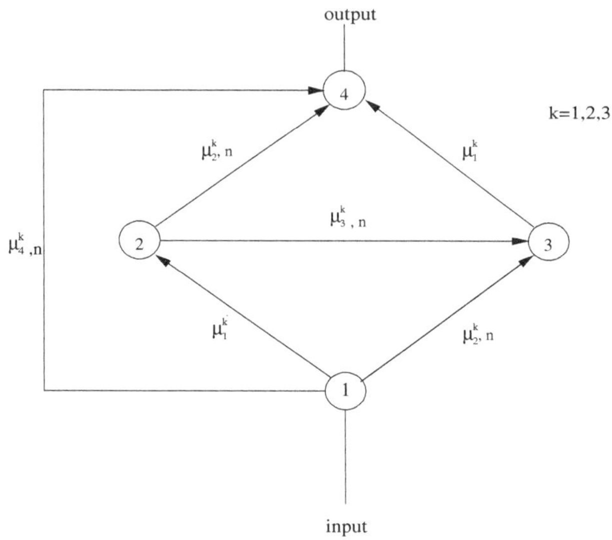

Figure 3.3. New network

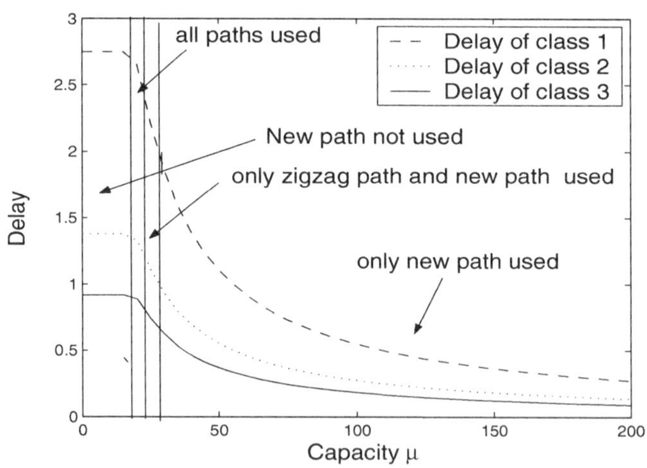

Figure 3.4. Delay as a function of the added capacity in path (1,4)

6.4 Multiplying the capacity of all links ($l \in \mathbb{L}$) by a constant factor $\alpha > 1$.

Now we use the method proposed in Theorem 3.5 for efficiently adding resources to this network.

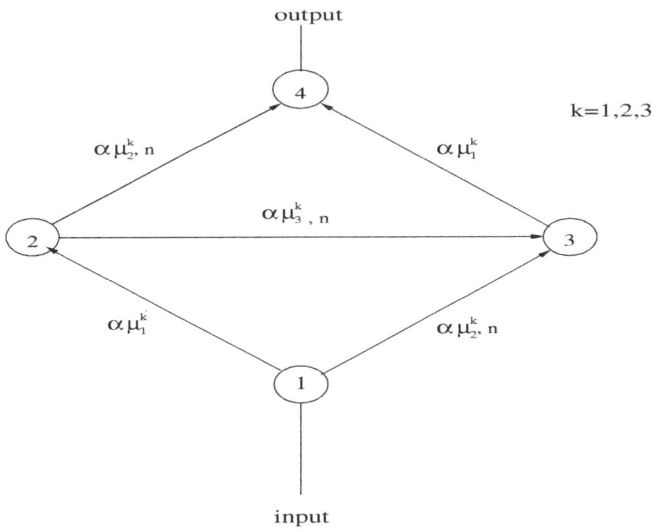

Figure 3.5. New network

Figure 3.6 shows the delay of each class as a function of the additional capacity μ where $\mu = (\alpha - 1)(2\mu_1 + 2\mu_2 + \mu 3)$ with $\mu_1 = 2.7$, $\mu_2 = 27$

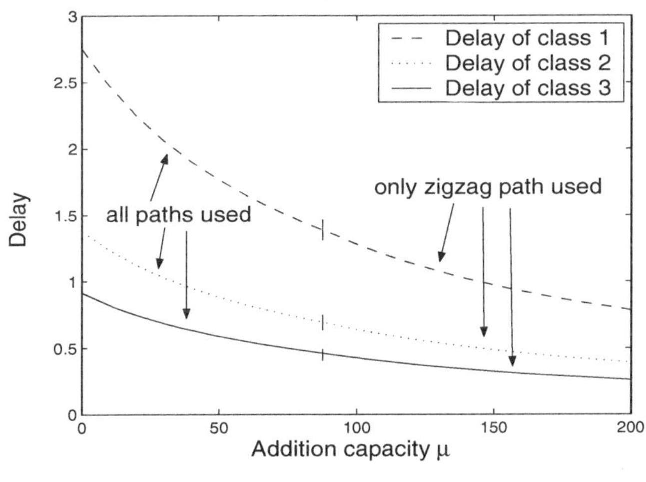

Figure 3.6. Delay of each class as a function of the added capacity in all links

and $\mu_3 = 40$. Figure 3.6 indicates that the delay of each class decreases when the additional capacity μ increases. Hence the Braess paradox is indeed avoided.

Acknowledgments. This work was partially supported by a research contract with France Telecom R&D No. 001B001.

References

Altman, E., El Azouzi, R., and Pourtallier, O. (2001). Avoiding paradoxes in routing games. *Proceedings of the 17th International Teletraffic Congress*, Salvador da Bahia, Brazil, September 24–28.

Baskett, F., Chandy, K.M., Muntz, R.R., and Palacios, F. (1975). Open, closed and mixed networks of queues with different classes of customers. *Journal of the ACM*, 22(2):248–260.

Beans, N.G., Kelly, F.P., and Taylor, P.G. (1997). Braess's paradox in a loss network. *Journal of Applied Probability*, 34:155—159.

Braess, D. (1968). Über ein Paradoxen aus der Verkehrsplanung. *Unternehmensforschung*, 12:258–268.

Calvert, B., Solomon, W., and Ziedins, I. (1997). Braess's paradox in a queueing network with state-dependent routing. *Journal of Applied Probability*, 34:134–154.

Cohen, J.E. and Jeffries, C. (1997). Congestion Resulting from Increased Capacity in Single-Server Queueing Networks. *IEEE/ACM Transactions on Networking*, 5(2):1220–1225.

Cohen, J.E. and Kelly, F.P. (1990). A Paradox of Congestion in a Queueing Network. *Journal of Applied Probability*, 27:730–734.

Dafermos and Nagurney, A. (1984a). On some traffic equilibrium theory paradoxes. *Transportation Research B*, 18:101–110.

Dafermos and Nagurney, A. (1984b). Sensitivity analysis for the asymmetric network equilibrium problem. *Mathematical Programming*, 28: 174–184.

Gupta, P. and Kumar, P.R. (1998). A system and traffic dependent adaptive routing algorithm for Ad Hoc networks. *Proceedings of the 37th IEEE Conference on Decision and Control*, Tampa, Florida, USA.

Kameda, H. and Pourtallier, O. (2002). Paradoxes in Distributed Decisions on Optimal Load Balancing for Networks of Homogeneous Computers. *Journal of the ACM*, 49(3):407-433.

Kameda, H. and Zhang, Y. (1995). Uniqueness of the solution for optimal static routing in open BCMP queueing networks. *Mathematical and Computer Modelling*, 22(10–12):119–130.

Kameda,H., Altman, E., and Kozawa, T. (1999). A case where a paradox like Braess's occurs in the Nash equilibrium but does not occur in the Wardrop equilibrium — a situation of load balancing in distributed computer systems. *Proceedings of the 38th IEEE Conference on Decision and Control*, Phoenix, Arizona, USA.

Kameda, H., Altman, E., Kozawa, T., and Hosokawa, Y. (2000). Braess-like paradoxes of Nash Equilibria for load balancing in distributed computer systems. *IEEE Transactions on Automatic Control*, 45(9): 1687–1691.

Kelly, F.P. (1979). *Reversibility and Stochastic Networks.* John Wiley & Sons, Ltd., New York.

Kobayashi, H. (1978). *Modelling and Analysis, an Introduction to system Performance Evaluation Methodology.* Addison Wesley.

Korilis, Y.A., Lazar, A.A., and Orda, A. (1995). Architecting noncooperative networks. *IEEE Journal on Selected Areas in Communications*, 13:1241–1251.

Korilis, Y.A., Lazar, A.A., and Orda, A. (1997). Capacity allocation under non-cooperative routing. *IEEE Transactions on Automatic Control*, 42(3):309–325.

Korilis, Y.A., Lazar, A.A., and Orda, A. (1999). Avoiding the Braess paradox in non-cooperative networks. *Journal of Applied Probability*, 36:211–222.

Massuda, Y. (1999). *Braess's Paradox and Capacity Management in Decentralised Network*, manuscript.

Orda, A., Rom, R., and Shimkin, N. (1993). Competitive routing in multi-user environments. *IEEE/ACM Transactions on Networking*, 1:510–521.

Patriksson, M. (1994). *The Traffic Assignment Problem: Models and Methods.* VSP BV, Topics in Transportation, The Netherlands.

Roughgarden, T. and Tardos, E. (2002). How bad is selfish routing. *Journal of the ACM*, 49(2):236–259.

Smith, M.J. (1979). The marginal cost taxation of a transportation network. *Transportation Research-B*, 13:237–242.

Wardrop, J.G. (1952). Some theoretical aspects of road traffic research. *Proceedings of the Institution of Civil Engineers*, Part 2, pages 325–378.

Chapter 4

PRODUCTION GAMES AND PRICE DYNAMICS

Sjur Didrik Flåm

Abstract This note considers production (or market) games with transferable utility. It brings out that, in many cases, explicit core solutions may be defined by shadow prices — and reached via quite natural dynamics.

1. Introduction

Noncooperative game theory has, during recent decades, come to occupy central ground in economics. It now unifies diverse fields and links many branches (Forgó, Szép and Szidarovszky (1999), Gintis (2000), Verga-Redondo (2003)). Much progress came with circumventing the strategic form, focusing instead on extensive games. Important in such games are the rules that prescribe who can do what, when, and on the basis of which information.

Cooperative game theory (Forgó, Szép and Szidarovszky (1999), Peyton Young (1994)) has, however, in the same period, seen less of expansion and new applications. This fact might mirror some dissatisfaction with the plethora of solution concepts — or with applying only the *characteristic function.* Notably, that function subsumes — and often conceals — underlying activities, choices and data; all indispensable for proper understanding of the situation at hand. By directing full attention to payoff (or costs) sharing, the said function presumes that each relevant input already be processed.

Such predilection to work only with reduced, essential data may entail several risks: One is to overlook prospective "devils hidden in the details." Others could come with ignoring particular features, crucial for formation of viable coalitions. Also, the timing of players' decisions, and the associated information flow, might easily escape into the background (Flåm (2002)).

All these risks apply, in strong measure, to so-called*production* (or *market*) games, and they are best mitigated by keeping data pretty much as is. The said games are of frequent occurrence and great importance. Each instance involves parties concerned with equitable sharing of efficient production costs. Given a nonempty finite player set I, *coalition* $S \subseteq I$ presumably incurs a *stand-alone cost* $C_S \in \mathbb{R} \cup \{+\infty\}$ that results from explicit planning and optimization.[1] Along that line I consider here the quite general format

$$C_S := \inf \{f_S(x) + h_S(g_S(x))\}. \tag{4.1}$$

In (4.1) the function f_S takes a prescribed set $\mathbb{X}_S$ into $\mathbb{R} \cup \{+\infty\}$; the operator g_S maps that same set $\mathbb{X}_S$ into a real vector space $\mathbb{E}$; and finally, $h_S : \mathbb{E} \to \mathbb{R} \cup \{+\infty\}$ is a sort of penalty function. Section 2 provides examples.

As customary, a cost profile $c = (c_i) \in \mathbb{R}^I$ is declared in the *core*, written $c \in core$, iff it embodies

full cost cover:	$\sum_{i \in I} c_i \geq C_I$	and
coalitional stability:	$\sum_{i \in S} c_i \leq C_S$	for each nonempty subset $S \subseteq I$.

Plainly, this solution concept makes good sense when *core* is neither empty, nor too large, nor very sensitive to data.

Given the characteristic function $S \mapsto C_S$, defined by (4.1), my objects below are three: *first*, to study duality without invoking integer programming[2]; *second*, to display explicit core solutions, generated by so-called shadow prices; and *third*, to elaborate how such entities might be reached.

Motivation stems from several sources: There is the recurrent need to reach beyond instances with convex preferences and production sets. Notably, some room should be given to discrete activity (decision) sets $\mathbb{X}_S$ — as well as to non-convex objectives f_S and constraint functions g_S. Besides, the well known bridge connecting competitive equilibrium to core outcomes (Ellickson (1993)), while central in welfare economics, deserves easier and more frequent crossings — in both directions. Also, it merits emphasis that Lagrangian duality, the main vehicle here, often invites more tractable computations than might commonly be expected. And, not the least, like in microeconomic theory (Mas-Colell, Whinston

[1] References include Dubey and Shapley (1984), Evstigneev and Flåm (2001), Flåm (2002), Flåm and Jourani (2003), Granot (1986), Kalai and Zemel (1982), Owen (1975), Samet and Zemel (1994), Sandsmark (1999) and Shapley and Shubik (1969).

[2] Important issues concern indivisibilities and mathematical programming, but these will be avoided here; see Gomory and Baumol (1960), Scarf (1990), Scarf (1994), Wolsey (1981).

and Green (1995)), one wonders about the emergence and stability of equilibrium prices.

What imports in the sequel is that key resources — seen as private endowments, and construed as vectors in $\mathbb{E}$ — be perfectly divisible and transferable. Resource scarcity generates common willingness to pay for appropriation. Thus emerge endogenous shadow prices that equilibrate intrinsic exchange markets. Regarding the grand coalition, Section 3 argues that — absent a duality gap, and granted attainment of optimal dual values — these prices determine specific core imputations. Section 4 brings out that equilibrating prices can be reached via repeated play.

2. Production games

As said, coalition S, if it were to form, would attempt to solve problem (4.1). For interpretation construe $f_S : \mathbb{X}_S \to \mathbb{R} \cup \{+\infty\}$ as an aggregate cost function. Further, let $g_S : \mathbb{X}_S \to \mathbb{E}$ govern — and account for — resource consumption or technological features. Finally, $h_S : \mathbb{E} \to \mathbb{R} \cup \{+\infty\}$ should be seen as a penalty mechanism meant to enforce feasibility.

Coalition S has $\mathbb{X}_S$ as activity (decision) space. In the sequel no linear or topological structure will be imposed on the latter. Note though that $\mathbb{E}$, the vector space of "resource endowments", is common to all agents and coalitions. By tacit assumption $h_S(g_S(x)) = +\infty$ when $x \notin \mathbb{X}_S$.

To my knowledge *TU* production (or market) games have rarely been defined in such generality. Format (P_S) can accommodate a wide variety of instances. To wit, consider

EXAMPLE 4.1 (*Nonlinear constrained, cooperative programming.*) For each $i \in I$ there is a nonempty set $\mathbb{X}_i$, two functions $f_i : \mathbb{X}_i \to \mathbb{R} \cup \{+\infty\}$, $g_i : \mathbb{X}_i \to \mathbb{E}$, and a constraint $g_i(x_i) \in K_i \subset \mathbb{E}$. Let then $\mathbb{X}_S := \Pi_{i \in S} \mathbb{X}_i$. Further, posit $f_S(x) := \sum_{i \in S} f_i(x_i)$ and $g_S(x) := \sum_{i \in S} g_i(x_i)$. Finally, define $h_S(e) = 0$ if $e \in \sum_{i \in S} K_i$, and let $h_S(e) = +\infty$ otherwise.

EXAMPLE 4.2 (*Inf-convolution.*) Of particular importance is the special case of the preceding example where $\mathbb{X}_i = \mathbb{E}$, $g_i(x_i) = x_i - e_i$, and $K_i = \{0\}$. Coalition cost is then defined by the so-called *infimal convolution*

$$C_S := \inf \left\{ \sum_{i \in S} f_i(x_i) : \sum_{i \in S} x_i = \sum_{i \in S} e_i \right\}.$$

In Example 4.2 only convexity is needed to have a nonempty core. This is brought out by the following

PROPOSITION 4.1 (Convex separable cost yields nonempty core.) *Suppose* $\mathbb{X}_S = \Pi_{i\in S}\mathbb{X}_i$ *with each* $\mathbb{X}_i$ *convex. Also suppose*

$$C_S \geq \inf\left\{\sum_{i\in S} f_i(x_i) : \sum_{i\in S} \mathcal{A}_i x_i = 0, x_i \in \mathbb{X}_i\right\} \quad \forall S \subset I,$$

with each f_i *convex, each* $\mathcal{A}_i : \mathbb{X}_i \to \mathbb{E}$ *affine, and equality when* $S = I$. *Then the core is nonempty.*

Proof. Let $S \mapsto w_S \geq 0$ be any balanced collection of weights. That is, assume $\sum_{S:i\in S} w_S = 1$ for all i. Pick any positive ε and for each coalition S a profile $x_S = (x_{iS}) \in \mathbb{X}_S$ such that $\sum_{i\in S} f_i(x_{iS}) \leq C_S + \varepsilon$ and $\sum_{i\in S} \mathcal{A}_i x_{iS} = 0$. Posit $x_i := \sum_{S:i\in S} w_S x_{iS}$. Then $x_i \in \mathbb{X}_i$, $\sum_{i\in I} \mathcal{A}_i x_i = 0$, and

$$C_I \leq \sum_{i\in I} f_i\Big(\sum_{S:i\in S} w_S x_{iS}\Big) \leq \sum_{i\in I}\sum_{S:i\in S} w_S f_i(x_{iS}) = \sum_S w_S \sum_{i\in S} f_i(x_{iS})$$
$$\leq \sum_S w_S \left[C_S + \varepsilon\right].$$

Since $\varepsilon > 0$ was arbitrary, it follows that $C_I \leq \sum_S w_S C_S$. The Bondareva-Shapley Shapley (1967) result now certifies that the core is nonempty. □

Proposition 4.1 indicates good prospects for finding nonempty cores, but it provides less than full satisfaction: No explicit solution is listed. "Too much" convexity is required in the activity sets $\mathbb{X}_i$ and cost functions F_i. Resource aggregation is "too linear." And the original data do not enter explicitly. Together these drawbacks motivate next a closer look at the *grand coalition* $S = I$.

3. Lagrange multipliers, subgradients and min-max

This section contains auxiliary, quite useful material. It takes out time and space to consider the problem and cost

$$(P_I) \qquad C_I := \inf\left\{f_I(x) + h_I(g_I(x))\right\}$$

of the grand coalition. For easier notations, write simply P, C, f, g, h, $\mathbb{X}$ instead of P_I, C_I, f_I, g_I, h_I, $\mathbb{X}_I$, respectively. Much analysis revolves hereafter around the perturbed function

$$(x, e, y) \in \mathbb{X} \times \mathbb{E} \times \mathbb{Y} \mapsto f(x) + h(g(x) + e) - \langle y, e\rangle . \qquad (4.2)$$

Here $\mathbb{Y}$ is a judiciously chosen, convex, nonempty set of linear functionals $y : \mathbb{E} \to \mathbb{R}$. The appropriate nature of $\mathbb{Y}$ is made precise later. This means that additional properties of the functionals y (besides linearity) will be invoked only when needed. As customary, the expression $\langle y, e \rangle$ stands for $y(e)$. Objective (4.2) features a *perturbation* e available at a premium $\langle y, e \rangle$. Thus (4.2) relaxes and imbeds problem (P) into a competitive market where any endowment $e \in \mathbb{E}$ is evaluated at constant "unit price" y. To mirror this situation, associate the *Fenchel conjugate*

$$h^*(y) := \sup \{\langle y, e \rangle - h(e) : e \in \mathbb{E}\}$$

to h. If $h(e)$ denotes the cost of an enterprise that produces output e at revenues $\langle y, e \rangle$, then $h^*(y)$ is the corresponding profit. In economic jargon the firm at hand is a *price-taker* in the output market. Clearly, $h^* : \mathbb{Y} \to \mathbb{R} \cup \{\pm\infty\}$ is convex, and the *biconjugate function*

$$h^{**}(e) := \sup \{\langle y, e \rangle - h^*(y) : y \in \mathbb{Y}\}$$

satisfies $h^{**} \le h$. Anyway, the relaxed objective (4.2) naturally generates a *Lagrangian*

$$\begin{aligned} L(x, y) &:= \inf \quad \{f(x) + h(g(x) + e) - \langle y, e \rangle : e \in \mathbb{E}\} \\ &= \qquad f(x) + \langle y, g(x) \rangle - h^*(y), \end{aligned}$$

defined on $\mathbb{X} \times \mathbb{Y}$. Call now $\bar{y} \in \mathbb{Y}$ a *Lagrange multiplier* iff it belongs to the set

$$M := \left\{ \bar{y} \in \mathbb{Y} : \inf_x L(x, \bar{y}) \ge C =: \inf(P) \right\}.$$

Note that M is convex, but maybe empty. Intimately related to the problem (P) is also the marginal (optimal value) function

$$e \in \mathbb{E} \mapsto V(e) := \inf \{f(x) + h(g(x) + e) : x \in \mathbb{X}\}.$$

Of prime interest are differential properties of $V(\cdot)$ at the distinguished point $e = 0$. The functional $\bar{y} \in \mathbb{Y}$ is called a *subgradient* of $V : \mathbb{E} \to \mathbb{R} \cup \{\pm\infty\}$ at 0, written $\bar{y} \in \partial V(0)$, iff

$$V(e) \ge V(0) + \langle \bar{y}, e \rangle \text{ for all } e.$$

And V is declared *subdifferentiable* at 0 iff $\partial V(0)$ is nonempty. The following two propositions are, at least in parts, well known. They are included and proven here for completeness:

PROPOSITION 4.2 (Subgradient = Lagrange Multiplier.) *Suppose* $C = \inf(P) = V(0)$ *is finite. Then*

$$\partial V(0) = M.$$

Proof. Letting $\bar{e} := g(x) + e$ we get

$$\begin{aligned}
&\bar{y} \in \partial V(0) && \Leftrightarrow \\
&f(x) + h(\bar{e}) = f(x) + h(g(x) + e) \geq V(e) \geq V(0) + \langle \bar{y}, e \rangle \\
&\qquad\qquad\qquad\qquad \forall (x, e) \in \mathbb{X} \times \mathbb{E} && \Leftrightarrow \\
&f(x) + \langle \bar{y}, g(x) \rangle + h(\bar{e}) - \langle \bar{y}, \bar{e} \rangle \geq V(0) \quad \forall (x, \bar{e}) \in \mathbb{X} \times \mathbb{E} && \Leftrightarrow \\
&f(x) + \langle \bar{y}, g(x) \rangle - h^*(\bar{y}) \geq V(0) \quad \forall x \in \mathbb{X} && \Leftrightarrow \\
&\inf_x L(x, \bar{y}) \geq \inf(P) \Leftrightarrow \bar{y} \in M.
\end{aligned}$$

□

PROPOSITION 4.3 (Strong stability, min-max, biconjugacy and value attainment.) *The value function V is subdifferentiable at 0 — and problem (P) is then called* ***strongly stable*** *— iff $\inf(P)$ is finite and equals the saddle value*

$$\inf_x L(x, \bar{y}) = \inf_x \sup_y L(x, y) \quad \text{for each Lagrange multiplier } \bar{y}.$$

In that case $V^{**}(0) = V(0)$. Moreover, if problem (P) is strongly stable, then

- *for each optimal solution $\bar{x}$ to (P) and Lagrange multiplier $\bar{y}$ it holds that $\bar{y} \in \partial h(g(\bar{x}))$ and*

$$f(\bar{x}) + \langle \bar{y}, g(\bar{x}) \rangle = \min \{ f(x) + \langle \bar{y}, g(x) \rangle : x \in \mathbb{X} \}; \tag{4.3}$$

- *for each pair $(\bar{x}, \bar{y}) \in \mathbb{X} \times \mathbb{Y}$ that satisfies (4.3) with $\bar{y} \in \partial h(g(\bar{x}))$, the point $\bar{x}$ solves (P) optimally.*

Proof. By Proposition 4.2 $\partial V(0) \neq \emptyset \Leftrightarrow M \neq \varnothing \Leftrightarrow \exists y \in \mathbb{Y}$ such that $\inf_x L(x, y) \geq V(0)$. In this string, any $\bar{y} \in \partial V(0) = M$ applies to yield

$$\sup_y \inf_x L(x, y) \geq \inf_x L(x, \bar{y}) \geq \inf(P).$$

In addition, the inequality

$$f(x) + h(g(x)) \geq f(x) + \langle y, g(x) \rangle - h^*(y)$$

is valid for all $(x, y) \in \mathbb{X} \times \mathbb{Y}$. Consequently, $\inf(P) \geq \inf_x \sup_y L(x, y)$. Thus the inequality $\inf_x L(x, \bar{y}) \geq \inf(P)$ is squeezed in sandwich:

$$\sup_y \inf_x L(x, y) \geq \inf_x L(x, \bar{y}) \geq \inf(P) \geq \inf_x \sup_y L(x, y). \tag{4.4}$$

Equalities hold in (4.4) because $\sup_y \inf_x L(x,y) \le \inf_x \sup_y L(x,y)$.

From $V^*(y) = -\inf_x L(x,y)$ it follows that $V^{**}(0) = \sup_y \inf_x L(x,y)$. If (P) is strongly stable, then — by the preceding argument — the last entity equals $\inf(P) = V(0)$.

Finally, given any minimizer $\bar{x}$ of (P), pick an arbitrary $\bar{y} \in M = \partial V(0)$. It holds for each $x \in \mathbb{X}$ that

$$f(\bar{x}) + h(g(\bar{x})) = \inf(P) \le L(x,\bar{y}) = f(x) + \langle \bar{y}, g(x)\rangle - h^*(\bar{y}).$$

Insert $x = \bar{x}$ on the right hand side to have $h(g(\bar{x})) \le \langle \bar{y}, g(\bar{x})\rangle - h^*(\bar{y})$ whence

$$h(g(\bar{x})) + h^*(\bar{y}) = \langle \bar{y}, g(\bar{x})\rangle . \tag{4.5}$$

This implies first, $\bar{y} \in \partial h(g(\bar{x}))$ and second, (4.3). For the last bullet, (4.3) and $\bar{y} \in h(g(\bar{x}))$ ($\Leftrightarrow$ (4.5)) yield

$$f(\bar{x}) + h(g(\bar{x})) = \min_x \{f(x) + \langle \bar{y}, g(x)\rangle - h^*(\bar{y})\} = \min_x L(x,\bar{y}).$$

This tells, in view of (4.4), that $\bar{x}$ minimizes (P) — and that $\bar{y} \in M$. □

So far, using only algebra and numerical ordering, Lagrange multipliers — or equivalently, subgradients — were proven expedient for Lagrangian duality. It remains to be argued next that such multipliers do indeed exist in common circumstances. For that purpose recall that a point c in a subset C of real vector space is declared *absorbing* if for each non-zero direction d in that space there exists a real $r > 0$ such that $c +]0,r[d \subset C$. Also recall that $convC$ denotes the *convex hull* of C, and $epiV := \{(e,r) \in \mathbb{E} \times \mathbb{R} : V(e) \le r\}$ is short hand for the *epigraph* of V.

Proposition 4.4 (Linear support of V at 0.) *Suppose 0 is absorbing in $domV$. Also suppose $conv(epiV)$ contains an absorbing point, but that $(0, V(0))$ is not of such sort. Then, letting $\mathbb{Y}$ consist of all linear $y : \mathbb{E} \to \mathbb{R}$, the subdifferential $\partial V(0)$ is nonempty.*

Proof. By the Hahn-Banach separation theorem there is a hyperplane that supports $C := conv(epiV)$ in the point $(0, V(0))$. That hyperplane is defined in terms of a linear functional $(e^*, r^*) \neq 0$ by

$$\langle e^*, e\rangle + r^* r \ge r^* V(0) \text{ for all } (e,r) \in C. \tag{4.6}$$

Plainly, $(e,r) \in C \; \& \; \bar{r} > r \Rightarrow (e,\bar{r}) \in C$. Consequently, $r^* \ge 0$. If $r^* = 0$, then, since 0 is absorbing in $domV$, it holds that $\langle e^*, e\rangle \ge 0$ for all $e \in \mathbb{E}$, whence $e^* = 0$, and the contradiction $(e^*, r^*) = 0$ obtains. So, divide through (4.6) with $r^* > 0$, define $y := -e^*/r^*$, and put $r = V(e)$ to have

$$V(e) \geq V(0) + \langle y, e \rangle \text{ for all } e \in \mathbb{E}.$$

That is, $y \in \partial V(0)$. □

PROPOSITION 4.5 (Continuous linear support of V at 0.) *Let $\mathbb{E}$ be a topological, locally convex, separated, real vector space. Denote by $\check{V}$ the largest convex function $\leq V$. Suppose V is finite-valued, bounded above on a neighborhood of 0 and $\check{V}(0) = V(0)$. Then, letting $\mathbb{Y}$ consist of all continuous linear functionals $y : \mathbb{E} \to \mathbb{R}$, the subdifferential $\partial V(0)$ is nonempty.*

Propositions 4.4 and 4.5 emphasize the convenience of $(0, V(0))$ being "non-interior" to $conv(epiV)$. In particular, it simplifies things to have $epiV$ — or equivalently, V itself — convex.

4. Saddle-points and core solutions

After so much preparation it is time to reconsider the production game having coalition costs C_S defined by (4.1). Quite reasonably, suppose henceforth that a weak form of subadditivity holds, namely: $C_I \leq \sum_{i \in I} C_i < +\infty$. As announced, the objective is to find *explicit* cost allocations $c = (c_i) \in$ *core*. For that purpose, in view of Example 4.2, recall that

$$L_S(x, y) = f_S(x) + \langle y, g_S(x) \rangle - h_S^*(y),$$

and introduce a standing

Hypothesis on additive estimates: *Let henceforth $\mathbb{X}_S := \Pi_{i \in S} \mathbb{X}_i$ and suppose there exist for each i, functions $f_i : \mathbb{X}_i \to \mathbb{R} \cup \{+\infty\}$, $g_i : \mathbb{X}_i \to \mathbb{E}$; and $h_i : \mathbb{E} \to \mathbb{R} \cup \{+\infty\}$ such that for all $S \subseteq I$ and $y \in \mathbb{Y}$,*

$$\inf_x L_S(x, y) \geq \inf \left\{ \sum_{i \in S} [f_i(x_i) + \langle y, g_i(x_i) \rangle - h_i^*(y)] : x_i \in \mathbb{X}_i \right\}. \quad (4.7)$$

Further, for the grand coalition $S = I$ it should hold that

$$\inf_x L_I(x, y) \leq \sup_y \inf \left\{ \sum_{i \in I} [f_i(x_i) + \langle y, g_i(x_i) \rangle - h_i^*(y)] : x_i \in \mathbb{X}_i \right\}.$$

PROPOSITION 4.6 *The standing hypothesis holds if for all $S \subseteq I$, $x \in \mathbb{X}_S$, $y \in \mathbb{Y}$*

$$f_S(x) + \langle y, g_S(x) \rangle \geq \sum_{i \in S} \{ f_i(x_i) + \langle y, g_i(x_i) \rangle \}, \quad (4.8)$$

and for all $e \in \mathbb{E}$,

$$h_S(e) \geq \inf\left\{\sum_{i\in S} h_i(e_i) : \sum_{i\in S} e_i = e\right\}, \tag{4.9}$$

with equalities when $S = I$.

Proof. (4.9) implies $h_S^*(y) \leq \sum_{i\in S} h_i^*(y)$. □

EXAMPLE 4.3 (*Positive homogenous penalty.*) Let $h : \mathbb{E} \to \mathbb{R} \cup \{+\infty\}$ be positively homogeneous. For example, h could be the support function of some nonempty subset of a pre-dual to $\mathbb{E}$. Then h^*, restricted to $\mathbb{Y}$, is the extended indicator δ_Y of some convex set $Y \subseteq \mathbb{Y}$. That is, $h^*(y) = 0$ if $y \in Y$, $+\infty$ otherwise. Suppose $h^* = h_S^*$ for all $S \subseteq I$. Then (4.8) implies (4.7).

EXAMPLE 4.4 (*Cone constraints.*) Of special notice is the instance when h, as described in Example 4.3, equals the extended indicator δ_K of a convex cone $K \subset \mathbb{E}$. Then $h^* = \delta_{K^*}$ where $K^* := \{y : \langle y, K\rangle \leq 0\}$ is the negative dual cone. In Example 4.1 let all K_i be the same convex cone $K \subset \mathbb{E}$ and posit $h_S^* := h^*$ for all $S \subseteq I$. Then the standing hypothesis is satisfied, and coalition S incurs cost

$$C_S := \inf\left\{\sum_{i\in S} f_i(x_i) : \sum_{i\in S} g_i(x_i) \in K, x_i \in \mathbb{X}_i\right\}.$$

Observe that costs and constraints are here pooled additively. However, no activity set can be transferred from any agent to another.

EXAMPLE 4.5 (*Inf-convolution of penalties.*) When

$$h_S(e) := \inf\left\{\sum_{i\in S} h_i(x_i) : \sum_{i\in S} x_i = e\right\},$$

we get $h_S^*(y) = \sum_{i\in S} h_i^*(y)$.

THEOREM 4.1 (Nonemptiness of the core and explicit allocations.)

- *Suppose* $V_I^{**}(0) = V_I(0)$. *Then, under the standing hypothesis, core* $\neq \emptyset$.
- *If moreover,* $V_I(\cdot)$ *is subdifferentiable at* 0 — *that is, if* (P_I) *is strongly stable* — *then each Lagrange multiplier* $\bar{y}$ *for problem* (P_I) *generates a cost allocation* $c = (c_i) \in$ *core by the formula*

$$c_i = c_i(\bar{y}) := \inf\left\{f_i(x_i) + \langle \bar{y}, g_i(x_i)\rangle - h_i^*(\bar{y}) : x_i \in \mathbb{X}_i\right\}. \tag{4.10}$$

Proof. By the standing assumption it holds for *any price* $y \in \mathbb{Y}$ and *each coalition* S that

$$\sum_{i\in S} c_i(y) \le \inf_x L_S(x,y) \le \sup_y \inf_x L_S(x,y) \le \inf_x \sup_y L_S(x,y)$$
$$= \inf_x \{f_S(x) + h_S^{**}(g_S(x))\} \le \inf_x \{f_S(x) + h_S(g_S(x))\} = C_S.$$

So, no coalition ought reasonably block a payment scheme of the said sort $i \mapsto c_i(y)$. In addition, if $\bar{y}$ is a Lagrange multiplier, then

$$\sum_{i\in I} c_i(\bar{y}) = \inf_x L_I(x,\bar{y}) \ge C_I.$$

In lack of strong stability, when merely $V_I^{**}(0) = V_I(0)$, choose for each integer n a "price" $y^n \in \mathbb{Y}$ such that the numbers $c_i^n = c_i(y^n), i \in I$, satisfy

$$\sum_{i\in I} c_i^n = \inf_x L_I(x,y^n) \ge C_I - 1/n.$$

As argued above, $\sum_{i\in S} c_i^n \le C_S$ for all $S \subseteq I$. In particular, $c_i^n \le C_i < +\infty$. From $c_i^n \ge C_I - 1/n - \sum_{j\neq i} C_j$ it follows that the sequence (c^n) is bounded. Clearly, any accumulation point c belongs to *core*. □

EXAMPLE 4.6 (*Cooperative linear programming.*) A special and important version of Example 4.1 — and Example 4.2 — has $\mathbb{X}_i := \mathbb{R}_+^{n_i}$ with linear cost $k_i^T x_i$, $k_i \in \mathbb{R}^{n_i}$, and linear constraints $g_i(x_i) := A_i x_i - e_i$, $e_i \in \mathbb{R}^m$, A_i being a $m \times n_i$ matrix. Posit $K_i := \{0\}$ for all i to get, for coalition S, cost given by the standard linear program

$$(P_S) \qquad C_S := \inf\left\{\sum_{i\in S} k_i^T x_i : \sum_{i\in S} A_i x_i = \sum_{i\in S} e_i \text{ with } x_i \ge 0 \text{ for all } i\right\}$$

Suppose that the primal problem (P_I), as just defined, and its *dual*

$$(D_I) \qquad \sup\left\{y^T \sum_{i\in I} e_i : A_i^T y \le k_i \text{ for all } i\right\}$$

are both feasible. Then $\inf(D_I)$ attained and, by Theorem 4.1, for any optimal solution $\bar{y}$ to (D_I), the payment scheme $c_i := \bar{y}^T e_i$ yields $(c_i) \in$ *core*. Thus, regarding e_i as the production target of "factory" or corporate division i, those targets are evaluated by a common price $\bar{y}$ generated endogenously.

EXAMPLE 4.7 (*Inf-convolution continued.*) Each Lagrange multiplier $\bar{y}$ that applies to Example 4.2, generates a cost allocation $(c_i) \in core$ via

$$c_i := \langle \bar{y}, e_i \rangle - f_i^*(\bar{y}).$$

This formula is quite telling: each agent is charged for his "production target" less the price-taking profit he can generate, both entities calculated at shadow prices.

5. Price dynamics

Suppose henceforth that there exists at least one Lagrange multiplier. That is, suppose $M \neq \emptyset$. Further, for simplicity, let the endowment space $\mathbb{E}$, and its topological dual $\mathbb{Y}$, be real (finite-dimensional) Euclidean.[3] Denote by $D(y) := \inf_{x \in \mathbb{X}_I} L_I(x, y)$ the so-called *dual objective.* Most importantly, the function $y \in \mathbb{Y} \mapsto D(y) \in \mathbb{R} \cup \{-\infty\}$ so defined, is upper semicontinuous concave. And M, the optimal solution set of the *dual problem*

$$\sup \{D(y) : y \in \mathbb{Y}\},$$

is nonempty closed convex. Consequently, each $y \in \mathbb{Y}$ has a unique, orthogonal projection (closest approximation) $\bar{y} = P_M y$ in M. Let $Y := domD$ and suppose Y is closed. Denote by $Ty := cl\mathbb{R}_+ \{Y - y\}$ the *tangent cone* of Y at the point $y \in Y$ and by $Ny := \{y^* : \langle y^*, Ty \rangle \leq 0\}$ the associated negative dual or *normal cone.*

PROPOSITION 4.7 (Continuous-time price convergence to Lagrange multipliers.) *For any initial point $y(0) \in Y$, at which D is superdifferentiable, the two differential inclusions*

$$\dot{y} \in P_{Ty} \partial D(y) \quad and \quad \dot{y} \in \partial D(y) - Ny \tag{4.11}$$

admit the same, unique, absolutely continuous, infinitely extendable trajectory $0 \leq t \mapsto y(t) \in Y$. Moreover, $\|y(t) - P_M y(t)\| \to 0$.

Proof. Let $\Delta(y) := \min \{\|y - \bar{y}\| : \bar{y} \in M\}$ denote the Euclidean distance from y to the set M of optimal dual solutions. System $\dot{y} \in \partial D(y) - Ny$ has a unique, infinitely extendable solution $y(\cdot)$ along which

$$\begin{aligned}\frac{d}{dt}\Delta(y(t))^2/2 &= \langle y - P_M y, \dot{y} \rangle \in \langle y - P_M y, \partial D(y) - Ny \rangle \\ &\leq \langle y - P_M y, \partial D(y) \rangle \leq D(y) - D(P_M y).\end{aligned}$$

[3]Real Hilbert spaces $\mathbb{E}$ can also be accommodated.

Concavity explains the last inequality. Because $D(y) - D(P_M y) \leq 0$, it follows that $\|y(t) - P_M y(t)\| \to 0$. System $\dot{y} \in P_{Ty}\partial D(y)$ also has an infinitely extendable solution by Nagumo's theorem Aubin (1991). Since $P_{Ty}\partial D(y) \subseteq \partial D(y) - Ny$ that solution is unique as well. □

PROPOSITION 4.8 (Discrete-time price convergence to Lagrange multipliers) *Suppose D is superdifferentiable on Y with $\|\partial D(y)\|^2 \leq \kappa(1 + \|y - P_M y\|^2)$ for some constant $\kappa > 0$. Select step sizes $s_k > 0$ such that $\sum s_k = +\infty$ and $\sum s_k^2 < +\infty$. Then, for any initial point $y_0 \in Y$, the sequence $\{y_k\}$ generated iteratively by the difference inclusion*

$$y_{k+1} \in P_Y \left[y_k + s_k \partial D(y_k)\right], \tag{4.12}$$

is bounded, and every accumulation point $\bar{y}$ must be a Lagrange multiplier.

Proof. This result is well known, but its proof is outlined for completeness. Let $\bar{y}_k := P_M y_k$ and $\alpha_k = \|y_k - \bar{y}_k\|^2$. Then (4.12) implies

$$\begin{aligned}
\alpha_{k+1} = \|y_{k+1} - \bar{y}_{k+1}\|^2 &\leq \|y_{k+1} - \bar{y}_k\|^2 \in \|P_Y\left[y_k + s_k \partial D(y_k)\right] - P_Y \bar{y}_k\|^2 \\
&\leq \|y_k + s_k \partial D(y_k) - \bar{y}_k\|^2 \\
&\leq \|y_k - \bar{y}_k\|^2 + s_k^2 \|\partial D(y_k)\|^2 \\
&\quad + 2 s_k \langle y_k - \bar{y}_k, \partial D(y_k)\rangle \\
&\leq \alpha_k(1 + \beta_k) + \gamma_k - \delta_k
\end{aligned}$$

with $\beta_k := s_k^2 \kappa$, $\gamma_k := s_k^2 \kappa$, and $\delta_k := -2 s_k \langle y_k - \bar{y}_k, \partial D(y_k)\rangle$. The demonstration of Proposition 4.7 brought out that

$$\|y - P_M y\| > 0 \Rightarrow \sup \langle y - P_M y, \partial D(y)\rangle < 0.$$

Thus $\delta_k \geq 0$. Since $\sum \beta_k < +\infty$ and $\sum \gamma_k < +\infty$, it must be the case that α_k converges, and $\sum \delta_k < +\infty$; see Benveniste, Métivier, and Priouret (1990) Chapter 5, Lemma 2. If $\lim \alpha_k > 0$, the property $\sum s_k = +\infty$ would imply the contradiction $\sum \delta_k = +\infty$. Thus $\alpha_k \to 0$, and the proof is complete. □

Acknowledgments. I thank INDAM for generous support, Universita degli Studi di Trento for great hospitality, and Gabriele H. Greco for stimulating discussions.

References

Aubin, J-P. (1991). *Viability Theory.* Birkhäuser, Boston.

Benveniste, A., Métivier, M., and Priouret, P. (1990). *Adaptive Algorithms and Stochastic Approximations.* Springer, Berlin.

Bourass, A. and Giner, E. (2001). Kuhn-Tucker conditions and integral functionals. *Journal of Convex Analysis* 8, 2, 533-553 (2001).

Dubey, P. and Shapley, L.S. (1984). Totally balanced games arising from controlled programming problems. *Mathematical Programming,* 29:245–276.

Ellickson, B. (1993). *Competitive Equilibrium.* Cambridge University Press.

Evstigneev, I.V. and Flåm, S.D. (2001). Sharing nonconvex cost. *Journal of Global Optimization,* 20:257–271.

Flåm, S.D. (2002). Stochastic programming, cooperation, and risk exchange. *Optimization Methods and Software,* 17:493–504.

Flåm, S.D., and Jourani, A. (2003). Strategic behavior and partial cost sharing. *Games and Economic Behavior,* 43:44–56.

Forgó, F., Szép, J., and Szidarovszky, F. (1999). *Introduction to the Theory of Games.* Kluwer Academic Publishers, Dordrecht.

Gintis, H. (2000). *Game Theory Evolving.* Princeton University Press.

Gomory, R.E. and Baumol, W.J. (1960). Integer programming and pricing. *Econometrica,* 28(3):521–550.

Granot, D. (1986). A generalized linear production model: A unifying model. *Mathematical Programming,* 43:212–222.

Kalai, E. and Zemel, E. (1982). Generalized network problems yielding totally balanced games. *Operations Research,* 30(5):998–1008.

Mas-Colell, A., Whinston, M.D., and Green, J.R. (1995). *Microeconomic Theory.* Oxford University Press.

Owen, G. (1975). On the core of linear production games. *Mathematical Programming,* 9:358–370.

Peyton Young, H. (1994). *Equity.* Princeton University Press.

Samet, D. and Zemel, S. (1994). On the core and dual set of linear programming games. *Mathematics of Operations Research,* 9(2): 309–316.

Sandsmark, M. (1999). Production games under uncertainty. *Computational Economics,* 3:237–253.

Scarf, H.E. (1990). Mathematical programming and economic theory. *Operations Research,* 38:377–385.

Scarf, H.E. (1994). The allocation of resources in the presence of indivisibilities. *Journal of Economic Perspectives,* 8(4):111–128.

Shapley, L.S. (1967). On balanced sets and cores. *Naval Research Logistics Quarterly,* 14:453–461.

Shapley, L.S. and Shubik, M. (1969). On market games. *Journal of Economic Theory,* 1:9–25.

Verga-Redondo, F. (2003). *Economics and the Theory of Games.* Cambridge University Press.

Wolsey, L.A. (1981). Integer programming duality: Price functions and sensitivity analysis. *Mathematical Programming,* 20:173–195.

Chapter 5

CONSISTENT CONJECTURES, EQUILIBRIA AND DYNAMIC GAMES

Alain Jean-Marie
Mabel Tidball

Abstract We discuss in this paper the relationships between conjectures, conjectural equilibria, consistency and Nash equilibria in the classical theory of discrete-time dynamic games. We propose a theoretical framework in which we define conjectural equilibria with several degrees of consistency. In particular, we introduce feedback-consistency, and we prove that the corresponding equilibria and Nash-feedback equilibria of the game coincide. We discuss the relationship between these results and previous studies based on differential games and supergames.

1. Introduction

This paper discusses the relationships between the concept of *conjectures* and the classical theory of equilibria in dynamic games.

The idea of introducing conjectures in games has a long history, which goes back to the work of Bowley (1924) and Frisch (1933). There are, at least, two related reasons for this. One is the wish to capture the idea that economic agents seem to have a tendency, in practice, to anticipate the move of their opponents. The other one is the necessity to cope with the lack or the imprecision of the information available to players.

The first notion of conjectures has been developed for static games and has led to the theory of *conjectural variations* equilibria. The principle is that each player i assumes that her opponent j will "respond" to (infinitesimal) variations of her strategy δe_i by a proportional variation $\delta e_j = r_{ij} \delta e_i$. Considering this, player i is faced with an optimization problem in which her payoff Π^i is perceived as depending only on her strategy e_i. A set of conjectural variations r_{ij} and a strategy profile

$(e_1, \ldots, e_n)$ is a conjectural variations equilibrium if it solves simultaneously all players' optimization problems. The first order conditions for those are:

$$\frac{\partial}{\partial e_i}\Pi^i(e_1, \ldots, e_n) + \sum_{j \neq i} r_{ij} \frac{\partial}{\partial e_j}\Pi^i(e_1, \ldots, e_n) = 0. \qquad (5.1)$$

Conjectural variations equilibria generalize Nash equilibria, which corresponds to "zero conjectures" $r_{ij} = 0$.

The concept of conjectural variations equilibria has received numerous criticisms. First, there is a problem of rationality. Under the assumptions of complete knowledge, and common knowledge of rationality, the process of elimination of dominated strategies usually rules out anything but the Nash equilibrium. Second, the choice of the conjectural variations r_{ij} is, *a priori*, arbitrary, and without a way to point out particularly reasonable conjectures, the theory seems to be able to explain any observed outcome. Bresnahan (1981) has proposed to select conjectures that are *consistent*, in the sense that reaction functions and conjectured actions mutually coincide. Yet, the principal criticisms persist.

These criticisms are based on the assumption of complete knowledge, and on the fact that conjectural variation games are static. Yet, from the onset, it was clear to the various authors discussing conjectural variations in static games, that the proper (but less tractable) way of modeling agents would be a dynamic setting. Only the presence of a dynamic structure, with repeated interactions, and the observation of what rivals have actually played, gives substance to the idea that players have *responses*. In a dynamic game, the structure of information is clearly made precise, specifying in particular what is the information observed and available to agents, based on which they can choose their decisions. This allows to state the principle of consistency as the coincidence of prior assumptions and observed facts. Making precise how conjectures and consistency can be defined is the main purpose of this paper.

Before turning to this point, let us note that there exists a second approach linking conjectures and dynamic games. Several authors have pointed out the fact that computing stationary Nash-feedback equilibria in certain dynamic games, leads to steady-state solutions which are identifiable to the conjectural variations equilibria of the associated static game. In that case, the value of the conjectural variation r_{ij} is precisely defined from the parameters of the dynamic game. This correspondence illustrates the idea that *static* conjectural variations equilibria are worthwhile being studied, if not as a factual description of interactions between players, but at least as a technical "shortcut" for studying more complex dynamic interactions. This principle has been applied in dynamic games, in particular by Driskill and McCafferty (1989), Wildasin (1991), Dockner (1992), Cabral (1995), Itaya and Shimomura (2001), Itaya and Okamura (2003) and Figuières (2002). These results are reported in Figuières et al. (2004), Chapter 2.

The rest of this paper is organized as follows. In Section 2, we present a model in which the notion of consistent conjecture is embedded in the definition of the game. We show in this context that consistent conjectures and *feedback* strategies are deeply related. In particular, we prove that, in contrast with the static case, Nash equilibria can be seen as consistent conjectural equilibria. This part is a development on ideas sketched in Chapter 3 of Figuières et al. (2004).

These results are the discrete-time analogy of results of Fershtman and Kamien (1985) who have first incorporated the notion of consistent conjectural equilibria into the theory of *differential* games. Section 3 is devoted to this case, in which conjectural equilibria provide a new interpretation of open-loop and closed-loop equilibria.

We finish with a description of the model of Friedman (1968), who was the first to develop the idea of consistent conjectures, in the case of *supergames* and with an infinite time horizon. We show how Friedman's proposal of *reaction function* equilibria fits in our general framework. We also review existence results obtained for such equilibria, in particular for Cournot's duopoly in a linear-quadratic setting (Section 4).

We conclude the discussion in Section 5.

2. Conjectures for dynamic games, equilibria and consistency

The purpose of this section is to present a theoretical framework for defining *consistent* conjectures in discrete-time dynamic games, based essentially on the ideas of Fershtman and Kamien (1985) and Friedman (1968). The general principle is that *i)* players form conjectures on how the other players react (or would react) to their actions, *ii)* they optimize their payoff based on this assumption, and *iii)* conjectures should be *consistent* with facts, in the sense that the evolution of the game should be the same as what was believed before implementing the decisions. The idea of individual optimization based on some assumed "reaction" of other players is the heart of the conjectural principle, as we have seen in the introduction when discussing conjectural variations equilibria.

We shall see however that differences appear in the way consistency is enforced. This depends on the information which is assumed available to players, in a way very similar to the definition of different equilibrium concepts in dynamic games. Among the possibilities, one postulates that consistency exists if conjectures and *best responses* coincide. This requirement is in the spirit of the definition of Bresnahan (1981) for consistency in static conjectural variations games.

We review the general framework and different variants for the notion of consistency in Section 2.1. The principal contribution of this survey is a terminology distinguishing between the different variants of conjectures and conjectural equilibria used so far in the literature. Next, in Section 2.2, we establish the links which exist between a particular concept of conjectural equilibria ("feedback-consistent" equilibria) and Nash-feedback equilibria for discrete-time dynamic games.

2.1 Principle

Consider a dynamic game with n players and time horizon T, finite or infinite. The state of the game at time t is described by a vector $\mathbf{x}(t) \in \mathbb{S}$. Player i has an influence on the evolution of the state, through a control variable. Let $e_i(t)$ be the control performed by player i in the t-th period of the game, that is, between time t and $t+1$. Let E^i be the set in which player i chooses her control, $\mathbb{E} = E^1 \times \cdots \times E^n$ and $\mathbb{E}^{-i} = E^1 \times \cdots \times E^{i-1} \times E^{i+1} \times \cdots \times E^n$ be the space of controls of player i's opponents. Let also $\mathbf{e}(t) = (e_1(t), \ldots, e_n(t)) \in \mathbb{E}$ denote the vector of controls applied by each player. An element of $\mathbb{E}^{-i}$ will be denoted by $\mathbf{e}_{-i}$. The state of the game at time $t+1$ results from the combination of the controls $\mathbf{e}(t)$ and the state $\mathbf{x}(t)$. Formally, the state evolves according to some dynamics

$$\mathbf{x}(t+1) = f_t(\mathbf{x}(t), \mathbf{e}(t)), \qquad \mathbf{x}(0) = \mathbf{x}_0, \tag{5.2}$$

for some sequence of functions $f_t : \mathbb{S} \times \mathbb{E} \to \mathbb{S}$. The instantaneous payoff of player i is a function $\Pi^i_t : \mathbb{S} \times \mathbb{E} \to \mathbb{R}$ of the state and controls, and her total payoff is given by:

$$V^i(\mathbf{x}_0; \mathbf{e}(0), \mathbf{e}(1), \mathbf{e}(2), \ldots, \mathbf{e}(T)) = \sum_{t=0}^{T} \Pi^i_t(\mathbf{x}(t), \mathbf{e}(t)), \tag{5.3}$$

for some sequence of payoff functions Π^i_t. The definition of conjectural equilibria involves the definition of conjectures, and the resolution of individual optimization problems.

Conjectures. Each player has a belief on the behavior of the other ones. More precisely player i *thinks* that player j chooses her control by applying some function ϕ^{ij}_t to observed quantities. Several degrees of behavioral complexity are possible here. We identify some of those found in the current literature in the following definition. In the sequel, we shall use the superscript "$i\dagger$" as a shorthand for "believed by player i".

DEFINITION 5.1 *The conjecture of player i about player j is a sequence of functions ϕ_t^{ij}, $t = 0, 1, \ldots, T$, which define the conjectured value of player j's control, $e_j^{i\dagger}(t)$. Depending on the situation, we may have:*

$$\phi_t^{ij} : \mathbb{S} \to E^j, \qquad \textit{with} \qquad e_j^{i\dagger}(t) = \phi_t^{ij}(\mathbf{x}(t)), \tag{5.4}$$

(state-based conjectures) or

$$\phi_t^{ij} : \mathbb{S} \times \mathbb{E} \to E^j, \qquad \textit{with} \qquad e_j^{i\dagger}(t) = \phi_t^{ij}(\mathbf{x}(t), \mathbf{e}(t-1)), \tag{5.5}$$

(state and control-based conjectures), or

$$\phi_t^{ij} : \mathbb{S} \times \mathbb{E}^{-i} \to E^j, \qquad \textit{with} \qquad e_j^{i\dagger}(t) = \phi_t^{ij}(\mathbf{x}(t), \mathbf{e}_{-j}(t)) \tag{5.6}$$

("complete" conjectures).

The first form (5.4) is the basic one, and is used by Fershtman and Kamien (1985) with differential games. The second one (5.5) is inspired by the supergame[1] model of Friedman (1968), in which the conjecture involves the last observed *move* of the opponents. We have generalized the idea in the definition, and we come back to the specific situation of supergames in the sequel. Conjectures that are based on the state *and* the control need the specification of an initial control $\mathbf{e}_{-1}$, observed at the beginning of the game, in addition to the initial state $\mathbf{x}_0$.

The third form was also introduced by Fershtman and Kamien (1985) who termed it "complete". In discrete time, endowing players with such conjectures brings forth the problem of justifying how all players can simultaneously think that their opponents observe their moves and react, as if they were all leaders in a Stackelberg game. Indeed, the conjecture involves the quantity $\mathbf{e}_{-j}(t)$, that is, the control that players i's opponents are *about* to choose, and which is not *a priori* observable at the moment player i's decision is done. In that sense, this form is more in the spirit of *conjectural variations*. We shall see indeed in Section 2.2 that the two first forms are related to Nash-feedback equilibria, while the third is more related to static conjectural variations equilibria.

Laitner (1980) has proposed a related form in a discrete-time supergame. He assumes a conjecture of the form $e_j^{i\dagger}(t) = \phi_t^{ij}(\mathbf{e}_{-j}(t),$

[1] We adopt here the terminology of Myerson (1991), according to which a supergame is a dynamic game with a constant state and complete information.

$\mathbf{e}(t-1))$, which could be generalized to $e_j^{i\dagger}(t) = \phi_t^{ij}(\mathbf{x}(t), \mathbf{e}_{-j}(t), \mathbf{e}(t-1))$ in a game with a non-trivial state evolution.

Individual optimization. Consider first state-based conjectures, of the form (5.4). Given her conjectures, player i is faced with a classical dynamic control problem. Indeed, she is led to conclude by replacing $e_j(t)$, for $j \neq i$, by $\phi_t^{ij}(.)$ in (5.2), that the state actually evolves according to some dynamics depending only on her own control sequence $e_i(t)$, $t = 0, \ldots, T$. Likewise, her payoff depends only on her own control. More precisely: the conjectured dynamics and payoff are:

$$\begin{aligned} \mathbf{x}^{i\dagger}(t+1) &= f_t\big(\mathbf{x}^{i\dagger}(t), \phi_t^{i1}(\mathbf{x}^{i\dagger}(t)), \ldots, \phi_t^{i,i-1}(\mathbf{x}^{i\dagger}(t)), e_i(t), \\ &\qquad \phi_t^{i,i+1}(\mathbf{x}^{i\dagger}(t)), \ldots \phi_t^{in}(\mathbf{x}^{i\dagger}(t))\big), \qquad (5.7) \\ \Pi_t^{i\dagger}(\mathbf{x}, e_i) &= \Pi_t^{i}\big(\mathbf{x}, \phi_t^{i1}(\mathbf{x}), \ldots, \phi_t^{i,i-1}(\mathbf{x}), e_i, \phi_t^{i,i+1}(\mathbf{x}), \ldots \phi_t^{in}(\mathbf{x})\big). \qquad (5.8) \end{aligned}$$

For player i, the system evolves therefore according to some dynamics of the form:

$$\mathbf{x}^{i\dagger}(t+1) = \tilde{f}_t^i(\mathbf{x}^{i\dagger}(t), e_i(t)), \qquad \mathbf{x}^{i\dagger}(0) = \mathbf{x}_0. \qquad (5.9)$$

If conjectures are of the form (5.5), a difficulty arises. Replacing the conjectures in the state dynamics (5.2), we obtain:

$$\begin{aligned} \mathbf{x}^{i\dagger}(t+1) &= f_t\big(\mathbf{x}^{i\dagger}(t), \phi_t^{i1}(\mathbf{x}^{i\dagger}(t), \mathbf{e}(t-1)), \ldots, \phi_t^{i,i-1}(\mathbf{x}^{i\dagger}(t), \mathbf{e}(t-1)), \\ &\qquad e_i(t), \phi_t^{i,i+1}(\mathbf{x}^{i\dagger}(t), \mathbf{e}(t-1)), \ldots, \phi_t^{in}(\mathbf{x}^{i\dagger}(t), \mathbf{e}(t-1))\big). \end{aligned}$$

This equation involves the $e_j(t-1)$. Replacing them by their conjectured values $\phi_{t-1}^{ij}(\mathbf{x}^{i\dagger}(t-1), \mathbf{e}(t-2))$ makes appear the previous state $\mathbf{x}^{i\dagger}(t-1)$ and still involves unresolved quantities $e_k(t-2)$. Unless there are only two players, this elimination process necessitates going backwards in time until $t = 0$. The resulting formula for $\mathbf{x}^{i\dagger}(t+1)$ will therefore involve all previous states as well as all previous controls $e_i(s)$, $s \leq t$. Such an evolution is improper for setting up a classical control problem with Markovian dynamics. In order to circumvent this difficulty, it is possible to define a proper control problem for player i in an *enlarged* state space. Indeed, define $\mathbf{y}(t) = \mathbf{e}(t-1)$. Then the previous equation rewrites as:

$$\begin{aligned} \mathbf{x}^{i\dagger}(t+1) &= f_t\big(\mathbf{x}^{i\dagger}(t), \phi_t^{i1}(\mathbf{x}^{i\dagger}(t), \mathbf{y}(t)), \ldots, \phi_t^{i,i-1}(\mathbf{x}^{i\dagger}(t), \mathbf{y}(t)), e_i(t), \\ &\qquad \phi_t^{i,i+1}(\mathbf{x}^{i\dagger}(t), \mathbf{y}(t)), \ldots, \phi_t^{in}(\mathbf{x}^{i\dagger}(t), \mathbf{y}(t))\big), \qquad (5.10) \\ y_j(t+1) &= \phi_t^{ij}(\mathbf{x}(t), \mathbf{y}(t)) \qquad j \neq i \qquad (5.11) \end{aligned}$$

$$y_i(t+1) \quad = \quad e_i(t). \tag{5.12}$$

The initial conditions are $\mathbf{x}(0) = \mathbf{x}_0$ and $\mathbf{y}(0) = \mathbf{e}_{-1}$. Similarly, the conjectured cost function can be written as:

$$\begin{aligned}&\Pi_t^{i\dagger}(\mathbf{x}, \mathbf{y}, e_i)\\ &= \Pi_t^i(\mathbf{x}, \phi_t^{i1}(\mathbf{x}, \mathbf{y}), \ldots, \phi_t^{i,i-1}(\mathbf{x}, \mathbf{y}), e_i, \phi_t^{i,i+1}(\mathbf{x}, \mathbf{y}), \ldots, \phi_t^{in}(\mathbf{x}, \mathbf{y})).\end{aligned}$$

With this cost function and the state dynamics (5.10)–(5.12), player i faces a well-defined control problem.

Consistency. In general terms, consistency of a conjectural mechanism is the requirement that the outcome of each player's individual optimization problem corresponds to what has been conjectured about her. But different interpretations of this general rule are possible, depending on what kind of "outcome" is selected.

It is well-known that solving a deterministic dynamic control problem can be done in an "open-loop" or in a "feedback" perspective. In the first case, player i will obtain an optimal action profile $\{e_i^{i\dagger}(t), t = 0, \ldots, T\}$, assumed unique. In the second case, player i will obtain an optimal feedback $\{\gamma_t^{i\dagger}(\cdot), t = 0, \ldots, T\}$, where each $\gamma_t^{i\dagger}$ is a function of the state space into the action space E^i. Select first the open-loop point of view. Starting from the computed optimal profile $e_i^{i\dagger}(t)$, player i deduces the conjectured actions of her opponents using her conjecture scheme ϕ_t^{ij}. She therefore obtains $\{e_j^{i\dagger}(t)\}$, $j \neq i$. Replacing in turn these values in the dynamics, she obtains a conjectured state path $\{\mathbf{x}^{i\dagger}(t)\}$.

If all players j actually implements their decision rule $e_j^{j\dagger}$, the evolution of the state will follow the real dynamics (5.2), and result in some actual trajectory $\{\mathbf{x}^a(t)\}$. Specifically, the actual evolution of the state is:

$$\mathbf{x}^a(t+1) = f_t(\mathbf{x}^a(t), e_1^{1\dagger}(t), \ldots, e_n^{n\dagger}(t)), \quad \mathbf{x}^a(0) = \mathbf{x}_0. \tag{5.13}$$

Players will observe a discrepancy with their beliefs unless the actual path coincides with their conjectured path. If it does, no player will have a reason to challenge their conjectures and deviate from the "optimal" control they have computed. This leads to the following definition of a *state-consistent* equilibrium. Denote by $\boldsymbol{\phi}_t^i$ the vector of functions $(\phi_t^{i1}, \ldots, \phi_t^{i,i-1}, \phi_t^{i,i+1}, \ldots, \phi_t^{in})$.

DEFINITION 5.2 (State-consistent conjectural equilibrium) *The vector of conjectures* $(\boldsymbol{\phi}_t^1, \ldots, \boldsymbol{\phi}_t^n)$ *is a state-consistent conjectural equilibrium if*

$$\mathbf{x}^{i\dagger}(t) = \mathbf{x}^a(t), \tag{5.14}$$

for all i and t, and for all initial state $\mathbf{x}_0$.

An alternative definition, proposed by Fershtman and Kamien (1985), requires the coincidence of control paths, given the initial condition of the state:

DEFINITION 5.3 (Weak control-consistent conjectural equilibrium) *The vector of conjectures $(\phi_t^1, \ldots, \phi_t^n)$ is a weak control-consistent conjectural equilibrium if*

$$\mathbf{e}^{i\dagger}(t) = \mathbf{e}^{j\dagger}(t), \tag{5.15}$$

for all $i \neq j$ and t, given the initial state $\mathbf{x}_0$.

The stronger notion proposed by Fershtman and Kamien (1985) (where it is termed the "perfect" equilibrium) requires the coincidence for all possible initial states:

DEFINITION 5.4 (Control-consistent conjectural equilibrium) *The vector of conjectures $(\phi_t^1, \ldots, \phi_t^n)$ is a control-consistent conjectural equilibrium if the coincidence of controls* (5.15) *holds for all $i \neq j$, all t and all initial state $\mathbf{x}_0$.*

Clearly, any control-consistent conjectural equilibrium is state-consistent provided the dynamics (5.2) have a unique solution. It is of course possible to define a weak state-consistent equilibrium, where coincidence of trajectories is required only for some particular initial value. This concept does not seem to be used in the existing literature.

Now, consider that the solution of the deterministic control problems is expressed as a *state feedback*. Accordingly, when solving her (conjectured) optimization problem, player i concludes that there exists a sequence of functions $\gamma_t^{i\dagger} : \mathbb{S} \to E^i$ such that:

$$e_i^{i\dagger}(t) = \gamma^{i\dagger}(\mathbf{x}(t)).$$

Consistency can then be defined as the requirement that optimal feedbacks coincide with conjectures.

DEFINITION 5.5 (Feedback-consistent conjectural equilibrium) *The vector of state-based conjectures $(\phi_t^1, \ldots, \phi_t^n)$ is a feedback-consistent conjectural equilibrium if $\gamma_t^{i\dagger} = \phi_t^{ji}$ for all $i \neq j$ and all t.*

Obviously, consistency in this sense implies that the conjectures of two different players about some third player i coincide:

$$\phi_t^{ji} = \phi_t^{ki}, \quad \forall i \neq j \neq k, \quad \forall t. \tag{5.16}$$

In addition, if the time horizon T is infinite, and if there exists a *stationary* feedback $\gamma_{\infty}^{i\dagger}$, then a conjecture which is consistent with this stationary feedback should coincide with it at any time. This implies that the conjecture does not vary over time. For a simple equilibrium in the sense of Definition 5.2, none of these requirements are necessary *a priori*. It may happen that trajectories coincide in a "casual" way, resulting from discrepant conjectures of the different players.

2.2 Feedback consistency and Nash-feedback equilibria

We now turn to the principal result establishing the link between conjectural equilibria and the classical Nash-feedback equilibria of dynamic games. We assume in this section that T is finite.

THEOREM 5.1 *Consider a game with state-based conjectures $\phi_t^{ji} : \mathbb{S} \to E^i$. In such a game, Feedback-consistent equilibria and Nash-feedback equilibria coincide.*

Proof. The proof consists in identifying the value functions of both problems. According to the description above, looking for a feedback-consistent conjectural equilibrium involves the solution of the control problem:

$$\max_{e_i(.)} \left\{ \sum_{t=0}^{T} \Pi_t^{i\dagger}(\mathbf{x}^{i\dagger}(t), e_i(t)) \right\},$$

with the state evolution (5.9):

$$\mathbf{x}^{i\dagger}(t+1) = \tilde{f}_t^i(\mathbf{x}^{i\dagger}(t), e_i(t)), \qquad \mathbf{x}^{i\dagger}(0) = \mathbf{x}_0,$$

and where $\tilde{f}_t^i$ and $\Pi_t^{i\dagger}$ have been defined by Equations (5.7)–(5.9). The optimal feedback control in this case is given by the solution of the dynamic programming equation:

$$W_{t-1}^i(\mathbf{x}) = \max_{e_i \in E^i} \left\{ \Pi_t^{i\dagger}(\mathbf{x}, e_i) + W_t^i(\tilde{f}_t^i(\mathbf{x}, e_i)) \right\}$$

which defines recursively the sequence of value functions $W_t^i(\cdot)$, starting with $W_{T+1}^i \equiv 0$.

Consider now the Nash-feedback equilibria (NFBE) of the dynamic game: for each player i, maximize

$$\sum_{t=0}^{T} \Pi_t^i(\mathbf{x}(t), \mathbf{e}(t))$$

with the state dynamics (5.2):

$$\mathbf{x}(t+1) = f_t(\mathbf{x}(t), \mathbf{e}(t)), \qquad \mathbf{x}(0) = \mathbf{x}_0.$$

According to Theorem 6.6 of Başar and Olsder (1999), the set of feedback strategies $\{(\gamma_t^{1*}(\cdot), \dots, \gamma_t^{n*}(\cdot)), t = 0, \dots, T\}$, where each γ_t^{i*} is a function from $\mathbb{S}$ to E^i, is a NFBE, if and only if there exists a sequence of functions defined recursively by:

$$\begin{aligned} & V_{t-1}^i(\mathbf{x}) \qquad (5.17) \\ = & \max_{e_i \in E^i} \left\{ \Pi_t^i \left(\mathbf{x}, \gamma_t^{1*}(\mathbf{x}), \dots, e_i, \dots, \gamma_t^{n*}(\mathbf{x}) \right) + V_t^i(\hat{f}_t^i(\mathbf{x}, e_i)) \right\}, \end{aligned}$$

with $V_{T+1}^i \equiv 0$, and where

$$\hat{f}_t^i(\mathbf{x}) = f_t\big(\mathbf{x}, \gamma_t^{1*}(\mathbf{x}), \dots, \gamma_t^{(i-1)*}(\mathbf{x}), e_i, \gamma_t^{(i+1)*}(\mathbf{x}), \dots, \gamma_t^{n*}(\mathbf{x})\big). \quad (5.18)$$

Now, replacing γ^{j*} by ϕ^{ij} in Equations (5.17) and (5.18), we see that the dynamics $\hat{f}_t^i$ and $\tilde{f}_t^i$ coincide, as well as the sequences of value functions W_t^i and V_t^i. This means that every NFBE will provide a feedback consistent system of conjectures. Conversely, if feedback consistent conjectures ϕ_t^{ij} are found, then $\phi_t^{ki}(\mathbf{x})$ will solve the dynamic programming equation (5.17) in which γ^{j*} is set to ϕ^{kj} (recall that in a feedback-consistent system of conjectures, the functions ϕ_t^{ij} are actually independent from i). Therefore, such a conjecture will be a NFBE. □

Using the same arguments, we obtain a similar result for state and control-based conjectures.

THEOREM 5.2 *Consider a game with state and control-based conjectures* $\phi_t^{ij} : \mathbb{S} \times \mathbb{E} \to E^j$. *The Feedback-consistent equilibria of this game coincide with the Nash-feedback equilibria of the game with extended state space* $\mathbb{S} \times \mathbb{E}$ *and dynamics defined in Equations* (5.10)–(5.12).

Let us now turn to complete conjectures of the form (5.6). As in the case of state and control-based conjectures, using the conjectures does not allow player i to formulate a control problem, unless there are only two players. We therefore assume $n = 2$. The optimization problem of player i is then, with $j \neq i$,

$$\max_{e_i(\cdot)} \sum_{t=0}^{T} \Pi_t^i \left(\mathbf{x}(t), e_i(t), \phi_t^{ij}(\mathbf{x}(t), e_i(t)) \right),$$

with the conjectured evolution of the state:

$$\mathbf{x}(t+1) = f_t(\mathbf{x}(t), e_i(t), \phi_t^{ij}(\mathbf{x}(t), e_i(t))), \quad \mathbf{x}(0) = \mathbf{x}_0.$$

Accordingly, the optimal reaction satisfies the following necessary conditions (see Theorem 5.5 of Başar and Olsder (1999)):

THEOREM 5.3 *Consider a two-player game, with complete conjectures ϕ_t^{ij} which are differentiable. Let $e^{i\dagger}(t)$ be the conjectured optimal control path of player i, and*

$$\mathbf{x}^{i\dagger}(t+1) = f_t(\mathbf{x}^{i\dagger}(t), e_i^{i\dagger}(t), \phi_t^{ij}(\mathbf{x}^{i\dagger}(t), e_i^{i\dagger}(t))), \quad \mathbf{x}^{i\dagger}(0) = \mathbf{x}_0$$

be the conjectured optimal state path. Then there exist for each player a sequence $p^i(t)$ of costate vectors such that:

$$p^i(t) = \left(\frac{\partial \Pi_t^i}{\partial \mathbf{x}} + \frac{\partial \Pi_t^i}{\partial e_j}\frac{\partial \phi_t^{ij}}{\partial \mathbf{x}}\right) + (p^i(t+1))' \cdot \left(\frac{\partial f_t}{\partial \mathbf{x}} + \frac{\partial f_t}{\partial e_j}\frac{\partial \phi_t^{ij}}{\partial \mathbf{x}}\right)$$

with $p^i(T+1) = 0$, where functions f_t and Π_t^i are evaluated at $(\mathbf{x}^{i\dagger}(t)$, $e_i^{i\dagger}(t)$, $\phi_t^{ij}(\mathbf{x}^{i\dagger}(t), e_i^{i\dagger}(t)))$ and ϕ_t^{ij} is evaluated at $(\mathbf{x}^{i\dagger}(t), e_i^{i\dagger}(t)))$. Also, for all $i \neq j$,

$$\begin{aligned} e_i^{i\dagger}(t) \in \arg\max_{e_i \in E^i} \Big\{ & \Pi_t^i\left(\mathbf{x}^{i\dagger}(t), e_i, \phi_t^{ij}(\mathbf{x}^{i\dagger}(t), e_i)\right) \\ & + (p^i(t+1))' \cdot f_t(\mathbf{x}^{i\dagger}(t), e_i, \phi_t^{ij}(\mathbf{x}^{i\dagger}(t), e_i))\Big\}. \end{aligned}$$

For this type of conjectures, several remarks arise. First, the notions of consistency which can be appropriate are state consistency or control consistency. Clearly from the necessary conditions above, computing consistent equilibria will be more complicated than for state-based conjectures.

Next, the first order conditions of the maximization problem are:

$$0 = \left(\frac{\partial \Pi_t^i}{\partial e_i} + \frac{\partial \Pi_t^i}{\partial e_j}\frac{\partial \phi_t^{ij}}{\partial e_i}\right) + (p^i(t+1))' \cdot \left(\frac{\partial f_t}{\partial e_i} + \frac{\partial f_t}{\partial e_j}\frac{\partial \phi_t^{ij}}{\partial e_i}\right).$$

One recognizes in the first term of the right-hand side the formula for the conjectural variations equilibrium of a static game, see Equation (5.1).

Observe also that when the conjecture ϕ_t^{ij} is just a function of the state, we are back to state-based conjectures and the consistency conditions obtained with the theorem above are that of a Nash-feedback equilibrium. This was observed by Fershtman and Kamien (1985) for differential games.

Finally, if there are more than two players having complete conjectures, the decision problem at each instant in time is not a collection of individual control problems, but a game. We are not aware of results in the literature concerning this situation.

3. Consistent conjectures in differential games

The model of Fershtman and Kamien (1985) is a continuous-time, finite-horizon game. With respect to the general framework of the previous section, the equation of the dynamics (5.2) becomes

$$\dot{\mathbf{x}}(t) = f_t(\mathbf{x}(t), \mathbf{e}(t)),$$

and the total payoff is:

$$\int_0^T \Pi_t^i(\mathbf{x}(t), \mathbf{e}(t))\, dt.$$

Players have conjectures of the form (5.4), or "complete conjectures" of the form (5.6). Conjectures are assumed to be the same for all players (Condition (5.16)). Classically, the definition of a dynamic game must specify the space of strategies with which players can construct their action $e_i(t)$ at time t. The information potentially available being the initial state $\mathbf{x}_0$ and the current state $\mathbf{x}(t)$, three classes of strategies are considered by Fershtman and Kamien: *i)* closed-loop no-memory strategies, where $e_i(t) = \psi^i(\mathbf{x}_0, \mathbf{x}, t)$, *ii)* feedback Nash strategies where $e_i(t) = \psi^i(\mathbf{x}, t)$, and *iii)* open-loop strategies where $e_i(t) = \psi^i(\mathbf{x}_0, t)$.

The first concept of consistent conjectural equilibrium studied is the one of Definition 5.3 (weak control-consistent equilibrium). The following results are then obtained:

- Open-loop Nash equilibria are weak control-consistent conjectural equilibria.
- Weak control-consistent conjectural equilibria are closed-loop no-memory equilibria.

In other words, the class of weak control-consistent conjectural equilibria is situated between open-loop and closed-loop no-memory equilibria.

Fershtman and Kamien further define *perfect* conjectural equilibria as in Definition 5.4: control-consistent equilibria. The result is then:

- Control-consistent conjectural equilibria and feedback Nash equilibria coincide.

Further results of the paper include the statement of the problem of calculating complete conjectural equilibria (defined as Definition 5.3

with conjectures of the form (5.6)). The particular case of a duopoly market is studied. The price is the state variable $x(t)$ of this model; it evolves according to a differential equation depending on the quantities $(e_1(t), e_2(t))$ produced by both firms. The complete conjectures have here the form: $\phi^{ij}(x; e_j)$. The feedback Nash equilibrium is computed, as well as the complete conjectural equilibrium with affine conjectures. The two equilibria coincide when conjectures are actually constant. When $\partial\phi^{ij}(x; e_j)/\partial e_j = 1$, the stationary price is the monopoly price.

4. Consistent conjectures for supergames

In this section, we consider the problem set in Friedman (1968) (see also Friedman (1976) and Friedman (1977), Chapter 5). The model is a discrete-time, infinite-horizon supergame with n players. The total payoff of player i has the form:

$$V^i(\mathbf{x}_0; \mathbf{e}(0), \mathbf{e}(1), \dots) = \sum_{t=0}^{\infty} \theta_i^t \, \Pi^i(\mathbf{e}(t)),$$

for some discount factor θ_i, and where $\mathbf{e}(t) \in \mathbb{E}$ is the profile of strategies played at time t. Friedman assumes that players have (time-independent) conjectures of the form $e_j(t+1) = \phi^{ij}(\mathbf{e}(t))$, and proposes the following notion of equilibrium. Given this conjecture, player i is faced with an infinite-horizon control problem, and when solving it, she hopefully ends up with a stationary feedback policy $\gamma^i : \mathbb{E} \to E^i$. It can be seen as a *reaction function* to the vector of conjectures $\boldsymbol{\phi}^i$. This gives the name to the equilibrium advocated in Friedman (1968):

DEFINITION 5.6 (Reaction function equilibrium) *The vector of conjectures $(\boldsymbol{\phi}^1, \dots, \boldsymbol{\phi}^n)$ is a reaction function equilibrium if*

$$\gamma^i = \phi^{ki}, \forall k, \forall i.$$

In the terminology we have introduced in Section 2.1, the conjectures are "state and control-based" (Definition 5.1). Applying Theorem 5.2 to this special situation where the state space is reduced to a single element, we have:

THEOREM 5.4 *Reaction function equilibria coincide with the stationary Nash-Feedback equilibria of the game described by the dynamics:*

$$\mathbf{x}(t) = \mathbf{e}(t-1),$$

and the payoff functions: $\widetilde{\Pi}_0^i \equiv 0$ *and*

$$\widetilde{\Pi}_t^i(\mathbf{x}, \mathbf{e}) = \theta_i^{t-1} \, \Pi^i(\mathbf{x}), \quad t \geq 1.$$

In the process of finding reaction function equilibria, Friedman introduces a refinement. He suggests to solve the control problem with a finite horizon T, and then let T tend to infinity to obtain a stationary optimal feedback control. If the finite-horizon solutions converge to a stationary feedback control, this has the effect of selecting certain solutions among the possible solutions of the infinite-horizon problem. Once the stationary feedbacks γ^i are computed, the problem is to match them with the conjectures ϕ^{ji}.

No concrete example of such an equilibrium is known so far in the literature, except for the obvious one consisting in the repetition of the Nash equilibrium of the static game. Indeed, we have:

THEOREM 5.5 (The repeated static Nash equilibrium is a reaction function equilibrium.) *Assume there exists a unique Nash equilibrium $(e_1^N, \ldots, e_n^N)$ for the one-stage (static) game. If some player i conjectures that the other players will play the strategies $\mathbf{e}_{-i}^N$ at each stage, then her own optimal response is unique and is to play e_i^N at each stage.*

Proof. Let (e_i^N, e_{-i}^N) denote the unique Nash equilibrium of the static game. Since player i assumes that her opponents systematically play e_{-i}^N, we have $\mathbf{e}_{-i}(t) = e_{-i}^N$ for all t. Therefore, her perceived optimization problem is:

$$\max_{\{e_i(0), e_i(1), \ldots\}} \sum_{t=0}^{\infty} \theta_i^t \, \Pi^i \left(e_i(t), \mathbf{e}_{-i}^N\right).$$

Since e_i^N is the best response to $\mathbf{e}_{-i}^N$, the optimal control of player i is $e_i(t) = e_i^N$ for all $t \geq 0$. In other words, player i should respond to her "Nash conjecture" by playing Nash repeatedly. □

Based on Theorem 5.4 and tools of optimal control and games, it is possible to develop the calculation of Friedman's reaction function equilibria in the case of linear-quadratic games. Even for such simple games, finding Nash-feedback equilibria usually involves the solution of algebraic Ricatti equations, which cannot always be done in closed form. However, the games we have here have a special form, due to their simplified dynamics.

The detailed analysis, reported in Figuières et al. (2004), proceeds in several steps. First, we consider a finite time horizon game with stationary affine conjectures, and we construct the optimal control of each player. We obtain that the optimal control is also affine, and that its multiplicative coefficients are always proportional to those of the conjecture. We deduce conditions for the existence of a consistent equilibrium in the sense of Definition 5.5, in the case of symmetric players. Those conditions bear on the sign of quantities involving the coefficients of the conjectures and the parameters of the model.

Applied to Cournot's and Bertrand's duopoly models, we demonstrate that the repeated-Nash strategy of Theorem 5.5 is the unique reaction function equilibrium of the game. We therefore answer to Friedman's interrogation about the multiplicity of such equilibria in the duopoly.

5. Conclusion

In this paper, we have put together several concepts of consistent conjectural equilibria in a dynamic setting, collected from different papers. This allows us to draw a number of perspectives.

First, finding in which class of dynamic games the different definitions of Section 2.1 coincide is an interesting research direction.

As we have observed above, an equilibrium according to Definition 5.4 (control-consistent) is an equilibrium according to Definition 5.2 (state-consistent). We feel however that when conjectures are of the simple form (5.4), it may be that actions of the players are not observable, so that players should be happy with the coincidence of state paths. State-consistency is then the natural notion. The stronger control-consistency is however appropriate when conjectures are of the "complete" form. We have further observed that for repeated games, Definitions 5.2 and 5.4 coincide.

The problem disappears when consistency in feedback is considered, since the requirements of feedback-consistency (Definition 5.5) imply the coincidence of conjectures and actual values for both controls and states. Indeed, if the feedback functions of different players coincide, their conjectured state paths and control paths will coincide, since the initial state $\mathbf{x}_0$ is common knowledge.

Another issue is that of the information available to optimizing agents. In the models of Section 2, agents do not know the payoff functions of their opponents when they compute their optimal control, based on their own conjectures. Computing an equilibrium requires however the complete knowledge of the payoffs. This is not in accordance with the idea that players hold conjectures in order to compensate for the lack of information. On the other hand, *verifying* that a conjecture is consistent requires less information. Weak control-consistency (and the weak state-consistency that could have been defined in the same spirit) is verified by the observation of the equilibrium path. A possibility in this case is to develop learning models such as in Jean-Marie and Tidball (2004). The stronger state-consistency, control-consistency and feedback-consistency can be checked by computations based on the knowledge of one's payoff function and the conjectures of the opponents.

Finally, we point out that other interesting dynamic game models with conjectures and/or consistency have been left out of this survey. We have already mentioned the work of Laitner (1980). Other ideas are possible: for instance, it is possible to assume as in Başar, Turnovsky and d'Orey (1986) that players consider the game as a static conjectural variations game at each instant in time. Consistency in the sense of Bresnahan is then used. Also related is the paper of Kalai and Stanford (1985) in which a model similar to that of Friedman is analyzed. These papers illustrate the fact that some types of conjectures may lead to a multiplicity of consistent equilibria.

References

Başar, T. and Olsder, G.J. (1999). *Dynamic Noncooperative Game Theory.* Second Ed., SIAM, Philadelphia.

Başar, T., Turnovsky, S.J., and d'Orey, V. (1986). Optimal strategic monetary policies in dynamic interdependent economies. *Springer Lecture Notes on Economics and Management Science*, 265:134–178.

Bowley, A.L. (1924). *The Mathematical Groundwork of Economics.* Oxford University Press, Oxford.

Bresnahan, T.F. (1981). Duopoly models with consistent conjectures. *American Economic Review*, 71(5):934–945.

Cabral, L.M.B. (1995). Conjectural variations as a reduced form. *Economics Letters*, 49:397–402.

Dockner, E.J. (1992). A dynamic theory of conjectural variations. *The Journal of Industrial Economics*, XL(4):377–395.

Driskill, R.A. and McCafferty, S. (1989). Dynamic duopoly with adjustment costs: A differential game approach. *Journal of Economic Theory*, 49:324–338.

Fershtman, C. and Kamien, M.I. (1985). Conjectural equilibrium and strategy spaces in differential games. *Optimal Control Theory and Economic Analysis*, 2:569–579.

Figuières, C. (2002). Complementarity, substitutability and the strategic accumulation of capital. *International Game Theory Review*, 4(4): 371–390.

Figuières, C., Jean-Marie, A., Quérou, N., and Tidball, M. (2004). *The Theory of Conjectural Variations.* World Scientific Publishing.

Friedman, J.W. (1968). Reaction functions and the theory of duopoly. *Review of Economic Studies*, 35:257–272.

Friedman, J.W. (1976). Reaction functions as Nash equilibria. *Journal of Economic Studies*, 43:83–90.

Friedman, J.W. (1977). *Oligopoly and the Theory of Games*. North-Holland, Amsterdam.

Frisch, R. (1933). Monopole, Polypole – La notion de force en économie. *Nationaløkonomisk Tidsskrift*, 71:241 259. (reprinted "Monopoly, Polypoly: The concept of force in the economy" *International Economic Papers*, 1:23–36, (1951).

Itaya, J.-I. and Okamura, M. (2003). Conjectural variations and voluntary public good provision in a repeated game setting. *Journal of Public Economic Theory*, 5(1):51–66.

Itaya, J.-I. and Shimomura, K. (2001). A dynamic conjectural variations model in the private provision of public goods: A differential game approach. *Journal of Public Economics*, 81:153–172.

Jean-Marie, A. and Tidball, M. (2004). Adapting behaviors through a learning process. To appear in *Journal of Economic Behavior and Organization.*

Kalai, E. and Stanford, W. (1985). Conjectural variations strategies in accelerated Cournot games. *Intl. Journal of Industrial Organization*, 3:133–152.

Laitner, J. (1980). Rational duopoly equilibria. *The Quarterly Journal of Economics*, 641–662.

Myerson, R.B. (1991). *Game Theory*, Harvard University Press.

Wildasin, D.E. (1991). Some rudimentary 'duopolity theory'. *Regional Science and Urban Economics*, 21:393–421.

Chapter 6

COOPERATIVE DYNAMIC GAMES WITH INCOMPLETE INFORMATION

Leon A. Petrosjan

Abstract The definition of cooperative game in characteristic function form with incomplete information on a game tree is given. The notions of optimality principle and based on it solution concepts are introduced. The new concept of "imputation distribution procedure" is defined connected with the basic definitions of time-consistency and strongly time-consistency. Sufficient conditions of the existence of time-consistent solutions are derived. For a large class of games where these conditions cannot be satisfied the regularization procedure is developed and new characteristic function is constructed. The "regularized" core is defined and strongly time-consistency proved. The special case of stochastic games is also investigated in details.

1. Introduction

In n-persons games in extensive form as in classical simultaneous game theory different solution concepts are used. The most common approach is a noncooperative setting where as solution the Nash Equilibrium is considered. In the same time not much attention is given to the problem of time-consistency of solution considered in each specific case. This may follow from the fact that in most cases the Nash Equilibrium turns to be time-consistent, but also not always as it was shown in Petrosjan (1996).

The problem becomes more serious when cooperation in games in extensive form is considered. Usually in cooperative settings players agree to use such strategies which maximize the sum of their payoffs. As a result the game then develops along the cooperative trajectory (conditionally optimal trajectory). The corresponding maximal total payoff satisfies Bellman's Equation and thus is time-consistent. But the values of characteristic function for each subcoalition of players naturally did not satisfy this property along conditionally optimal trajectory. The characteristic function plays key role in construction of solution concepts in cooperative game theory. And impossibility to satisfy Bellman's Equation for values of characteristic function for subcoalitions implies the time-inconsistency of cooperative solution concepts. This was seen first for n-person differential games in papers (Filar and Petrosjan (2000); Haurie (1975); Kaitala and Pohjola (1988)) and in the papers (Petrosjan (1995); Petrosjan (1993); Petrosjan and Danilov (1985)) it was purposed to introduce a special rule of distribution of the players gain under cooperative behavior over time interval in such a way that time-consistency of the solution could be restored in given sense.

In this paper we formalize the notion of time-consistency and strongly time-consistency for cooperative games in extensive form with incomplete information, propose the regularization method which makes possible to restore classical simultaneous solution concepts in a way they became useful in the games in extensive form. We prove theorems concerning strongly time-consistency of regularized solutions and give a constructive method of computing of such solutions.

2. Definition of the multistage game with incomplete information in characteristic function form

To define the multistage cooperative game (in characteristic function form) with incomplete information we have to define first the multistage game in extensive form. In this definition we follow H. Kuhn (1953), with the only difference that in our definition we shall not allow chance moves, and the payoffs of players will be defined at each vertex of the game tree.

DEFINITION 6.1 *The n-person multistage game in extensive form is defined by*

1. *Specifying the finite graph tree $G = (X, F)$ with initial vertex x_0 referred to as the initial position of the game (here X is the set of vertexes and $F : X \to 2^X$ is a point-to-set mapping, and let $F_x = F(x)$).*

2. *Partition of the set of all vertices X into $n+1$ sets $X_1, X_2, \ldots, X_n$, X_{n+1} — called players partition, where X_i is interpreted as the set of vertices (positions) where player i "makes a move", $i = 1, \ldots, n$, and $X_{n+1} = \{x : F_x = \emptyset\}$ is called the set of final positions.*

3. *For each $x \in X$ specifying the vector function $h(x) = (h_1(x), \ldots, h_n(x))$, $h_i(x) \geq 0$, $i = 1, \ldots, n$; the vector function $h_i(x)$ is called the instantaneous payoff of the ith player.*

4. Subpartition of the set X_i, $i = 1, \dots, n$, into non-overlapping subsets X_i^j referred to as information sets of the ith player. In this case, for any position of one and the same information set the set of its subsequent vertices should contain one and the same number of vertices, i.e., for any x, $y \in X_i^j$ $|F_x| = |F_y|$ ($|F_x|$ is the number of elements of the set F_x), and no vertex of the information set should follow another vertex of this set, i.e., if $x \in X_i^j$ then there is no other vertex $y \in X_i^j$ such that

$$y \in \hat{F}_x = \{x \cup F_x \cup F_x^2 \cup \cdots \cup F_x^r \cup \cdots\}. \tag{6.1}$$

(here F_x^k is defined by induction $F_x^1 = F_x$, $F_x^1 = F(F_x^{k-1}) = \bigcup_{y \in F_x^{k-1}} F_y$).

The conceptual meaning of the informational partition is that when a player makes his move in position $x \in X_i$ in terms of incomplete information he does not know the position x itself, but knows that this position is in certain set $X_i^j \subset X_i$ ($x \in X_i^j$). Some restrictions are imposed by condition 4 on the players information sets. the requirement $|F_x| = |F_y|$ for any two vertices of the same information set are introduced to make vertices $x, y \in X_i^j$ indistinguishable. In fact, with $|F_x| \neq |F_y|$ player i could distinguish among the vertices $x, y \in X_i^j$ by the number of arcs emanating therefrom. If one information set could have two vertices x, y such that $y \in \hat{F}_x$ this would mean that a play of game can intersect twice an information set, but this in turn is equivalent to the fact that player j has no memory of the number of the moves he made before given stage which can hardly be concerned in the actual play of the game.

Denote the multistage game in extensive form starting from the vertex $x_0 \in X$ by $\Gamma(x_0)$.

For purpose of further discussion we need to introduce some additional notions.

DEFINITION 6.2 *The arcs incidental with x, i.e., $\{(x, y) : y \in F_x\}$, are called alternatives at the vertex $x \in X$.*

If $|F_x| = k$ then there are k alternatives at vertex x. We assume that if at the vertex x there are k alternatives then they are designated by integers $1, \dots, k$ with the vertex x bypassed in clockwise sense. The first alternative at the vertex x_0 is indicated in an arbitrary way. If some vertex $x \neq x_0$ is bypassed in a clockwise sense, then an alternative which follows a single arc (F_x^{-1}, x) entering into x is called the first alternative at x. Suppose that in the game all alternatives are enumerated as above. Let A_k be the set of all vertices $x \in X$ having exactly k alternatives, i.e., $A_k = \{x : |F_x| = k\}$.

Let $I_i = \{X_i^j : X_i^j \subset X_i\}$ be the set of all information sets of player i. By definition the pure strategy of player i means the function u_i mapping I_i into the set of positive numbers so that $u_i(X_i^j) \leq k$ if $X_i^j \subset A_k$. We say that the strategy u_i chooses alternative l in position $x \in X_i^j$ if $u_i(X_i^j) = l$, where l is the number of the alternative.

One may see that to each n-tuple in pure strategies $u(\cdot) = (u_1(\cdot), \ldots, u_n(\cdot))$ uniquely corresponds a path (trajectory) $w = x_0, \ldots, x_l$, $x_l \in X_{n+1}$, and hence the payoff

$$K_i(x_0; u_1(\cdot), \ldots, u_n(\cdot)) = \textstyle\sum_{k=0}^{l} h_i(x_k), \tag{6.2}$$

$$h_i \geq 0. \tag{6.3}$$

Here $x_l \in X_{n+1}$ is a final position (vertex) and $w = \{x_0, x_1, \ldots, x_l\}$ is the only path leading (F is a tree) from x_0 to x_l. The condition that the position (vertex) y belongs to w will be written as $y \in w$.

Consider the cooperative form of the game $\Gamma(x_0)$. In this formalization we suppose, that the players before starting the game agree to play $u_1^*, \ldots, u_n^*$ such that the corresponding path (trajectory) $w^* = \{x_0^*, \ldots, x_k^*, \ldots, x_l^*\}$ ($x_l^* \in X_{n+1}$) maximizes the sum of the payoffs

$$\max_u \sum_{i=1}^{n} K_i(x_0; u_1(\cdot), \ldots, u_n(\cdot)) = \sum_{i=1}^{n} K_i(x_0; u_1^*(\cdot), \ldots, u_n^*(\cdot))$$

$$= \sum_{i=1}^{n} \left[\sum_{k=0}^{l} h_i(x_k^*) \right] = v(N; x_0),$$

where x_0 is the initial vertex of the game $\Gamma(x_0)$ and N is the set of all players in $\Gamma(x_0)$. The trajectory w^* is called *conditionally optimal.* To define the cooperative game one has to introduce the characteristic function. The values of characteristic function for each coalition are defined in a classical way as values of associated zero-sum games. Consider a zero-sum game defined over the structure of the game $\Gamma(x_0)$ between the coalition S as first player and the coalition $N \setminus S$ as second player, and suppose that the payoff of S is equal to the sum of payoffs of players from S. Denote this game as $\Gamma_S(x_0)$. Suppose that $v(S; x_0)$ is the value of such game. The characteristic function is defined for each $S \subset N$ as value $v(S; x_0)$ of $\Gamma_S(x_0)$. From the definition of $v(S; x_0)$ it follows that $v(S; x_0)$ is superadditive (see Owen (1968)). It follows from the superadditivity condition that it is advantageous for the players to form a maximal coalition N and obtain a maximal total payoff $v(N; x_0)$ that is possible in the game. Purposefully, the quantity $v(S; x_0)$ ($S \neq N$) is equal to a maximal guaranteed payoff of the coalition S obtained irrespective of the behavior of other players, even the other form a coalition $N \setminus S$ against S.

Note that the positiveness of payoff functions h_i, $i = 1, \dots, n$ implies that of characteristic function. From the superadditivity of v it follows that

$$v(S'; x_0) \geq v(S; x_0)$$

for any $S, S' \subset N$ such that $S \subset S'$, i.e., the superadditivity of the function v in S implies that this function is monotone in S.

The pair $\langle N, v(\cdot, x_0) \rangle$, where N is the set of players, and v the characteristic function, is called the *cooperative game with incomplete information in the form of characteristic function* v. For short, it will be denoted by $\Gamma_v(x_0)$.

Various methods for "equitable" allocation of the total profit among players are treated as solutions in cooperative games. The set of such allocations satisfying an optimality principle is called a solution of the cooperative game (in the sense of this optimality principle). We will now define solutions of the game $\Gamma_v(N; x_0)$.

Denote by ξ_i a share of the player $i \in N$ in the total gain $v(N; x_0)$.

DEFINITION 6.3 *The vector* $\xi = (\xi_1, \dots, \xi_n)$, *whose components satisfy the conditions:*

1. $\xi_i \geq v(\{i\}; x_0), \quad i \in N,$

2. $\sum_{i \in N} \xi_i = v(N; x_0),$

is called an imputation in the game $\Gamma_v(x_0)$.

Denote the set of all imputations in $\Gamma_v(x_0)$ by $L_v(x_0)$.

Under the solution of $\Gamma_v(x_0)$ we will understand a subset $W_v(x_0) \subset L_v(x_0)$ of imputation set which satisfies additional "optimality" conditions.

The equity of the allocation $\xi = (\xi_1, \dots, \xi_n)$ representing an imputation is that each player receives at least maximal guaranteed payoff and the entire maximal payoff is distributed evenly without a remainder.

3. Principle of time-consistency (dynamic stability)

Formalization of the notion of optimal behavior constitutes one of fundamental problems in the theory of n-person games. At present, for the various classes of games different solution concepts are constructed. Recall that the players' behavior (strategies in noncooperative games or imputations in cooperative games) satisfying some given optimality principle is called a solution of the game in the sense of this principle and must possess two properties. On the one hand, it must be feasible under conditions of the game where it is applied. On the other hand, it must adequately reflect the conceptual notion of optimality providing special features of the class of games for which it is defined.

In dynamic games, one more requirement is naturally added to the mentioned requirements, viz. the purposefulness and feasibility of an optimality principle are to be preserved throughout the game. This requirement is called the *time-consistency of a solution of the game (dynamic stability)*.

The time-consistency of a solution of dynamic game is the property that, when the game proceeds along a "conditionally optimal" trajectory, at each instant of time the players are to be guided by the same optimality principle, and hence do not have any ground for deviation from the previously adopted "optimal" behavior throughout the game. When the time-consistency is betrayed, at some instant of time there are conditions under which the continuation of the initial behavior becomes non-optimal and hence initially chosen solution proves to be unfeasible.

Assume that at the start of the game the players adopt an optimality principle and construct a solution based on it (an imputation set satisfying the chosen principle of optimality, say the core, nucleolus, NM–solution, etc.). From the definition of cooperative game it follows that the evolution of the game is to be along the trajectory providing a maximal total payoff for the players. When moving along this "conditionally optimal" trajectory, the players pass through subgames with current initial states and current duration. In due course, not only the conditions of the game and the players opportunities, but even the players' interests may change. Therefore, at some stage (at some vertex y on the conditionally optimal trajectory) the initially optimal solution of the current game may not exist or satisfy players at this stage. Then, at this stage (starting from vertex y) players will have no ground to keep to the initially chosen "conditionally optimal" trajectory. The latter exactly means the time-inconsistency of the chosen optimality principle and, as a result, the instability of the motion itself.

We now focus our attention on time-consistent solutions in the cooperative games with incomplete information.

Let an optimality principle be chosen in the game $\Gamma_v(x_0)$. The solution of this game constructed in the initial state x_0 based on the chosen principle of optimality is denoted by $W_v(x_0)$. The set $W_v(x_0)$ is a subset of the imputation set $L_v(x_0)$ in the game $\Gamma_v(x_0)$. Assume that $W_v(x_0) \neq \emptyset$. Let $w^* = \{x_0^*, \ldots, x_k^*, \ldots, x_l^*\}$ be the conditionally optimal trajectory.

The definition suggests that along the conditionally optimal trajectory players obtain the largest total payoff.

For further consideration an important assumption is needed.

Assumption A. The n-tuple $u^*(\cdot) = (u_1^*(\cdot), \ldots, u_n^*(\cdot))$ and the corresponding trajectory $w^* = \{x_0^*, \ldots, x_k^*, \ldots, x_l^*\}$ are common knowledge in $\Gamma_v(x_0)$.

This assumption means that being at vertex $x_k^* \in X_i$ player i knows that he is in x_k^*. This changes the informational structures of subgames $\Gamma(x_k^*)$ along w^* in the following natural way.

Denote by $G(x_k^*)$ the subtree of tree G corresponding to the subgame $\Gamma(x_k^*)$ with initial vertex x_k^*. The information sets in $\Gamma(x_k^*)$ coincide with the intersections $G(x_k^*) \cap X_i^j = X_i^j(k)$ for all i, j where X_i^j is the information set in $\Gamma(x_0)$. The informational structure of $\Gamma(x_k^*)$ consists from the sets $X_i^j(k)$, for all i, j.

As before we can define the current cooperative subgame $\Gamma_v(x_k^*)$ of the subgame $\Gamma(x_k^*)$.

We will now consider the behavior of the set $W_v(x_0)$ along the conditionally optimal trajectory w^*. Towards this end, in each current state x_k^* current subgame $\Gamma_v(x_k^*)$ is defined as follows. In the state x_k^*, we define the characteristic function $v(S; x_k^*)$ as the value of the zero-sum game $\Gamma_S(x_k^*)$ between coalitions S and $N \setminus S$ from the initial state x_k^* (as it was done already for the game $\Gamma(x_0)$).

The *current cooperative subgame* $\Gamma_v(x_k^*)$ is defined as $\langle N, v(S, x_k^*)\rangle$. The imputation set in the game $\Gamma_v(x_k^*)$ is of the form:

$$L_v(x_k^*) = \left\{ \xi \in R^n \mid \xi_i \geq v(\{i\}; x_k^*), i = 1, \ldots, n; \sum_{i \in N} \xi_i = v(N; x_k^*) \right\},$$

where

$$v(N; x_k^*) = v(N; x_0^*) - \sum_{m=0}^{k-1} \sum_{i \in N} h_i(x_m^*).$$

The quantity

$$\sum_{m=0}^{k-1} \sum_{i \in N} h_i(x_m^*)$$

is interpreted as the total gain of the players on the first $k-1$ steps when the motion is carried out along the trajectory w^*.

Consider the family of current games

$$\{\Gamma_v(x_k^*) = \langle N, v(S; x_k^*)\rangle,\ 0 \leq k \leq l\},$$

determined along the conditionally optimal trajectory w^* and their solutions $W_v(x_k^*) \subset L_v(x_k^*)$ generated by the same principle of optimality as the initial solution $W_v(x_0^*)$.

It is obvious that the set $W_v(x_l^*)$ is a solution of terminal game $\Gamma_v(x_l^*)$ and is composed of the only imputation $h(x_l^*) = \{h_i(x_l^*),\ i = 1, \dots, n\}$, where $h_i(x_l^*)$ is the terminal part of player i's payoff along the trajectory w^*.

4. Time-consistency of the solution

Let the conditionally optimal trajectory w^* be such that $W_v(x_k^*) \neq \emptyset$, $0 \leq k \leq l$. If this condition is not satisfied, it is impossible for players to adhere to the chosen principle of optimality, since at the very first stage k, when $W_v(x_k^*) = \emptyset$, the players have no possibility to follow this principle. Assume that in the initial state x_0 the players agree upon the imputation $\xi^0 \in W_v(x_0)$. This means that in the state x_0 the players agree upon such an allocation of the total maximal gain that (when the game terminates at x_l^*) the share of ith player is equal to ξ_i^0, i.e., the ith component of the imputation ξ^0. Suppose the player i's payoff (his share) on the first k stages $x_0^*, x_1^*, \dots, x_{k-1}^*$ is $\xi_i(x_{k-1}^*)$. Then, on the remaining stages $x_k^*, \dots, x_l^*$ according to the ξ^0 he has to receive the gain $\eta_i^k = \xi_i^0 - \xi_i(x_{k-1}^*)$. For the original agreement (the imputation ξ^0) to remain in force at the instant k, it is essential that the vector $\eta^k = (\eta_1^k, \dots, \eta_n^k)$ belongs to the set $W_v(x_k^*)$, i.e., a solution of the current subgame $\Gamma_v(x_k^*)$. If such a condition is satisfied at each stage along the trajectory w^*, then the imputation ξ^0 is realized. Such is the conceptual meaning of the time-consistency of the imputation.

Along the trajectory w^*, the coalition N obtains the payoff

$$v(N; x_k^*) = \sum_{i \in N} \left[\sum_{m=k}^{l} h_i(x_m^*) \right].$$

Then the difference

$$v(N; x_0) - v(N; x_k^*) = \sum_{m=0}^{k-1} \sum_{i \in N} h_i(x_m^*)$$

is equal to the payoff the coalition N obtains on the first k stages $(0, \dots, k-1)$. The share of the ith player in this payoff, considering the transferability of payoffs, may be represented as

$$\gamma_i(k-1) = \sum_{m=0}^{k-1} \beta_i(m) \sum_{i=1}^{n} h_i(x_m^*) = \gamma_i(x_{k-1}^*, \beta), \tag{6.4}$$

where $\beta_i(m)$ satisfies the condition

$$\sum_{i=1}^{n} \beta_i(m) = 1, \ \beta_i(m) \geq 0, m = 0, 1, \ldots, l, \quad i \in N. \tag{6.5}$$

From (6.4) we necessarily get

$$\gamma_i(k) - \gamma_i(k-1) = \beta_i(k) \sum_{i=1}^{n} h_i(x_k^*).$$

This quantity may be interpreted as an instantaneous gain of the player i at the stage k. Hence it is clear the vector $\beta(k) = (\beta_1(k), \ldots, \beta_n(k))$ prescribes distribution of the total gain among the members of coalition N. By properly choosing $\beta(k)$, the players can ensure the desirable outcome, i.e., to regulate the players' gain receipt with respect to time, so that at each stage k there will be no objection against realization of original agreement (the imputation ξ^0).

DEFINITION 6.4 *The imputation $\xi^0 \in W_v(x_0)$ is called time-consistent in the game $\Gamma_v(x_0)$ if the following conditions are satisfied:*

1. *there exists a conditionally optimal trajectory $w^* = \{x_0^*, \ldots, x_k^*, \ldots, x_l^*\}$ along which $W_v(x_k^*) \neq \emptyset, \quad k = 0, 1, \ldots, l,$*
2. *there exists such vectors $\beta(k) = (\beta_1(k), \ldots, \beta_n(k))$ that for each $k = 0, 1, \ldots, l, \quad \beta_i(k) \geq 0, \quad \sum_{i=1}^{n} \beta_i(k) = 1$ and*

$$\xi^0 \in \bigcap_{k=0}^{l} [\gamma(x_{k-1}^*, \beta) \oplus W_v(x_k^*)], \tag{6.6}$$

where $\gamma(x_k^*, \beta) = (\gamma_1(x_k^*, \beta), \ldots, \gamma_n(x_k^*, \beta))$, and $W_v(x_k^*)$ is a solution of the current game $\Gamma_v(x_k^*)$.

The sum $\oplus$ in the above definition has the following meaning: for $\eta \in R^n$ and $A \subset R^n$ $\eta \oplus A = \{\eta + a \mid a \in A\}$.

The game $\Gamma_v(x_0)$ has a time-consistent solution $W_v(x_0)$ if all the imputations $\xi \in W_v(x_0)$ are time-consistent.

The conditionally optimal trajectory along which there exists a time-consistent solution of the game $\Gamma_v(x_0)$ is called an *optimal trajectory.*

The time-consistent imputation $\xi^0 \in W_v(x_0)$ may be realized as follows. From (6.6) at any stage k we have

$$\xi^0 \in [\gamma(x_{k-1}^*, \beta) \oplus W_v(x_k^*)], \tag{6.7}$$

where

$$\gamma(x^*_{k-1}, \beta) = \sum_{m=0}^{k-1} \beta(m) \sum_{i \in N} h_i(x^*_m)$$

is the payoff vector on the first k stages, the player i's share in the gain on the same interval being

$$\gamma_i(x^*_{k-1}, \beta) = \sum_{m=0}^{k-1} \beta_i(m) \sum_{i \in N} h_i(x^*_m).$$

When the game proceeds along the optimal trajectory, the players on the first k stages share the total gain

$$\sum_{m=0}^{k-1} \sum_{i \in N} h_i(x^*_m)$$

among themselves,

$$\xi^0 - \gamma(x^*_{k-1}, \beta) \in W_y(x^*_k) \tag{6.8}$$

so that the inclusion (6.8) is satisfied. Furthermore, (6.8) implies the existence of such vector $\xi^k \in W_v(x^*_k)$ that $\xi^0 = \gamma(x^*_{k-1}, \beta) + \xi^k$. That is, in the description of the above method of choosing $\beta(m)$, the vector of the gains to be obtained by the players at the remaining stages of the game

$$\xi^k = \xi^0 - \gamma(x^*_{k-1}, \beta) = \sum_{m=k}^{l} \beta_i(m) h(x^*_m)$$

belongs to the set $W_v(x^*_k)$.

We also have

$$\xi^0 = \sum_{m=0}^{l} \beta_i(m) \sum_{i \in N} h(x^*_m).$$

The vector

$$\alpha_i(m) = \beta_i(m) \sum_{i \in N} h(x^*_m), \quad i \in N, \quad m = 0, 1, \ldots, l,$$

is called the imputation distribution procedure (IDP)

In general, it is fairly easy to see that there may exist an infinite number of vectors $\beta(m)$ satisfying conditions (6.4), (6.5). Therefore the sharing method proposed here seems to lack true uniqueness. However, for any vector $\beta(m)$ satisfying conditions (6.4), (6.5) at each stage k

the players are guided by the imputation $\xi^k \in W_v(x_k^*)$ and the same optimality principle throughout the game, and hence have no reason to violate the previously achieved agreement.

Let us make the following additional assumption

Assumption B. The vectors $\xi^k \in W_v(x_k^*)$ may be chosen as monotone nonincreasing sequence of the argument k.

Show that by properly choosing $\beta(m)$ we may always ensure time-consistency of the imputation $\xi^0 \in W_v(x_0)$ under assumption B. and the first condition of definition (i.e., along the conditionally optimal trajectory at each stage k $W_v(x_k^*) \neq \emptyset$).

Choose $\xi^k \in W_v(x_k^*)$ to be a monotone nonincreasing sequence. Construct the difference $\xi^0 - \xi^k = \gamma(k-1)$ then we get $\xi^k + \gamma(k-1) \in W_v(x_k^*)$. Let $\beta(k) = (\beta_1(k), \ldots, \beta_n(k))$ be vectors satisfying conditions (6.4), (6.5). Instead of writing $\gamma(x_k^*, \beta)$ we will write for simplicity $\gamma(k)$. Rewriting (6.4) in vector form we get

$$\sum_{m=0}^{k-1} \beta(m) \sum_{i \in N} h_i(x_m^*) = \gamma(k-1)$$

and we get the following expression for $\beta(k)$

$$\beta(k) = \frac{\gamma(k) - \gamma(k-1)}{\sum_{i \in N} h_i(x_k^*)} = -\frac{\xi^k - \xi^{k-1}}{\sum_{i \in N} h_i(x_k^*)} \geq 0. \tag{6.9}$$

Here the last expression follows from equality

$$\xi^0 = \gamma(k) + \xi^k.$$

THEOREM 6.1 *If the assumption B is satisfied and*

$$W(x_k^*) \neq \emptyset, \; k = 0, 1, \ldots, l \tag{6.10}$$

solution $W(x_0)$ is time-consistent.

Theoretically, the main problem is to study conditions imposed on the vector function $\beta(m)$ in order to ensure time-consistency of specific forms of solutions $W_v(x_0)$ in various classes of games.

Consider the new concept of strongly time-consistency and define time-consistent solutions for cooperative games with terminal payoffs.

5. Strongly time-consistent solutions

For the time consistent imputation $\xi^0 \in W_v(x_0)$, as follows from the definition, there exists sequence of vectors $\beta(m)$ and imputation

ξ^k (generally nonunique) from the solution $W_v(x_k^*)$ of the current game $\Gamma_v(x_k^*)$ that $\xi^0 = \gamma(x_{k-1}^*, \beta) + \xi^k$. The conditions of time-consistency do not affect the imputation from the set $W_v(x_k^*)$ which fail to satisfy this equation. Furthermore, of interest is the case when any imputation from current solution $W_v(x_k^*)$ may provide a "good" continuation for the original agreement, i.e., for a time-consistent imputation $\xi^0 \in W_v(x_0)$ at any stage k and for every $\xi^k \in W_v(x_k^*)$ the condition $\gamma(x_{k-1}^*, \beta) + \xi^k \in W_v(x_0)$, where $\gamma(x_l^*, \beta) = \xi^0$, be satisfied. By slightly strengthening this requirement, we obtain a qualitatively new time-consistency concept of the solution $W_v(x_0)$ of the game $\Gamma_v(x_0)$ and call it a strongly time-consistency.

DEFINITION 6.5 *The imputation $\xi^0 \in W_v(x_0)$ is called strongly time-consistent (STC) in the game $\Gamma_v(x_0)$, if the following conditions are satisfied:*

1. *the imputation ξ^0 is time-consistent;*
2. *for any $0 \leq q \leq r \leq l$ and β^0 corresponding to the imputation ξ^0 we have,*

$$\gamma(x_r^*, \beta^0) \oplus W_v(x_r^*) \subset \gamma(x_q^*, \beta^0) \oplus W_v(x_r^*). \tag{6.11}$$

The game $\Gamma_v(x_0)$ has a strongly time-consistent solution $W_v(x_0)$ if all the imputations from $W_v(x_0)$ are strongly time-consistent.

6. Terminal payoffs

In (6.2) let $h_i(x_k) \equiv 0$, $i = 1, \ldots, n$, $k = 1, \ldots, l-1$. The cooperative game with terminal payoffs is denoted by the same symbol $\Gamma_v(x_0)$. In such games the payoffs are payed when the game terminates (at terminal vertex $x_l \in X_{n+1}$).

THEOREM 6.2 *In the cooperative game $\Gamma_v(x_0)$ with terminal payoffs $h_i(x_l)$, $i = 1, \ldots, n$, only the vector $h(x_l^*) = \{h_i(x_l^*), i = 1, \ldots, n\}$ whose components are equal to the players payoffs at the terminal point of the conditionally optimal trajectory may be time-consistent.*

Proof. It follows from the time-consistency of the imputation $\xi^0 \in W_v(x_0)$ that

$$\xi^0 \in \bigcap_{0 \leq k \leq l} W_v(x_k^*).$$

But since the current game $\Gamma_v(x_l^*)$ is of zero duration, then therein $L_v(x_l^*) = W_v(x_l^*) = h(x_l^*)$. Hence

$$\bigcap_{0 \leq k \leq l} W_v(x_k^*) = h(x_l^*),$$

i.e., $\xi^0 = h(x_l^*)$ and there are no other imputations. □

THEOREM 6.3 *For the existence of the time-consistent solution in the game with terminal payoff it is necessary and sufficient that for all* $0 \leq k \leq l$

$$h(x_l^*) \in W_v(x_k^*),$$

where $h(x_l^*)$ *is the players payoff vector at the terminal point of the conditionally optimal trajectory* $w^* = \{x_0^*, \ldots, x_k^*, \ldots, x_l^*\}$, *with* $W_v(x_k^*)$, $0 \leq k \leq l$ *being the solutions of the current games along the conditionally optimal trajectory generated by the chosen principle of optimality.*

This theorem is a corollary of the previous one.

Thus, if in the game with terminal payoffs there is a time consistent imputation, then the players in the initial state x_0 have to agree upon realization of the vector (imputation) $h(x_l^*) \in W_v(x_0)$ and, with the motion along the optimal trajectory $w^* = \{x_0^*, \ldots, x_k^*, \ldots, x_l^*\}$, at each stage $0 \leq k \leq l$ this imputation $h(x_l^*)$ belongs to the solution of the current games $\Gamma_v(x_k^*)$.

As the theorem shows, in the game with terminal payoffs only a unique imputation from the set $W_v(x_0)$ may be time-consistent. Which is a highly improbable event since this means,that imputation $h(x_l^*)$ belongs to the solutions of all subgames along the conditionally optimal trajectory. Therefore, in such games there is no point in discussing both the time-consistency of the solution $W_v(x_0)$ as a whole and its strongly time-consistency.

7. Regularization

For some economic applications it is necessary that the instantaneous gain of player i at the stage k, which by properly choosing $\beta(k)$ regulates ith player's gain receipt with respect to time

$$\beta_i(k) \sum_{i \in N} h_i(x_k^*) = \alpha_i(k)$$

be nonnegative (IDP, $\alpha_i \geq 0$). Unfortunately this condition cannot be always guaranteed. In the same time we shall purpose a new characteristic function (c. f.) based on classical one defined earlier, such that solution defined in games with this new c. f. would be strongly time-consistent and would guarantee nonnegative instantaneous gain of player i at each stage k.

Let $v(S; x_k^*)$ $\quad S \subset N$ be the c. f. defined in subgame $\Gamma(x_k^*)$ in Section 2 using classical maxmin approach.

For the function $V(N; x_k^*)$ $(S = N)$ the Bellman's equation along $w^* = \{x_0^*, \ldots, x_k^*, \ldots, x_l^*\}$ is satisfied, i.e.,

$$V(N; x_0) = \sum_{m=0}^{k-1} \sum_{i=1}^{n} h_i(x_m^*) + V(N; x_k^*). \tag{6.12}$$

Define the new "regularized" function $\overline{v}(S; x_0), S \subset N$ by formula

$$\overline{v}(S; x_0) = \sum_{m=0}^{l} v(S; x_m^*) \frac{\sum_{i=1}^{n} h_i(x_m^*)}{v(N; x_m^*)} \tag{6.13}$$

And in the same manner for $0 \leq k \leq l$

$$\overline{v}(S; x_k^*) = \sum_{m=k}^{l} v(S, x_m^*) \frac{\sum_{i=1}^{n} h_i(x_m^*)}{v(N; x_m^*)} \tag{6.14}$$

It can be proved that $\overline{v}$ is superadditive and $\overline{v}(N; x_k^*) = v(N; x_k^*)$

Denote the set of imputations defined by characteristic functions $v(S; x_k^*)$, $\overline{v}(S; x_k^*)$, $k = 0, 1, \ldots, l$ by $L(x_k^*)$ and $\overline{L}(x_k^*)$ correspondingly. Let $\xi^k \in L(x_k^*)$ be a selector, $0 \leq k \leq l$, define

$$\overline{\xi} = \sum_{k=0}^{l} \xi^k \frac{\sum_{i=1}^{n} h_i(x_k^*)}{v(N; x_k^*)}, \tag{6.15}$$

$$\overline{\xi}^k = \sum_{m=k}^{l} \xi^m \frac{\sum_{i=1}^{n} h_i(x_m^*)}{v(N; x_m^*)}. \tag{6.16}$$

DEFINITION 6.6 *The set $\overline{L}(x_0)$ consists of vectors defined by (6.15) for all possible selectors ξ^k, $0 \leq k \leq l$ with values in $L(x_k^*)$.*

Let $\xi \in \overline{L}(x_0)$ and the functions $\alpha_i(k), i = 1, \ldots, n,\ 0 \leq k \leq l$ satisfy the condition

$$\sum_{k=0}^{l} \alpha_i(k) = \overline{\xi}_i, \quad \alpha_i \geq 0. \tag{6.17}$$

The vector function $\alpha(k) = \{\alpha_i(k)\}$ defined by the formula (6.17) is called "imputation distribution procedure" (IDP) (see Section 4). Define

$$\sum_{m=0}^{k-1} \alpha_i(m) = \overline{\xi}_i(k-1), \quad i = 1, \ldots, n.$$

The following formula connects α_i and β_i (see Section 4)

$$\alpha_i(k) = \beta_i(k) \sum_{i \in N} h_i(x_k^*).$$

Let $\overline{C}(x_0) \subset \overline{L}(x_0)$ be any of the known classical optimality principles from the cooperative game theory (core, nucleolus, NM-solution, Shapley value or any other OP). Consider $\overline{C}(x_0)$ as an optimality principle in $\Gamma(x_0)$. In the same manner let $\overline{C}(x_k^*)$ be an optimality principle in $\Gamma(x_k^*)$, $0 \leq k \leq l$.

The STC of the optimality principle means that if an imputation $\xi \in C(x_0)$ and an IDP $\alpha(k) = \{\alpha_i(k)\}$ of ξ are selected, then after getting by the players, on the first k stages, the amount

$$\overline{\xi}_i(k-1) = \sum_{m=0}^{k-1} \alpha_i(m), \; i = 1, \dots, n,$$

the optimal income (in the sense of the optimality principle $C(x_k^*)$) on the last $l-k$ stages in the subgame $\Gamma(x_k^*)$ together with $\overline{\xi}(k-1)$ constitutes the imputation belonging to the OP in the original game $\Gamma(x_0)$. The condition is stronger than time-consistency, which means only that the part of the previously considered "optimal" imputation belongs to the solution in the corresponding current subgame $\Gamma(x_{k-1}^*)$.

Suppose $\overline{C}(x_0) = \overline{L}(x_0)$ and $\overline{C}(x_k^*) = \overline{L}(x_k^*)$, then

$$\overline{L}(x_0) \supset \overline{\xi}(k-1) \oplus \overline{L}(x_k^*)$$

for all $0 \leq k \leq l$ and this implies that the set of all imputations $\overline{L}(x_0)$ if considered as solution in $\Gamma(x_0)$ is strongly time consistent (here $a \oplus B$, $a \in R^n$, $B \subset R^n$ is the set of vectors $a + b$, $b \in B$).

Suppose that the set $\overline{C}(x_0)$ consists of unique imputation — the Shapley value. In this case from time consistency the strong time-consistency follows immediately.

Suppose now that $C(x_0) \subset L(x_0)$, $C(x_k^*) \subset L(x_k^*)$, $0 \leq k \leq l$ are cores of $\Gamma(x_0)$ and correspondingly of subgames $\Gamma(x_k^*)$.

We suppose that the sets $C(x_k^*)$, $0 \leq k \leq l$ are nonempty. Let $\widehat{C}(x_0)$ and $\widehat{C}(x_k^*)$, $0 \leq k \leq l$, be the sets of all possible vectors $\overline{\xi}$, $\overline{\xi}_k$ from (6.15), (6.16) and $\xi^k \in C(x_k^*)$, $0 \leq k \leq l$. And let $\overline{C}(x_0)$ and $\overline{C}(x_k^*)$, $0 \leq k \leq l$, be cores of $\Gamma(x_0)$, $\Gamma(x_k^*)$ defined for c. f. $\overline{v}(S; x_0)$, $\overline{v}(S; x_k^*)$.

PROPOSITION 6.1 *The following inclusions hold*

$$\widehat{C}(x_0) \subset \overline{C}(x_0) \tag{6.18}$$

$$\widehat{C}(x_k^*) \subset \overline{C}(x_k^*), 0 \leq k \leq l. \tag{6.19}$$

Proof. The necessary and sufficient condition for imputation $\bar{\xi}$ belong to the core $\overline{C}(x_0)$ is the condition

$$\sum_{i \in S} \bar{\xi}_i \geq \bar{v}(S; x_0), \quad S \subset N.$$

If $\bar{\xi} \in \widehat{C}(x_0)$, then

$$\bar{\xi} = \sum_{m=0}^{l} \xi^m \frac{\sum_{i=1}^n h_i(x_m^*)}{v(N; x_m^*)},$$

where $\xi^m \in C(x_m^*)$. Thus

$$\sum_{i \in S} \xi_i^m \geq v(S; x_m^*), S \subset N, 0 \leq m \leq l.$$

And we get

$$\begin{aligned} \sum_{i \in S} \bar{\xi}_i &= \sum_{m=0}^{l} \sum_{i \in S} \xi^m \frac{\sum_{i=1}^n h_i(x_m^*)}{v(N; x_m^*)} \\ &\geq \sum_{m=0}^{l} v(S; x_m^*) \frac{\sum_{i=1}^n h_i(x_m^*)}{v(N; x_m^*)} = \bar{v}(S; x_0). \end{aligned}$$

the inclusion (6.18) is proved similarly.

Define a new solution in $\Gamma(x_0)$ as $\widehat{C}(x_0)$ which we will call "regularized" subcore. $\widehat{C}(x_0)$ is always time-consistent and strongly time-consistent

$$\widehat{C}(x_0) \supset \sum_{m=0}^{k-1} \xi^m \frac{\sum_{i=1}^n h_i(x_m^*)}{v(N; x_m^*)} \oplus \widehat{C}(x_m^*), 0 \leq k \leq m.$$

Here under $a \oplus A$, where $a \in R^n$, $A \subset R^n$ the set of all vector $a+b$, $b \in A$ is understood. The quantity

$$\alpha_i = \xi^m \frac{\sum_{i=1}^n h_i(x_m^*)}{v(N; x_m^*)} \geq 0$$

is IDP and is nonnegative. □

8. Cooperative stochastic games

Stochastic games (Shapley (1953b)) constitute a special subclass of extensive form games, but our previous construction cannot be used to create the cooperative theory for such games, since we considered only games in extensive form (and incomplete information) which do not contain chance moves. But chance move play an an essential role in stochastic games. Although the theory is very close to the discussed in previous sections one cannot derive from it immediately the results, for cooperative stochastic games, and we shall provide here the corresponding investigation in details.

8.1 Cooperative game

Consider a finite graph tree $G = (Z, L)$ where Z is the set of all vertexes and $L : Z \to 2^Z$ point to set mapping ($L_z = L(z) \subset Z, z \in Z$). In our setting each vertex $z \in Z$ is considered as an n-person simultaneous (one stage) game

$$\Gamma(z) = \langle N; X_1^z, \ldots, X_n^z; K_1^z, \ldots, K_n^z \rangle ,$$

where $N = \{1, \ldots, n\}$ is the set of players which is the same for all $z \in Z$, X_i^z — the set of strategies of player $i \in N$, $K_i^z(x_1^z, \ldots, x_n^z)$ (we suppose that $K_i^z \geq 0$) is the payoff of player i ($i \in N, x_i^z \in X_i^z$). The n-tuple $x^z = (x_1^z, \ldots, x_n^z)$ is called situation in the game $\Gamma(z)$. The game $\Gamma(z)$ is called a stage game. For each $z \in Z$ the transition probabilities

$$p(z, y; x_1^z, \ldots, x_n^z) = p(z, y; x^z) \geq 0,$$

$$\sum_{y \in L_z} p(z, y; x^z) = 1$$

are defined. $p(z, y; x^z)$ is the probability that the game $\Gamma(y)$, $y \in L_z$, will be played next after the game $\Gamma(z)$ under the condition that in $\Gamma(z)$ the situation $x^z = (x_1^z, \ldots, x_n^z)$ was realized.

Also $p(z, y; x^z) \equiv 0$ if $L_z = \emptyset$.

Suppose that in the game the path

$$z_0, z_1, \ldots, z_l \ (L_{z_l} = \emptyset)$$

is realized. Then the payoff of player $i \in N$ is defined as

$$K_i(z_0) = \sum_{j=0}^{l} K_i^{z_j}(x_1^{z_j}, \ldots, x_n^{z_j}) = \sum_{j=0}^{l} K_i^{z_j}(x^{z_j}).$$

But since the transition from one stage game to the other has stochastic character, one has to consider the mathematical expectation of player's payoff

$$E_i(z_0) = \exp K_i(z_0).$$

The following formula holds

$$E_i(z_0) = K_i^{z_0}(x^{z_0}) + \sum_{y \in L_{z_0}} p(z_0, y; x^{z_0}) E_i(y) \tag{6.20}$$

where $E_i(y)$ is the mathematical expectation of player ith payoff in the stochastic subgame starting from the stage game $\Gamma(y)$, $y \in L_{z_0}$.

The strategy $\pi_i(\cdot)$ of player $i \in N$ is a mapping which for each stage game $\Gamma(y)$ determines which local strategy x_i^y in this stage game is to be selected. Thus $\pi_i(y) = x_i^y$ for $y \in Z$.

We shall denote the described stochastic game as $\overline{G}(z_0)$. Denote by $\overline{G}(z)$ any subgame of $\overline{G}(z_0)$ starting from the stage game $\Gamma(z)$ (played on a subgraph of the graph G starting from vertex $z \in Z$).

If $\pi_i(\cdot)$ is a strategy of player $i \in N$ in $\overline{G}(z_0)$, then the trace of this strategy $\pi_i^y(\cdot)$ defined on a subtree $G(y)$ of G is a strategy in a subgame $\overline{G}(y)$ of the game $\overline{G}(z_0)$.

The following version of (6.20) holds for a subgame $\overline{G}(z)$ (for the mathematical expectation of player ith payoff in $\overline{G}(z)$)

$$E_i(z) = K_i^z(x^z) + \sum_{y \in L_z} p(z, y; x^z) E_i(y).$$

As mixed strategies in $\overline{G}(z_0)$ we consider behavior strategies (Kuhn (1953)). Denote them by $q_i(\cdot)$, $i \in N$, and the corresponding situation as

$$q^N(\cdot) = (q_1(\cdot), \ldots, q_n(\cdot)).$$

Here $q_i(z)$ for each $z \in Z$ is a mixed strategy of player i in a stage game $\Gamma(z)$.

Denote by $\bar{\pi}^N(\cdot) = (\bar{\pi}_1(\cdot), \ldots, \bar{\pi}_n(\cdot))$ the n-tuple of pure strategies in $\overline{G}(z_0)$ which maximizes the sum of expected players' payoffs (cooperative solution). Denote this maximal sum by $V(z_0)$

$$V(z_0) = \max E(z_0) = \max \left[\sum_{i \in N} E_i(z_0) \right].$$

It can be easily seen that the maximization of the sum of the expected payoffs of players over the set of behavior strategies does not exceed $V(z_0)$.

In the same way we can define then cooperative n-tuple of strategies for any subgame $\overline{G}(z)$, $z \in Z$, starting from the stage game $\Gamma(z)$. From Bellman's optimality principle it follows that each of these n-tuples is a trace of $\bar{\pi}^N(\cdot)$ in the subgame $\Gamma(z)$. The following Bellman equation holds (Bellman (1957))

$$V(z) = \max_{\substack{x_i^z \in X_i^z \\ i \in N}} \left\{ \sum_{i \in N} K_i^z(x_i^z) + \sum_{y \in L_z} p(z, y; x^z) V(y) \right\}$$
$$= \sum_{i \in N} K_i^z(\bar{x}^z) + \sum_{y \in L_z} p(z, y; \bar{x}^z) V(y) \tag{6.21}$$

with the initial condition

$$V(z) = \max_{x_i^z \in X_i^z i \in N} \sum_{i \in N} K_i^z(x^z), \ z \in \{z : L_z = \emptyset\}. \tag{6.22}$$

The maximizing n-tuple $\bar{\pi}^N(\cdot) = (\bar{\pi}_1(\cdot), \ldots, \bar{\pi}_n(\cdot))$ defines the probability measure over the game tree $G(z_0)$. Consider a subtree $\hat{G}(z_0)$ of $G(z_0)$ which consists of paths in $G(z_0)$ having the positive probability under the measure generated by $\bar{\pi}^N(\cdot)$. We shall call $\hat{G}(z_0)$ cooperative subtree and the vertexes in $\hat{G}(z_0)$ shall denote by $CZ \subset Z$.

For each $z \in CZ$ define a zero-sum game over the structure of the graph $\overline{G}(z)$ between the coalition $S \subset N$ as maximizing player and coalition $N \setminus S$ as minimizing. Let $V(S, z)$ be the value of this game in behavior strategies (the existence follows from Kuhn (1953)). Thus for each subgame $\overline{G}(z)$, $z \in CZ$, we can define a characteristic function $V(S, z)$,$S \subset N$, with $V(N, z) = V(z)$ defined by (6.21), (6.22). Consider the cooperative version $\overline{\overline{G}}(z)$, $z \in Z$, of a subgame $\overline{G}(z)$ with characteristic function $V(S, z)$.

Let $I(z)$ be the imputation set in $\overline{\overline{G}}(z)$

$$I(z) = \left\{ \alpha^z : \sum_{i \in N} \alpha_i^z = V(z) = V(N, z), \alpha_i^z \geq V(\{i\}, z) \right\}. \tag{6.23}$$

As solution in $\overline{\overline{G}}(z)$ we can understand any given subset $C(z) \subset I(z)$. This can be any of classical cooperative solution (nucleous, core, NM-solution, Shapley Value). For simplicity (in what follows) we suppose that $C(z)$ is a Shapley Value

$$C(z) = Sh(z) = \{Sh_1(z), \ldots, Sh_n(z)\} \subset I(z)$$

but all conclusions can be automatically applied for other cooperative solution concepts.

8.2 Cooperative Payoff Distribution Procedure (CPDP)

The vector function $\beta(z) = (\beta_1(z), \ldots, \beta_n(z))$ is called CPDP if

$$\sum_{i \in N} \beta_i(z) = \sum_{i \in N} K_i^z(\bar{x}_1^z, \ldots, \bar{x}_n^z), \tag{6.24}$$

where $\bar{x}^z = (\bar{x}_1^z, \ldots, \bar{x}_n^z)$ satisfies (6.21). In each subgame $\overline{\overline{G}}(z)$ with each path $\bar{z} = z, \ldots, z_m$ ($L_{z_m} = \emptyset$) in this subgame one can associate the random variable — the sum of $\beta_i(z)$ along this path $\bar{z}$. Denote the expected value of this sum in $\overline{\overline{G}}(z)$ as $B_i(z)$.

It can be easily seen that $B_i(z)$ satisfies the following functional equation

$$B_i(z) = \beta_i(z) + \sum_{y \in L_z} p(z, y; x^z) B_i(y). \tag{6.25}$$

Calculate Shapley Value (Shapley (1953a)) for each subgame $G(z)$ for $z \in CZ$

$$Sh_i(z) = \sum_{S \subset N i \in S} \frac{(|S|-1)!(n-|S|)!}{n!} (V(S,z) - V(S \setminus \{i\}, z)) \tag{6.26}$$

where $|S|$ is the number of elements in S.

Define $\gamma_i(z)$ by formula

$$Sh_i(z) = \gamma_i(z) + \sum_{y \in Z} p(z, y; x^z) Sh_i(y). \tag{6.27}$$

Since $Sh(z) \in I(z)$ we get from (6.27)

$$V(N; z) = \sum_{i \in N} \gamma_i(z) + \sum_{y \in L_z} p(z, y; x^z) V(N; y),$$

$$\text{and} V(N; z) =\leftrightarrow \sum_{i \in N} \gamma_i(z), \quad \text{for } z \in \{z : L_z = \emptyset\}. \tag{6.28}$$

Comparing (6.28) and (6.21) we get that

$$\sum_{i \in N} \gamma_i(z) = \sum_{i \in N} K_i^z(\tilde{x}^z) \tag{6.29}$$

for $\overline{x}^z = (\overline{x}_1^z, \ldots, \overline{x}_n^z)$, $\overline{x}_i^z \in X_i^z$, $i \in N$ and thus the following lemma holds

LEMMA 6.1 *$\gamma(z) = (\gamma_1(z), \ldots, \gamma_n(z))$ defined by (6.27) is CPDP.*

DEFINITION 6.7 *Shapley Value* $\{Sh(z_0)\}$ *is called time-consistent in* $\overline{\overline{G}}(z_0)$ *if there exists a nonnegative CPDP* ($\beta_i(z) \geq 0$) *such that the following condition holds*

$$Sh_i(z) = \beta_i(z) + \sum_{y \in L_z} p(z, y; x^z) Sh_i(y), i \in N, z \in Z. \tag{6.30}$$

From (6.30) we get

$$\beta_i(z) = Sh_i(z) - \sum_{y \in L_z} p(z, y; x^z) Sh_i(y)$$

and the nonnegativity of CPDP $\beta_i(z)$ can follow from the monotonicity of Shapley Value along the paths on cooperative subgame $\hat{\hat{G}}(z_0)$ ($Sh_i(y) \leq Sh_i(z)$ for $y \in L_z$). In the same time the nonnegativity of CPDP $\beta_i(z)$ from (6.30) in general does not hold.

Denote as before by $B_i(z)$ the expected value of the sums of $\beta_i(y)$ from (6.30), $y \in Z$ along the paths in the cooperative subgame $\hat{\hat{G}}(z)$ of the game $\hat{\hat{G}}(z_0)$.

LEMMA 6.2

$$B_i(z) = Sh_i(z), i \in N. \tag{6.31}$$

We have for $B_i(z)$ the equation (6.25)

$$B_i(z) = \beta_i(z) + \sum_{y \in L_z} p(z, y; x^z) B_i(y) \tag{6.32}$$

with initial condition

$$B_i(z) = Sh_i(z) \text{ for } z \in \{z: \ L_z = \emptyset\}, \tag{6.33}$$

and for the Shapley Value we have

$$Sh_i(z) = \beta_i(z) + \sum_{y \in L_z} p(z, y; x^z) Sh_i(y). \tag{6.34}$$

From (6.32), (6.33), (6.34) it follows that $B_i(z)$ and $Sh_i(z)$ satisfy the same functional equations with the same initial condition (6.33), and the proof follows from backward induction consideration.

Lemma 6.2 gives natural interpretation for CPDP $\beta_i(z)$, $\beta_i(z)$ can be interpreted as the instantaneous payoff which player has to get in a stage game $\Gamma(z)$ when this game actually occurs along the paths of cooperative subtree $\hat{\hat{G}}(z_0)$, if his payoff in the whole game equals to the i-th component of the Shapley Value. So the CPDP shows the distribution in time of the Shapley Value in such a way that the players in each subgame are oriented to get the current Shapley Value of this subgame.

8.3 Regularization

In this section we purpose the procedure similar to one used in differential cooperative games (Petrosjan (1993)) which will guarantee the existence of time-consistent Shapley Value in the cooperative stochastic game (nonnegative CPDP).

Introduce

$$\bar{\beta}_i(z) = \frac{\sum_{i \in N} K_i(\bar{x}_1^z, \ldots, \bar{x}_n^z)}{V(N, z)} Sh_i(z) \tag{6.35}$$

where $\bar{x}^z = (\bar{x}_1^z, \ldots, \bar{x}_n^z)$, $z \in Z$ is the realization of the n-tuple of strategies $\bar{\pi}(\cdot) = (\bar{\pi}_1(\cdot), \ldots, \bar{\pi}_n(\cdot))$ maximizing the mathematical expectation of the sum of players' payoffs in the game $\overline{G}(z_0)$ (cooperative solution) and $V(N, z)$ is the value of c. f. for the grand coalition N in a subgame $\overline{\overline{G}}(z)$. Since $\sum_{i \in N} Sh_i(z) = V(N, z)$ from (6.35) it follows that $\bar{\beta}_i(z)$, $i \in N$, $z \in Z$, is CPDP. From (6.35) it follows also that the instantaneous payoff of the player in a stage game $\Gamma(z)$ must be proportional to the Shapley Value in a subgame $\overline{\overline{G}}(z)$ of the game $\overline{\overline{G}}(z_0)$.

Define the regularized Shapley Value (RSV) in $\overline{\overline{G}}(z)$ by induction as follows

$$\hat{S}h_i(z) = \frac{\sum_{i \in N} K_i(\bar{x}^z)}{V(N, z)} Sh_i(z) + \sum_{y \in L_z} p(z, y; \bar{x}^z) \hat{S}h_i(y) \tag{6.36}$$

with the initial condition

$$\hat{S}h_i(z) = \frac{\sum_{i \in N} K_i(\bar{x}^z)}{V(N, z)} Sh_i(z) = Sh_i(z) \text{ for } z \in \{z : L_z = \emptyset\}. \tag{6.37}$$

Since $K_i(x) \geq 0$ from (6.35) it follows that $\bar{\beta}_i(z) \geq 0$, and the new regularized Shapley Value $\hat{S}h_i(z)$ is time-consistent by (6.36).

Introduce the new characteristic function $\hat{V}(S, z)$ in $\overline{\overline{G}}(z)$ by induction using the formula ($S \subset N$)

$$\hat{V}(S, z) = \frac{\sum_{i \in N} K_i(\bar{x}^z)}{V(N, z)} V(S, z) + \sum_{y \in L_z} p(z, y; \bar{x}^z) \hat{V}(S, y) \tag{6.38}$$

with initial condition

$$\hat{V}(S, z) = V(S, z) \text{ for } z \in \{z : L_z = \emptyset\}.$$

Here $V(S,z)$ is superadditive, so is $\hat{V}(S,z)$, and $\hat{V}(N,z) = V(N,z)$ since both functions $\hat{V}(N,z)$ and $V(N,z)$ satisfy the same functional equation (6.21) with the initial condition (6.22). Rewriting (6.38) for $\{S \setminus i\}$ we get

$$\hat{V}(S \setminus i, z) = \frac{\sum_{i \in N} K_i(\bar{x}^z)}{V(N,z)} V(S \setminus i, z) + \sum_{y \in L_z} p(z,y;\bar{x}^z)\hat{V}(S \setminus i, y). \quad (6.39)$$

Subtracting (6.39) from (6.38) and multiplying on $\frac{(|S|-1)!(n-|S|)!}{n!}$ and summing upon $S \subset N$, $S \ni i$ we get

$$\begin{aligned} &\sum_{\substack{S \subset N \\ S \ni i}} \frac{(|S|-1)!(n-|S|)!}{n!} \left[\hat{V}(S,z) - \hat{V}(S \setminus i, z)\right] \\ &= \left\{ \sum_{\substack{S \subset N \\ S \ni i}} \frac{(|S|-1)!(n-|S|)!}{n!} \left[V(S,z) - V(S \setminus i, z)\right] \right\} \frac{\sum_{i \in N} K_i(\bar{x}^z)}{V(N,z)} \\ &\quad + \sum_{y \in L_z} p(z,y;\bar{x}^z) \left\{ \sum_{\substack{S \subset N \\ S \ni i}} \frac{(|S|-1)!(n-|S|)!}{n!} \left[V(S,z) - V(S \setminus i, z)\right] \right\}. \end{aligned} \quad (6.40)$$

From (6.36), (6.37) and (6.40) it follows that (RSV) $\hat{Sh}(z)$ is a Shapley Value for the c. f. $\hat{V}(S,z)$, since we got that $\hat{Sh}_i(z)$ and the function

$$\sum_{S \subset N S \ni i} \frac{(|S|-1)!(n-|S|)!}{n!} \left[\hat{V}(S,z) - \hat{V}(S \setminus i, z)\right]$$

satisfy the same functional equations with the initial condition.

$$\begin{aligned} \sum_{\substack{S \subset N \\ S \ni i}} \frac{(|S|-1)!(n-|S|)!}{n!} \left[\hat{V}(S,z) - \hat{V}(S \setminus i, z)\right] &= \\ \sum_{\substack{S \subset N \\ S \ni i}} \frac{(|S|-1)!(n-|S|)!}{n!} \left[V(S,z) - V(S \setminus i, z)\right] &= Sh_i(z) \end{aligned}$$

which also coincides with (6.37).

Thus

$$\hat{Sh}_i = \sum_{\substack{S \subset N \\ S \ni i}} \frac{(|S|-1)!(n-|S|)!}{n!} \left[\hat{V}(S,z) - \hat{V}(S \setminus i, z)\right].$$

THEOREM 6.4 *The RSV is Time-consistent and is a Shapley Value for the regularized characteristic function $\hat{V}(S, z)$ defined for any subgame $\overline{\overline{G}}(z)$ of the stochastic game $\overline{\overline{G}}(z_0)$.*

References

Bellman, R.E. (1957). *Dynamic Programming.* Princeton University Press, Prinstone, NY.

Filar, J. and Petrosjan, L.A. (2000). Dynamic cooperative games. *International Game Theory Review*, 2(1):42–65.

Haurie, A. (1975). On some properties of the characteristic function and core of multistage game of coalitions. *IEEE Transaction on Automatic Control*, 20(2):238–241.

Haurie, A. (1976). A note on nonzero-sum differential games with bargaining solution. *Journal of Optimization Theory and Application*, 18(1):31–39.

Kaitala, V. and Pohjola, M. (1988). Optimal recovery of a shared resource stock. A differential game model with efficient memory equilibria. *Natural Resource Modeling*, 3(1):191–199.

Kuhn, H.W. (1953). Extensive games and the problem of information. *Annals of Mathematics Studies*, 28:193–216.

Owen, G. (1968). *Game Theory.* W.B. Saunders Co., Philadelphia.

Petrosjan, L.A. (1993). *Differential Games of Pursuit.* World Scientific, London.

Petrosjan, L.A. (1995). The Shapley value for differential games. In: Geert Olsder (eds.), *New Trends in Dynamic Games and Applications*, pages 409–417, Birkhauser.

Petrosjan, L.A. (1996). On the time-concistency of the Nash equilibrium in multistage games with discount payoffs. *Applied Mathematics and Mechanics (ZAMM)*, 76(3):535–536.

Petrosjan, L.A. and Danilov, N.N. (1985). *Cooperative Differential Games.* Tomsk University Press, Vol. 276.

Petrosjan, L.A. and Danilov, N.N. (1979). Stability of solution in nonzero-sum differential games with transferable payoffs. *Vestnik of Leningrad University*, 1:52–59.

Petrosjan, L.A. and Zaccour, G. (2003). Time-consistent Shapley value allocation of pollution cost reduction. *Journal of Economic Dynamics and Control*, 27:381–398.

Shapley, L.S. (1953a). A value for n-person games. *Annals of Mathematics Studies*, 28:307–317.

Shapley, L.S. (1953b). Stochastic games. *Proceedings of the National Academy of U.S.A.*, 39:1095–1100.

Chapter 7

ELECTRICITY PRICES IN A GAME THEORY CONTEXT

Mireille Bossy
Nadia Maïzi
Geert Jan Olsder
Odile Pourtallier
Etienne Tanré

Abstract We consider a model of an electricity market in which $\mathcal{S}$ suppliers offer electricity: each supplier S_i offers a maximum quantity q_i at a fixed price p_i. The response of the market to these offers is the quantities bought from the suppliers. The objective of the market is to satisfy its demand at minimal price.

We investigate two cases. In the first case, each of the suppliers strives to maximize its market share on the market; in the second case each supplier strives to maximize its profit.

We show that in both cases some Nash equilibrium exists. Nevertheless a close analysis of the equilibrium for profit maximization shows that it is not realistic. This raises the difficulty to predict the behavior of a market where the suppliers are known to be mainly interested by profit maximization.

1. Introduction

Since the deregulation process of electricity exchanges has been initiated in European countries, many different market structures have appeared (*see e.g. Stoft (2002)*). Among them are the so called *day ahead markets* where suppliers face a decision process that relies on a centralized auction mechanism. It consists in submitting bids, more or less complicated, depending on the design of the *day ahead market* (power pools, power exchanges, ...). The problem is the determination of the quantity and price that will win the process of selection on the market. Our aim in this paper is to describe the behavior of the participants (suppliers) through a static game approach. We consider a market where $\mathcal{S}$ suppliers are involved. Each supplier offers on the market a maximal quantity of electricity, q, that it is ready to deliver at a fixed price p. The response of the market to these offers is the quantities bought from each supplier. The objective of the market is to satisfy its demand at minimal price.

Closely related papers are Supatchiat, Zhang and Birge (2001) and Madrigal and Quintana (2001). They also consider optimal bids on electricity markets. Nevertheless, in Supatchiat, Zhang and Birge (2001), the authors take the quantity of electricity proposed on the market as exogenous, whereas here we consider the quantity as part of the bid. In Madrigal and Quintana (2001), the authors do not consider exactly the same kind of market mechanism, in particular they consider open bids and fix the market clearing price as the highest price among the accepted bids. They consider fixed demand but also stochastic demand.

The paper is organized as follows. The model is described in Section 2, together with the proposed solution concept. In Section 3 we consider the case where the suppliers strive to maximize their market share, while in Section 4 we analyze the case where the goal is profit maximization. We conclude in Section 5 with some comparison remarks on the two criteria used, and some possible directions for future work.

2. Problem statement

2.1 The agents and their choices

We consider a single market, that has an inelastic demand for d units of electricity that is provided by $\mathcal{S}$ local suppliers called S_j, $j = 1, 2, \ldots, \mathcal{S}$.

2.1.1 The suppliers. Each supplier S_j sends an offer to the market that consists in a price function $p_j(\cdot)$, that associates to any quantity of electricity q, the unit price $p_j(q)$ at which it is ready to sell this quantity.

We shall use the following special form of the price function:

DEFINITION 7.1 *For supplier S_j, a* quantity-price *strategy, referred to as the pair (q_j, p_j), is a price function $p_j(\cdot)$ defined by*

$$p_j(q) = \begin{cases} p_j \geq 0, & \text{for } q \leq q_j, \\ +\infty, & \text{for } q > q_j. \end{cases} \tag{7.1}$$

q_j is the maximal quantity S_j offers to sell at the finite price p_j. For higher quantities the price becomes infinity.

Note that we use the same notation, (p_j) for the price function and for the fixed price. This should not cause any confusion.

2.1.2 The market. The market collects the offers made by the suppliers, i.e., the price functions $p_1(\cdot), p_2(\cdot), \ldots, p_{\mathcal{S}}(\cdot)$, and has to choose the quantities $\overline{q}_j$ to buy from each supplier S_j, $j = 1, \ldots, \mathcal{S}$. The unit price paid to S_j is $p_j(\overline{q}_j)$.

We suppose that an admissible choice of the market is such that the demand is fully satisfied at finite price, i.e., such that,

$$\sum_{j=1}^{\mathcal{S}} \overline{q}_j = d, \; \overline{q}_j \geq 0, \text{ and } p_j(\overline{q}_j) < +\infty, \; \forall j. \tag{7.2}$$

When the set of admissible choices is empty, i.e., when the demand cannot be satisfied at finite cost (for example when the demand is too large with respect to some finite production capacity), then the market buys the maximal quantity of electricity it can at finite price, though the full demand is not satisfied.

2.2 Evaluation functions and objective

2.2.1 The market. We suppose that the objective of the market is to choose an admissible strategy (i.e., satisfying (7.2)), $(\overline{q}_1, \ldots, \overline{q}_{\mathcal{S}})$ in response to the offers $p_1(\cdot), \ldots, p_{\mathcal{S}}(\cdot)$ of the suppliers, so as to minimize the total cost.

More precisely the market problem is:

$$\min_{\{\overline{q}_j\}_{j=1\cdots\mathcal{S}}} \varphi_M(p_1(\cdot), \ldots, p_{\mathcal{S}}(\cdot), \overline{q}_1, \ldots, \overline{q}_{\mathcal{S}}), \tag{7.3}$$

with

$$\varphi_M(p_1(\cdot), \ldots, p_{\mathcal{S}}(\cdot), \overline{q}_1, \ldots, \overline{q}_{\mathcal{S}}) \stackrel{\text{def}}{=} \sum_{j=1}^{\mathcal{S}} p_j(\overline{q}_j)\overline{q}_j, \tag{7.4}$$

subject to constraints (7.2).

2.2.2 The suppliers. The two criteria, profit and market share, will be studied for the suppliers:

- **The profit** — When the market buys quantities $\overline{q}_j$, $j = 1, \ldots, \mathcal{S}$, supplier S_j's profit to be maximized is

$$\varphi_{S_j}(p_1(\cdot), \ldots, p_{\mathcal{S}}(\cdot), \overline{q}_j) \stackrel{\text{def}}{=} p_j(\overline{q}_j)\overline{q}_j - C_j(\overline{q}_j), \tag{7.5}$$

  where $C_j(\cdot)$ is the production cost function.

ASSUMPTION 7.1 *We suppose that, for each supplier S_j, the production cost $C_j(\cdot)$ is a piecewise $\mathcal{C}^1$ and convex function.*

When C_j is not differentiable we define the marginal cost $C'(q)$ as $\lim_{\epsilon \to 0^+} \frac{dC_j}{dq}(q - \epsilon)$.

Because of the assumption made on C_j, the marginal cost C'_j is monotonic and nondecreasing. In particular it can be a piecewise constant increasing function.

A typical special case in electricity production is when the marginal costs are piecewise constant. It corresponds to the fact that the producers starts producing in its cheapest production facility. If the market asks more electricity, the producers start up the one but cheapest production facility, etc.

- **The market share** — for supplier S_j, $\overline{q}_j$ is the quantity bought from him by the market, i.e., we define this criterion as

$$\varphi_{S_j}(p_1(\cdot), \ldots, p_{\mathcal{S}}(\cdot), \overline{q}_j) \stackrel{\text{def}}{=} \overline{q}_j. \tag{7.6}$$

For this criterion, it is necessary to introduce a price constraint. As a matter of fact, the obvious, but unrealistic, solution without price constraint would be to set the price to zero whatever the quantity bought is.

We need a constraint such that, for example, the profit is non-negative, or such that the unit price is always above the marginal cost, C'_j.

For the sake of generality we suppose the existence of a minimal unit price function $\mathcal{L}_j$ for each supplier. Supplier S_j is not allowed to sell the quantity q at a unit price lower than $\mathcal{L}_j(q)$.

A natural choice for $\mathcal{L}_j$ is C'_j, which expresses the usual constraint that the unit price is above the marginal cost.

2.3 Equilibria

From a game theoretical point of view, a two time step problem with $\mathcal{S}+1$ players will be formulated. At a first time step the suppliers announce their offers (the price functions) to the market, and at the second time step the market reacts to these offers by choosing the quantities $\overline{q}_j$ of electricity to buy from each supplier. Each player strives to optimize (i.e., maximize for the suppliers, minimize for the market) his own criterion function (φ_{S_j}, $j = 1, \ldots, \mathcal{S}$, φ_M) by properly choosing his own decision variable(s). The numerical outcome of each criterion function will in general depend on all decision variables involved. In contrast to conventional optimization problems, in which there is only one decision maker, and where the word "optimum" has an unambiguous meaning, the notion of "optimality" in games is open to discussion and must be defined properly. Various notions of "optimality" exist (see Başar and Olsder (1999)).

Here the structure of the problem leads us to use a combined *Nash Stackelberg* equilibrium. Please note that the "leaders", i.e., suppliers, choose and announce functions $p_j(\cdot)$. In Başar and Olsder (1999) the corresponding equilibrium is referred to as *inverse Stackelberg.*

More precisely, define $\{\overline{q}_j(p_1(\cdot), \ldots, p_{\mathcal{S}}(\cdot)), j = 1, \ldots, \mathcal{S}\}$, the best response of the market to the offers $(p_1(\cdot), \ldots, p_{\mathcal{S}}(\cdot))$ of the suppliers, i.e., a solution of the problem ((7.2)-(7.3)). The choices $(\{p_j^*(\cdot)\}, \{\overline{q}_j^*\}, \ j = 1, \ldots, \mathcal{S})$ will be said optimal if the following holds true,

$$\overline{q}_j^* \stackrel{\text{def}}{=} \overline{q}_j^*(p_1^*(\cdot), \ldots, p_{\mathcal{S}}^*(\cdot)), \tag{7.7}$$

For every supplier S_j, $j = 1, \ldots, \mathcal{S}$ and any admissible price function $\tilde{p}_j(\cdot)$ we have

$$\varphi_{S_j}(p_1^*(\cdot), \ldots, p_{\mathcal{S}}^*(\cdot), \overline{q}_j^*) \geq \varphi_{S_j}(p_1^*(\cdot), \ldots, \tilde{p}_j(\cdot), \ldots, p_{\mathcal{S}}^*(\cdot), \tilde{q}_j), \tag{7.8}$$

where

$$\tilde{q}_j \stackrel{\text{def}}{=} \overline{q}_j^*(p_1^*(\cdot), \ldots, \tilde{p}_j(\cdot), \ldots, p_{\mathcal{S}}^*(\cdot)). \tag{7.9}$$

The Nash equilibrium Equation (7.8) tells us that supplier S_j cannot increase its outcome by deviating unilaterally from its equilibrium choice $(p_j^*(\cdot))$. Note that in the second term of Equation (7.8), the action of the market is given by (7.9): if S_j deviates from $p_j^*(\cdot)$ by offering the price function $\tilde{p}_j(\cdot)$, the market reacts by buying from S_j the quantity $\tilde{q}_j$ instead of $\overline{q}_j^*$.

REMARK 7.1 As already noticed the minimization problem ((7.2)–(7.3)) defining the behavior of the market may not have any solution. In that case the market reacts by buying the maximal quantity of electricity it can at finite price.

At the other extreme, it may have infinitely many solutions (for example when several suppliers use the same price function). In that case $\overline{q}_j^*(\cdot)$ is not uniquely defined by Equation (7.7), nor consequently is the Nash equilibrium defined by Equation (7.8).

We would need an additional rule that says how the market reacts when its minimization problem has several (possibly infinitely many) solutions. Such an additional rule could be, for example, that the market first buys from supplier S_1 then from supplier S_2, etc. or that the market prefers the offers with larger quantities, etc. Nevertheless, it is not necessary to make this additional rule explicit in this paper. So we do assume that there is an additional rule, known by all the suppliers that insures that the reaction of the market is unique.

3. Suppliers maximize market share

In this section we analyze the case where the suppliers strive to maximize their market shares by appropriately choosing the price functions $p_j(\cdot)$ at which they offer their electricity on the market. We restrict our attention to price functions $p_j(\cdot)$ given in Definition 7.1 and referred to as the *quantity-price* pair (q_j, p_j).

For supplier S_j we denote $\mathcal{L}_j(\cdot)$ its minimal unit price function that we suppose nondecreasing with respect to the quantity sold. Classically this minimal unit price function may represent the marginal production cost.

Using a *quantity-price* pair (q_j, p_j) for each supplier, the market problem (7.3) can be written as

$$\min_{\{\overline{q}_j, j=1,\ldots,\mathcal{S}\}} \sum_{j=1}^{\mathcal{S}} p_j \overline{q}_j,$$

$$\text{under} \qquad 0 \le \overline{q}_j \le q_j, \ \sum_{j=1}^{\mathcal{S}} \overline{q}_j = d.$$

To define a unique reaction of the market we use Remark 7.1, when Problem (7.9) does not have any solution (i.e., when $\sum_{j=1}^{\mathcal{S}} q_j < d$) or at the other extreme when Problem (7.9) has possibly infinitely many solutions.

Hence we can define the evaluation function of the suppliers by

$$J_{S_j}((q_1, p_1), \ldots, (q_{\mathcal{S}}, p_{\mathcal{S}})) \stackrel{\text{def}}{=} \varphi_{S_j}(p_1(\cdot), \ldots, p_{\mathcal{S}}(\cdot), \overline{q}_j^*(p_1(\cdot), \ldots, p_{\mathcal{S}}(\cdot)),$$

where the price function $p_j(\cdot)$ is the pair (q_j, p_j) and $\overline{q}_j^*(p_1(\cdot), \ldots, p_{\mathcal{S}}(\cdot))$ is the unique optimal reaction of the market.

Now the Nash Stackelberg solution can be simply expressed as a Nash solution, i.e., find $\mathbf{u}^* \stackrel{\text{def}}{=} (u_1^*, \ldots, u_{\mathcal{S}}^*)$, $u_j^* \stackrel{\text{def}}{=} (q_j^*, p_j^*)$, so that for any supplier S_j and any pair $\tilde{u}_j = (\tilde{q}_j, \tilde{p}_j)$ we have

$$J_{S_j}(\mathbf{u}^*) \ge J_{S_j}(\mathbf{u}_{-j}^*, \tilde{u}_j), \tag{7.10}$$

where $(\mathbf{u}_{-j}^*, \tilde{u}_j)$ denotes the vector $(u_1^*, \ldots, u_{j-1}^*, \tilde{u}_j, u_{j+1}^*, \ldots, u_{\mathcal{S}}^*)$.

ASSUMPTION 7.2 *We suppose that there exist quantities* Q_j *for* $j = 1, \ldots, \mathcal{S}$, *such that*

$$\sum_{j=1}^{\mathcal{S}} Q_j \geq d, \tag{7.11}$$

and such that the minimal price functions $\mathcal{L}_j$ *are defined for the set* $[0, Q_j]$ *to* $\mathbb{R}^+$, *with finite values for any* q *in* $[0, Q_j]$. *The quantities* Q_j *represent the maximal quantities of electricity supplier* S_j *can offer to the market. It may reflect maximal production capacity for producers or more generally any other constraints such that transportation constraints.*

REMARK 7.2 The condition (7.11) insures that shortage can be avoided even if this implies high, but finite, prices.

We consider successively in the next subsections the cases where the minimal price functions $\mathcal{L}_j$ are continuous (Subsection 3.1) or discontinuous (Subsection 3.2). This last case is the most important from the application point of view, since we often take $\mathcal{L}_j = C'_j$ which is not in general continuous.

3.1 Continuous strictly increasing minimal price

We suppose the following assumption holds,

ASSUMPTION 7.3 *For any supplier* S_j, $j \in \{1, \ldots, \mathcal{S}\}$ *the minimal price function* $\mathcal{L}_j$ *is continuous and strictly increasing from* $[0, Q_j]$ *to* $\mathbb{R}^+$.

PROPOSITION 7.1

1. *Suppose that Assumption 7.3 holds. Then any strategy profile* $\mathbf{u}^* = (u_1^*, u_2^*, \ldots, u_{\mathcal{S}}^*)$ *with* $u_j^* = (q_j^*, p^*)$ *such that*

$$\begin{array}{l} \mathcal{L}_j(q_j^*) = p^*, \ \ \forall j \in \{1, \ldots, \mathcal{S}\} \text{ such that } q_j^* > 0 \\ \mathcal{L}_j(0) \geq p^*, \ \ \forall j \in \{1, \ldots, \mathcal{S}\} \text{ such that } q_j^* = 0 \\ \sum_{j \in \{1,2 \cdots \mathcal{S}\}} q_j^* = d, \end{array} \tag{7.12}$$

   is a Nash equilibrium.

2. *Suppose furthermore that Assumption 7.2 holds, then the equilibrium exists and is unique.*

We omit the proof of this proposition which is in the same vein as the proof of the following Proposition 7.2. Nevertheless it can be found in Bossy et al. (2004).

3.2 Discontinuous nondecreasing minimal price

We now address the problem where the minimal price functions $\mathcal{L}_j$ are not necessarily continuous and not necessarily strictly increasing. Nevertheless we assume that they are non decreasing. We set the following assumption,

ASSUMPTION 7.4 *We suppose that the minimal price functions $\mathcal{L}_j$ are nondecreasing, piecewise continuous, and that $\lim_{y \to x^-} \mathcal{L}_j(y) = \mathcal{L}(x)$ for any $x \geq 0$.*

Replacing Assumption 7.3 by Assumption 7.4, there may not be any strategy profile or at the other extreme there may be possibly infinitely many strategy profiles that satisfy Equations (7.12). Proposition 7.1 fail to characterize the Nash equilibria.

For any $p \geq 0$, we define $\rho_j(p)$, the maximal quantity supplier S_j can offer at price p, i.e.,

$$\rho_j(p) = \begin{cases} \max\{q \geq 0, \mathcal{L}_j(q) \leq p\}, & \text{if } j \text{ is such that } \mathcal{L}_j(0) \leq p, \\ 0, & \text{otherwise}. \end{cases} \tag{7.13}$$

Hence $\rho_j(p)$ is only determined by the structure of the minimal price function $\mathcal{L}_j$. In particular it is not dependent on any choice of the suppliers.

As a consequence of Assumption 7.4, $\rho_j(p)$ increases with p, and for any $p \geq 0$, $\lim_{y \to p^+} \rho_j(y) = \rho_j(p)$.

Denote by $O(\cdot)$ the function from $\mathbb{R}^+$ to $\mathbb{R}^+$ defined by

$$O(p) = \sum_{j=1}^{\mathcal{S}} \rho_j(p). \tag{7.14}$$

$O(p)$ is the maximal total offer that can be achieved at price p by the suppliers respecting the price constraints. The function O is possibly discontinuous, non decreasing (but not necessarily strictly increasing) and satisfies $\lim_{y \to p^+} O(y) = O(p)$. Assumption 7.2 implies that

$$O(\sup_j \mathcal{L}_j(Q_j)) \geq \sum_{j=1}^{\mathcal{S}} \rho_j(\mathcal{L}_j(Q_j)) \geq \sum_{j=1}^{\mathcal{S}} Q_j \geq d,$$

hence there exists a unique $p^* \leq \sup_j \mathcal{L}_j(Q_j) < +\infty$ such that

$$\begin{aligned} O(p^*) &= \sum_{j=1}^{\mathcal{S}} \rho_j(p^*) \geq d, \\ \forall \epsilon > 0, \quad & O(p^* - \epsilon) < d. \end{aligned} \tag{7.15}$$

The price p^* represents the minimal price at which the demand could be fully satisfied taking into account the minimal price constraint.

ASSUMPTION 7.5 *For p^* defined by (7.15), one of the following two condition holds:*

1. *We suppose that there exists a unique $\bar{j} \in \{1, \ldots, \mathcal{S}\}$ such that $\mathcal{L}_{\bar{j}}^{-1}(p^*) \neq \emptyset$, where $\mathcal{L}_{\bar{j}}^{-1}(p) \stackrel{\text{def}}{=} \{q \in [0, d],\ \ \mathcal{L}_j(q) = p\}$. In particular, there exists a unique $\bar{j} \in \{1, \ldots, \mathcal{S}\}$ such that $\mathcal{L}_{\bar{j}}(\rho_{\bar{j}}(p^*)) = p^*$, and such that for $j \neq \bar{j}$, we have $\mathcal{L}_j(\rho_j(p^*)) < p^*$.*

2. *At price p^* the maximal total quantity suppliers are allowed to propose is exactly d, i.e., $\sum_{j=1}^{\mathcal{S}} \rho_j(p^*) = d$.*

PROPOSITION 7.2 *Suppose Assumptions 7.4 and 7.5 hold. Consider the strategy profile $\mathbf{u}^* = (u_1^*, \ldots, u_{\mathcal{S}}^*)$, $u_j^* = (q_j^*, p^*)$ such that,*

- *p^* is defined by Equation (7.15),*
- *for $j \neq \bar{j}$,* i.e. *such that $\mathcal{L}_j(\rho_j(p^*)) < p^*$ (see Assumption 7.5), we have $q_j^* = \rho_j(p^*)$ and $p_j^* \in [\mathcal{L}_j(q_j^*), p^*[$,*
- *for $j = \bar{j}$,* i.e. *such that $\mathcal{L}_{\bar{j}}(\rho_{\bar{j}}(p^*)) = p^*$ (see Assumption 7.5), we have $q_{\bar{j}}^* \in [\min((d - \sum_{k \neq \bar{j}} q_k^*)\ ,\ \rho_{\bar{j}}(p^*))\ ,\ \rho_{\bar{j}}(p^*)]$, and $p_{\bar{j}}^* \in [p^*, \overline{p}[$, where $\overline{p}$ is defined by*

$$\overline{p} \stackrel{\text{def}}{=} \min\{\mathcal{L}_k(q_k^{*+}),\ k \neq \bar{j}\} \tag{7.16}$$

then, $\mathbf{u}^*$ is a Nash equilibrium.

REMARK 7.3 There exists an infinite number of strategy profiles that satisfy the conditions of Proposition 7.2 (the prices p_j^* are defined as elements of some intervals). Nevertheless, we can observe that there is no need for any coordination among the suppliers to get a Nash equilibrium. Each supplier can choose independently a strategy as described in Proposition 7.2, the resulting strategy profile is a Nash equilibrium. Note that this property does not hold in general for non-zero sum games (see the classical "battle of the sexes" game Luce and Raifa (1957)). We can also observe that for each supplier the outcome is the same whatever the Nash equilibrium set. In that sense we can say that all these Nash equilibria are equivalent.

A reasonable manner to select a particular Nash equilibrium is to suppose that the suppliers may strive for the maximization of their profits as an auxiliary criteria. More precisely, among the equilibria with market share maximization as criteria, they choose the equilibrium that brings them the maximal income. Because the equilibria we have found are independent, it is possible for each supplier to choose its preferred equilibrium. More precisely, with this auxiliary criterion, the equilibrium selected will be,

$$q_j^* = \rho_j(p^*),\ p_j^* = p^* - \epsilon,\quad \text{for } j \neq \bar{j} \text{ (i.e. such that } \mathcal{L}_j(\rho_j(p^*)) < p^*),$$
$$q_{\bar{j}}^* = \rho_{\bar{j}}(p^*),\ p_{\bar{j}}^* = \overline{p} - \epsilon,$$

where ϵ can be defined as the smallest monetary unit.

REMARK 7.4 Assumption 7.5 is necessary for the solution of the market problem (7.9) to have a unique solution for the strategies described in Proposition 7.2, which are consequently well defined.

If Assumption 7.5 does not hold, we would need to make the additional decision rule of the market explicit (see Remark 7.1). This is shown in the following example (Figure 7.1), with $\mathcal{S} = 2$. The Nash equilibrium may depend upon the additional decision rule of the market. In Figure 7.1, we have $\mathcal{L}_1(\rho_1(p^*)) = \mathcal{L}_2(\rho_2(p^*)) = p^*$ and $\rho_1(p^*) + \rho_2(p^*) > d$, where p^* is the price defined at (7.15). This means that Assumption 7.5 does not hold. Suppose the additional decision rule of the market

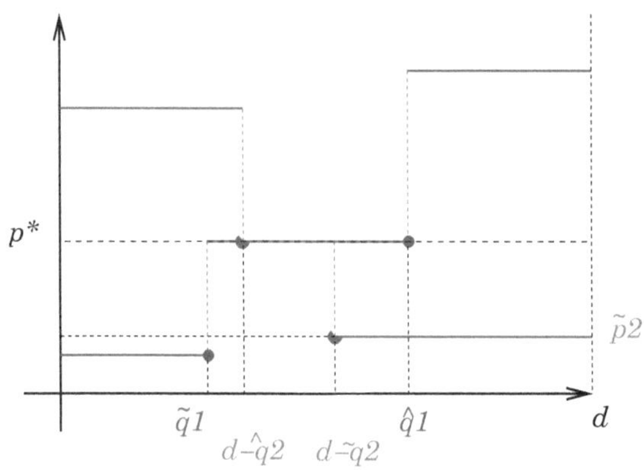

Figure 7.1. Example

is to give the preference to supplier S_1, i.e., for a pair of strategies $((q_1, p), (q_2, p))$ such that $q_1 + q_2 > d$ the market reacts by buying the respective quantities q_1 and $d - q_1$ respectively to supplier S_1 and to supplier S_2. The Nash equilibria for market share maximization are,

$$u_1^* = (q_1^* \in [d - \tilde{q}_2, \hat{q}_1], p^*), \quad u_2^* = (\tilde{q}_2, p_2^* \in [\tilde{p}_2, p^*[),$$

where $\hat{q}_i = \rho_i(p^*)$, $\tilde{q}_i = \rho_i(p^* - \epsilon)$, and $\tilde{p}_2 = \mathcal{L}_2(\tilde{q}_2)$.

Suppose now the additional decision rule of the market is a preference for supplier S_2. The previous pair of strategies is not a Nash equilibrium any more. Indeed, supplier S_2 can increase its offer, at price p^*, to the quantity $\hat{q}_2$. The equilibrium in that case is

$$u_1^* = (q_1^* \in [d \quad \hat{q}_2, \hat{q}_1], p^*), \quad u_2^* = (\hat{q}_2, p^*).$$

REMARK 7.5 In Proposition 7.2 we see that at equilibrium, the maximal price $\bar{p}$ that can be proposed is given by (7.16). A sufficient condition for that price to be finite is that for any $j \in \{1, 2, \ldots, \mathcal{S}\}$ we have,

$$\sum_{k \neq j} Q_k > d. \tag{7.17}$$

Equation (7.17) means that with the withdrawal of an individual supplier, the demand can still be satisfied. This will insure that none of the suppliers can create a fictive shortage and then increase unlimitedly the price of electricity.

Proof of Proposition 7.2. We have to prove that for supplier S_j there is no profitable deviation of strategy, *i.e.* for any $u_j \neq u_j^*$, we have $J_{P_j}(\mathbf{u}_{-j}^*, u_j^*) \geq J_{P_j}(\mathbf{u}_{-j}^*, u_j)$.

- Suppose first that $j \notin \mathcal{S}(p^*)$ so that $\mathcal{L}_j(\rho_j(p^*)) < p^*$. Since for the proposed Nash strategy $u_j^* = (q_j^*, p_j^*)$, we have $p_j^* < p^*$, the total quantity proposed by S_j is bought by the market ($\overline{q}_j = q_j^*$). Hence $J_j(\mathbf{u}^*) = q_j^*$.
  - If the deviation $u_j = (q_j, p_j)$ is such that $q_j \leq q_j^*$, then clearly $J_{P_j}(\mathbf{u}^*) = q_j^* \geq q_j \geq J_{P_j}(\mathbf{u}_{-j}^*, u_j)$, whatever the price p_j is.
  - If the deviation $u_j = (q_j, p_j)$ is such that $q_j > \rho_j(p^*)$ then necessarily, by the minimal price constraint, Assumption 7.5 and the definition of $q_j^* = \rho_j(p^*)$, we have

    $$p_j \geq \mathcal{L}_j(q_j) \geq \mathcal{L}_j(q_j^{*\,+}) > \sup_{k \neq j} p_k^*.$$

    Hence now supplier S_j is the supplier with the highest price. Consequently the market first buys from the other suppliers and satisfies the demand, when necessary, with the electricity produced by supplier S_j (instead of the supplier $S_{\bar{j}}$). Hence the market share of S_j cannot increase with this deviation.
- Suppose now that $j = \bar{j}$, i.e., we have, $\mathcal{L}_{\bar{j}}(\rho_{\bar{j}}(p^*)) = p^*$.

- If the first item of Assumption 7.5 holds, then at the proposed Nash equilibrium, supplier $S_{\bar{j}}$ is the supplier that meets the demand since it proposes the highest price.

  Hence if supplier $S_{\bar{j}}$ wants to increases its market share, it has to sell a quantity $\tilde{q}_{\bar{j}} \geq d - \sum_{k \neq \bar{j}} q_k^*$. But we have,

$$\mathcal{L}_{\bar{j}}(\tilde{q}_{\bar{j}}) \geq \mathcal{L}_{\bar{j}}(d - \sum_{k \neq \bar{j}} q_k^*) = p^* > \max_{k \neq \bar{j}} p_k^*.$$

  This proves that the quantity $\tilde{q}_{\bar{j}}$ cannot be offered at a price such that the market would buy it.

- If the second item of Assumption 7.5 holds, then the proposition states that the quantity proposed, and bought by the market is $\rho_{\bar{j}}$. An increase in the quantity proposed would imply a higher price, which would not imply a higher quantity proposed by the market since now the supplier would have the highest price.

□

Now we suppose that Assumption 7.5 does not hold. So for the price p^* defined by (7.15) we have more than one supplier S_j such that $\mathcal{L}_j(\rho_j(p^*)) = p^*$.

As shown in the example of Remark 7.4 (see Figure 7.1), the Nash equilibria may depend upon the reaction of the market when two suppliers, S_i and S_j, have the same price $p_i = p_j = p^*$. It is clear that for a supplier S_j in such a way that $\mathcal{L}_j(\rho_j(p^*)) = p^*$, two possibilities may occur at equilibrium. Either, for some supplier S_j that fixes its price to $p_j = p^*$, the market reacts in such a way that $\overline{q}_j < \rho_j(p^* - \epsilon)$, in which case at equilibrium we will have $p_j^* = p^* - \epsilon$, or the market reacts such that $\overline{q}_j \geq \rho_j(p^* - \epsilon)$, and in that case we will have $p_j^* = p^*$.

Although the existence of Nash equilibria seems clear for any possible reaction of the market, we restrict our attention to the case where the market reacts by choosing quantities $(\overline{q}_j)_{j=1,\dots\mathcal{S}}$ that are monotonically nondecreasing with respect to the quantity q_j proposed by each supplier S_j. More precisely we have the following assumption,

ASSUMPTION 7.6 *Let $\mathbf{u} = (\mathbf{u}_1, \dots, \mathbf{u}_{\mathcal{S}})$ be a strategy profile of the suppliers with $u_i = (q_i, p)$ for $i \in \{1, \dots, \mathtt{k}\}$. Suppose the market has to use its additional rule to decide how to share the quantity $\tilde{d} \leq d$ among suppliers S_1 to $S_{\mathtt{k}}$ (the quantity $d - \tilde{d}$ has already been bought from suppliers with a price lower than p).*

Let $i \in \{1, \dots, \mathtt{k}\}$, we define the function that associates $\overline{q}_i$ to any $q_i \geq 0$, where $\overline{q}_i$ is the i-th component of the reaction $(\overline{q}_1, \dots, \overline{q}_{\mathcal{S}})$ of the

market to the strategy profile $(\mathbf{u}_{-i}, (q_i, p))$. *We suppose that this function is not decreasing with respect to the quantity* q_i.

The meaning of this assumption is that the market does not penalize an "over-offer" of a supplier. For fixed strategies $\mathbf{u}_{-i}$ of all the suppliers but S_i, if supplier S_i, such that $p_i = p$ increases its quantity q_i, then the quantity bought by the market from S_i cannot decrease. It can increase or stay constant. In particular, it encompasses the case where the market has a preference order between the suppliers (for example, it first buys from supplier S_{j_1}, then supplier S_{j_2} etc), or when the market buys some fixed proportion from each supplier. It does not encompass the case where the market prefers the smallest offer.

PROPOSITION 7.3 *Suppose Assumption 7.5 does not hold while Assumption 7.6 does. Let the strategy profile* $((q_1^*, p_1^*), \ldots, (q_S^*, p_S^*))$ *be defined by*

- *If* S_j *is such that* $\mathcal{L}_j(\rho_j(p^*)) < p^*$ *then*

$$p_j^* = p^* - \epsilon, \quad q_j^* = \rho_j(p^* - \epsilon).$$

- *If* S_j *is such that* $\mathcal{L}_j(\rho_j(p^*)) = p^*$, *then either*

$$p_j^* = p^*, \quad q_j^* = \rho_j(p^*), \tag{7.18}$$

when the reaction of the market is such that

$$\overline{q}_j \geq \rho_j(p^* - \epsilon), \quad \forall \epsilon > 0,$$

or

$$p_j^* = p^* - \epsilon, \quad q_j^* = \rho_j(p^* - \epsilon). \tag{7.19}$$

when the new reaction of the market, for a deviation $p_j = p^*$ *would be such that* $\overline{q}_j < \rho_j(p^* - \epsilon)$.

This strategy profile is a Nash equilibrium.

Proof. The proof follows directly from the discussion made before the proposition, and from the proof of Proposition 7.2. □

Example. We consider a market with 5 suppliers and a demand d equal to 10. We suppose that the minimal price functions $\mathcal{L}_j$ of suppliers are increasing staircase functions, given in the following table (the notation $(]a, b]; c)$ indicates that the value of the function in the interval $]a, b]$ is c),

supplier 1	$([0,1];10),(]1,3];15),(]3,4];25),(]4,10];50)$
supplier 2	$([0,5];20),(]5,6];23),(]6,7];40),(]7,10];70)$
supplier 3	$([0,2];15),(]2,6];25),(]6,7];30),(]7,10];50)$
supplier 4	$([0,1];10),(]1,4];15),(]4,5];20),(]5,10];50)$
supplier 5	$([0,4];30),(]4,8];90),(]8,10];100)$

We display in the following table the values for $\rho_j(p)$ and $O(p)$ respectively defined by equations (7.13) and (7.14).

p	$\rho_1(p)$	$\rho_2(p)$	$\rho_3(p)$	$\rho_4(p)$	$\rho_5(p)$	$O(p)$
$p \in [0,10[$	0	0	0	0	0	0
$p \in [10,15[$	1	0	0	1	0	2
$p \in [15,20[$	3	0	2	4	0	9
$p \in [20,23[$	3	5	2	5	0	15

The previous table shows that for a price p in $[15, 20[$, only suppliers S_1, S_3 and S_4 can bring some positive quantity of electricity. The total maximal quantity that can be provided is 9 which is strictly lower than the demand $d = 10$. For a price in $[20, 23[$, we see that supplier S_2 can also bring some positive quantity of electricity, the total maximal quantity is then 15 which is higher than the demand. Then we conclude that the price p^* defined by Equation (7.15) is $p^* = 20$. Moreover, $\mathcal{L}_2(\rho_2(p^*)) = \mathcal{L}_4(\rho_4(p^*)) = p^*$ which means that Assumption 7.5 is not satisfied. Notice that for supplier S_5, we have $\mathcal{L}_5(0) = 30 > p^*$. Supplier S_5 will not be able to sell anything to the market, hence, whatever its bid is, we have $\overline{q}_5 = 0$. We suppose that Assumption 7.6 holds. According to Proposition 7.3, we have the following equilibria.

$$u_1^* = (3, p_1^* \in [15, 20[), u_3^* = (2, p_3^* \in [15, 20[) \text{ and } u_5^* = (p^*, q^*),\ p^* \geq \mathcal{L}_5(q)$$

to which the market reacts by buying the respective quantities $\overline{q}_1(\mathbf{u}^*) = 3$, $\overline{q}_3(\mathbf{u}^*) = 2$ and $\overline{q}_5(\mathbf{u}^*) = 0$. The quantity 5 remains to be shared between S_2 and S_4 according to the additional rule of the market. For example, suppose that the market prefers S_2 to all other suppliers. Then

$$u_2^* = (q_2^* \in [1,5], p_2^* = 20) \text{ and } u_4^* = (4, p_4^* \in [15, 20[).$$

to which the market reacts by buying $\overline{q}_2(\mathbf{u}^*) = 1$ and $\overline{q}_4(\mathbf{u}^*) = 4$. If now the market prefers S_4 to any other, then

$$u_2^* = (q_2^* \in [1,5], p_2^* = 20) \text{ and } u_4^* = (5, p_4^* = 20).$$

to which the market reacts by buying $\overline{q}_2(\mathbf{u}^*) = 0$ and $\overline{q}_4(\mathbf{u}^*) = 5$.

4. Suppliers maximize profit

In this section, the objective of the suppliers is to maximize their profit, *i.e.* for a strategy profile $\mathbf{u} = (u_1, \ldots, u_S)$, $u_j = (q_j, p_j)$, their evaluation functions are

$$J_{S_j}(\mathbf{u}) = p_j \overline{q}_j - C_j(\overline{q}_j), \tag{7.20}$$

where $C_j(\cdot)$ denotes supplier S_j's production cost function, and $\overline{q}_j$ is the optimal reaction of the market, *i.e.* the solution of Problem (7.9) together with an additional decision rule, known by all the suppliers, in case of nonunique solutions (see Remark 7.1). As before, we do not need to make this rule explicit.

In contrast to the market share maximization, we do not need a minimal price functions $\mathcal{L}_j$. Nevertheless we need a maximal unit price $p_{\max}$ under which the suppliers are allowed to sell their electricity. This maximal price can either be finite and fixed by the market or be infinite.

From all the assumptions previously made, we only retain in this section Assumption 7.1.

LEMMA 7.1 *We define, for any finite price $p \geq 0$, $\widehat{Q}_j(p)$ as the set of quantities that maximizes the quantity $qp - C_j(q)$, i.e.,*

$$\widehat{Q}_j(p) \stackrel{\text{def}}{=} \arg \max_{q \in [0,d]} qp - C_j(q),$$

and for infinite price,

$$\widehat{Q}_j(+\infty) = \min(Q_j, d),$$

where Q_j is the maximal production capacity of S_j.

We have for finite p,

$$\begin{array}{l} \widehat{Q}_j(p) = \{0\} \text{ if } C'(0) > p, \\ \widehat{Q}_j(p) = \{d\} \text{ if } C'(d) < p, \\ \widehat{Q}_j(p) = \{q,\ C_j'(q^-) \leq p \leq C_j'(q^+)\}, \text{ otherwise.} \end{array} \tag{7.21}$$

Proof. We prove the last equality of (7.21). For any $q \in \widehat{Q}_j(p)$, we have for any $\epsilon > 0$,

$$pq - C_j(q) \geq p(q+\epsilon) - C_j(q+\epsilon),$$

from which we deduce that

$$\frac{C_j(q+\epsilon) - C_j(q)}{\epsilon} \geq p,$$

and letting ϵ tends to zero, it follows that $C_j'(q^+) \geq p$. The other equality is obtained with negative ϵ.

The first two equalities of (7.21) follow directly from the fact that C' is supposed to be non decreasing. □

Note that if $C'(\cdot)$ is a continuous and non decreasing function, then (7.21) is equivalent to the classical first order condition for the evaluation function of the supplier.

LEMMA 7.2 *The function* $p \to \max_{q\in[0,d]}(qp - C_i(q))$ *is continuous and strictly increasing.*

Proof. We recognize the Legendre-Fenchel transform of the convex function C_j. The continuity follows from classical properties of this transform.

The function is strictly increasing, since for $p > p'$, if we denote by $\tilde{q}$ a quantity in $\arg\max_{q\in[0,d]}(qp' - C_i(q))$, we have

$$\begin{aligned}\max_{q\in[0,d]}(qp - C_i(q)) &\geq \tilde{q}p - C_i(\tilde{q}) \\ &> \tilde{q}p' - C_i(\tilde{q}) = \max_{q\in[0,d]}(qp' - C_i(q)).\end{aligned}$$

□

We now restrict our attention to the two suppliers' case, *i.e.* $\mathcal{S} = 2$.

Our aim is to determine the Nash equilibrium if such an equilibrium exists. Hence we need to find a pair $((q_1^*, p_1^*), (q_2^*, p_2^*))$ such that (q_1^*, p_1^*) is the best strategy of supplier S_1 if supplier S_2 chooses (q_2^*, p_2^*), and conversely, (q_2^*, p_2^*) is the best strategy of supplier S_2 if supplier S_1 chooses (q_1^*, p_1^*). Equivalently we need to find a pair $((q_1^*, p_1^*), (q_2^*, p_2^*))$ such that there is no profitable deviation for any supplier S_i, $i = 1, 2$.

Let us determine the conditions which a pair $((q_1^*, p_1^*), (q_2^*, p_2^*))$ must satisfy in order to be a Nash equilibrium, i.e., no profitable deviation exists for any supplier. We will successively examine the case where we have an excess demand ($q_1^* + q_2^* \leq d$) and the case where we have an excess supply ($q_1^* + q_2^* > d$).

Excess demand: $q_1^* + q_2^* \leq d$. In that case the market buys all the quantities proposed by the suppliers, i.e., $\overline{q}_i^* = q_i^*$, $i = 1, 2$.

1. Suppose that for at least one supplier, say supplier S_1, we have $p_1^* < p_{\max}$. Then supplier S_1 can increase its profit by increasing its price to $p_{\max}$. Since $q_1^* + q_2^* \leq d$ the reaction of the market to the new pair of strategies $((q_1^*, p_{\max}), (q_2^*, p_2^*))$ is still q_1^*, q_2^*. Hence the new profit of S_1 is now $q_1^* p_{\max} - C_1(q_1^*) > q_1^* p_1^* - C_1(q_1^*)$.

We have exhibited a profitable deviation, $(q_1^*, p_{\max})$ for supplier S_1. This proves that a pair of strategies such that $q_1^* + q_2^* \leq d$ with at least one price $p_i^* < p_{\max}$ cannot be a Nash equilibrium.

2. Suppose that $p_1^* = p_2^* = p_{\max}$, and that there exists at least one supplier, say supplier S_1, such that $q_1^* - \overline{q}_1^* \notin \widehat{Q}_1(p_{\max})$, i.e., such that the reaction of the market does not maximize S_1's profit (see Lemma 7.1). Consequently, the profit for S_1, associated with the pair $((q_1^*, p_{\max}), (q_2^*, p_{\max}))$ is such that

$$\overline{q}_1^* p_{\max} - C_1(\overline{q}_1^*) < \max_{q\in[0,d]} (q p_{\max} - C_1(q)).$$

Since

$$\lim_{\epsilon\to 0^+} \max_{q\in[0,d]} (q(p_{\max} - \epsilon) - C_1(q)) = \max_{q\in[0,d]} (q p_{\max} - C_1(q)),$$

there exists some $\bar{\epsilon} > 0$ such that

$$\max_{q\in[0,d]} (q(p_{\max} - \bar{\epsilon}) - C_1(q)) > \overline{q}_1^* p_{\max} - C_1(\overline{q}_1^*).$$

This proves that any deviation $(\hat{q}_1, p_{\max} - \bar{\epsilon})$ of supplier S_1, such that $\hat{q}_1 \in \widehat{Q}_1(p_{\max} - \bar{\epsilon})$, is profitable for S_1.

Hence, a pair of strategies such that $q_1^* + q_2^* \leq d$, $p_1^* = p_2^* = p^*$, to which the market reacts with, for at least one supplier, a quantity $\overline{q}_i^* \notin \hat{Q}_i(p_{\max})$ cannot be a Nash equilibrium.

3. Suppose that $p_1^* = p_2^* = p_{\max}$, $q_1^* = \overline{q}_1^* \in \hat{Q}_1(p_{\max})$ and $q_2^* = \overline{q}_2^* \in \hat{Q}_2(p_{\max})$ (i.e., the market reacts optimally for both suppliers).

In that case the pair $((q_1^*, p_1^*), (q_2^*, p_2^*))$ is a Nash equilibrium. As a matter of fact no deviation by changing the quantity can be profitable: since $\overline{q}_i^*$ is optimal for $p_{\max}$, the price cannot be increased, and a decrease of the profit will follow from a decrease of the price of one supplier (Lemma 7.2).

Excess supply: $q_1^* + q_2^* > d$. Two possibilities occur depending on whether the prices p_j, $j = 1, 2$ differ or not.

1. **The prices are different**, i.e., $p_1^* < p_2^*$ for example.

In that case the market first buys from the supplier with lower price (hence $\overline{q}_1^* = \inf(q_1^*, d)$), and then completes its demand to the supplier with highest price, S_2 (hence $\overline{q}_2^* = d - \overline{q}_1^*$).

For $\bar{\epsilon} > 0$ such that $p_1^* + \bar{\epsilon} < p_2^*$, we have

$$q_1^* p_1^* - C_1(q_1^*) \leq \max_{q \in [0,d]} \{qp_1^* - C_1(q)\} < \max_{q \in [0,d]} \{q(p_1^* + \bar{\epsilon}) - C_1(q)\}.$$

Hence supplier S_1 is better off increasing its price to $p_1^* + \bar{\epsilon}$ and proposing quantity $\hat{q}_1 \in \widehat{Q}_1(p_1^* + \bar{\epsilon})$. As a matter of fact, since $p_1^* + \bar{\epsilon} < p_2^*$, the reaction of the market will be $\overline{q}_1 = \hat{q}_1$.

So a pair of strategies with $p_1 \neq p_2$ cannot be a Nash equilibrium.

2. **The prices are equal**, i.e., $p_1^* = p_2^* \stackrel{\text{def}}{=} p^*$.

   Now the market faces an optimization problem (7.9) with several solutions. Hence it has to use its additional rule in order to determine the quantities $\overline{q}_1^*, \overline{q}_2^*$ to buy from each supplier in response to their offers q_1^*, q_2^*.

   If this response is not optimal for any supplier, i.e., $\overline{q}_1^* \notin \widehat{Q}_1(p^*)$ and $\overline{q}_2^* \notin \widehat{Q}_2(p^*)$, the same line of reasoning as in Item 2 of the excess demand case proves that the pair $((q_1^*, p_1^*), (q_2^*, p_2^*))$ cannot be a Nash equilibrium.

   Suppose that the reaction of the market is optimal for both suppliers, i.e., $\overline{q}_1^* \in \widehat{Q}_1(p^*)$ and $\overline{q}_2^* \in \widehat{Q}_2(p^*)$. A necessary condition for a supplier, say S_1, to increase its profit is to increase its price and consequently to complete the offer of the other supplier S_2. We have two possibilities,

   (a) If for at least one supplier, say supplier S_1, we have,

   $$(d - q_2^*)p_{\max} - C_1(d - q_2^*) > \max_{q \in [0,d]} \{qp^* - C_1(q)\},$$

   then supplier S_1 is better off increasing its price to $p_{\max}$ and completing the market to sell the quantity $d - q_2^*$.

   (b) Conversely, if none of the suppliers can increase its profit by "completing the offer of the other", i.e., if

   $$(d - q_2^*)p_{\max} - C_1(d - q_2^*) \leq \max_{q \in [0,d]} \{qp^* - C_1(q)\}, \quad (7.22)$$

   $$(d - q_1^*)p_{\max} - C_2(d - q_1^*) \leq \max_{q \in [0,d]} \{qp^* - C_2(q)\}, \quad (7.23)$$

   then the pair $((q_1^*, p_1^*), (q_2^*, p_2^*))$ is a Nash equilibrium.

   As a matter of fact, for one supplier, say S_1, changing only the quantity is not profitable since $q_1^* \in \widehat{Q}_1$, decreasing the price is not profitable because of Lemma 7.2. Inequality (7.22) prevents S_1 from increasing its price.

REMARK 7.6 Note that a sufficient condition for Inequality (7.22) and (7.23) to be true, is that both suppliers choose $q_j^* = d,\ j = 1, 2$.

With this choice each supplier prevents the other supplier from completing its demand with maximal price. This can be interpreted as a wish for the suppliers to obtain a Nash equilibrium. Nevertheless, to do that, the suppliers have to propose to the market a quantity d at price p^* which may be very risky and hence may not be credible. As a matter of fact, suppose S_1 chooses the strategy $q_1 = d, p_1 = p^* < p_{\max}$. If for some reason supplier S_2 proposes a quantity q_2 at a price $p_2 > p_1$, then S_1 has to provide the market with the quantity d at price p_1 since $\overline{q}_1 = d$, which may be disastrous for S_1.

Note also that if $p_{\max}$ is not very high compared with p^* the inequalities (7.22) and (7.23) will not be satisfied. Hence these inequalities could be used for the market to choose a maximal price $p_{\max}$ such that the equilibrium may be possible.

The previous discussion shows that in case of excess supply, the only possibility to have a Nash equilibrium, is that both suppliers propose the same price p^* and quantities $q_1^*,\ q_2^*$, such that, together with an additional rule, the market can choose optimal quantities $\overline{q}_1^*,\ \overline{q}_2^*$ that satisfy its demand and such that $\overline{q}_j^* \in \widehat{Q}_j(p^*)$.

This is clearly not possible for any price p^*. If the price p^* is too small, then the optimal quantity the suppliers can bring to the market is small, and for any $q \in \widehat{Q}_1(p^*) + \widehat{Q}_2(p^*)$, we have $q < d$. If the price p^* is too high, then the optimal quantity the suppliers are willing to bring to the market are large, and for any $q \in \widehat{Q}_1(p) + \widehat{Q}_2(p)$, we have $q > d$. The following Lemma characterizes the possible values of p^* for which it is possible to find q_1 and q_2 that satisfy $q_1 \in \widehat{Q}_1(p^*)$, $q_2 \in \widehat{Q}_2(p^*)$ and $q_1 + q_2 = d$.

Let us first define the function $\mathcal{C}_j'$ from $[0, d]$ to the set of intervals of $\mathbb{R}^+$ as

$$\mathcal{C}_j'(q) = [C_j'(q^-), C_j'(q^+)],$$

for q smaller than the maximal production quantity Q_j, and $\mathcal{C}_j(q) = \emptyset$ for $q > Q_j$. Clearly $\mathcal{C}_j'(q) = \{C_j'(q)\}$ except when C_j' has a discontinuity in q. We now can state the lemma.

LEMMA 7.3 *It is possible to find q_1, q_2 such that $q_1 + q_2 = d$, $q_1 \in \widehat{Q}_1(p)$ and $q_2 \in \widehat{Q}_2(p)$ if and only if*

$$p \in \mathcal{I},$$

where,

$$\mathcal{I} \stackrel{\text{def}}{=} \bigcup_{q\in[0,d]} (\mathcal{C}_1'(q) \cap \mathcal{C}_2'(d-q)),$$

or, equivalently, when $Q_1 + Q_2 \geq d$, $\mathcal{I} \stackrel{\text{def}}{=} [\mathcal{I}^-, \mathcal{I}^+]$, *where*

$$\mathcal{I}^- \stackrel{\text{def}}{=} \min\{p, \ \max(q \in \widehat{Q}_1(p)) + \max(q \in \widehat{Q}_2(p)) \geq d\},$$
$$\mathcal{I}^+ \stackrel{\text{def}}{=} \max\{p, \ \min(q \in \widehat{Q}_1(p)) + \min(q \in \widehat{Q}_2(p)) \leq d\},$$

and $\mathcal{I} = \emptyset$ *when* $Q_1 + Q_2 < d$.

Proof. If $p \in \mathcal{I}$, then there exists $q \in [0, d]$ such that $p \in \mathcal{C}_1'(q)$ and $p \in \mathcal{C}_2'(d-q)$. We take $q_1 = q$, $q_2 = d - q$ and conclude by applying Lemma 7.1.

Conversely, if $p \notin \mathcal{I}$, then it is not possible to find q_1, q_2 such that $q_1 + q_2 = d$, and such that $p \in \mathcal{C}_1'(q_1) \cap \mathcal{C}_2'(q_2)$ i.e., such that according to Lemma 7.1 $q_1 \in \widehat{Q}_1(p)$ and $q_2 \in \widehat{Q}_2(p)$.

Straightforwardly, if $\mathcal{I}^- \leq p \leq \mathcal{I}^+$, then there exists $q_1 \in \widehat{Q}_1(p)$ and $q_2 \in \widehat{Q}_2(p)$ such that $q_1 + q_2 = d$. □

We sum up the previous analysis by the following proposition.

PROPOSITION 7.4 *In a market with maximal price* $p_{\max}$ *with two suppliers, each having to propose quantity and price to the market, and each one wanting to maximize its profit, we have the following Nash equilibrium:*

1. *If* $p_{\max} < \min\{p \in \mathcal{I}\}$-**Excess demand case**, *any strategy profile* $((q_1^*, p_{\max}), (q_2^*, p_{\max}))$, *with* $q_1^* \in \widehat{Q}_1(p_{\max})$ *and* $q_2^* \in \widehat{Q}_2(p_{\max})$ *is a Nash equilibrium. In that case we have* $q_1^* + q_2^* < d$.
2. *If* $p_{\max} = \min\{p \in \mathcal{I}\}$, *any pair* $((q_1^*, p_{\max}), (q_2^*, p_{\max}))$ *where* $q_1^* \in \widehat{Q}_1(p_{\max})$ *and* $q_2^* \in \widehat{Q}_2(p_{\max})$ *is a Nash equilibrium. In that case we may have* $q_1^* + q_2^* \geq d$ *or* $q_1^* + q_2^* < d$.
3. *If* $p_{\max} > \min\{p \in \mathcal{I}\}$- **Excess supply case**, *any pair* $((q_1^*, p^*), (q_2^*, p^*))$, *such that* $p^* \in \mathcal{I}$, $p^* \leq p_{\max}$ *and which induces a reaction* $(\overline{q}_1, \overline{q}_2)$, $\overline{q}_1 \geq q_1^*$, $\overline{q}_2 \geq q_2^*$, *such that*
   - *(a)* $\overline{q}_1 + \overline{q}_2 = d$,
   - *(b)* $\overline{q}_1 \in \widehat{Q}_1(p^*)$, $\overline{q}_2 \in \widehat{Q}_2(p^*)$,
   - *(c)*
   $$(d - q_2^*)p_{\max} - C_1(d - q_2^*) \leq \overline{q}_1 p^* - C_1(\overline{q}_1),$$
   $$d - q_1^*)p_{\max} - C_2(d - q_1^*) \leq \overline{q}_2 p^* - C_2(\overline{q}_2),$$

   is a Nash equilibrium.

We now want to generalize the previous result to a market with $\mathcal{S} \geq 2$ suppliers.

Let $((q_1^*, p_1^*), \ldots, (q_{\mathcal{S}}^*, p_{\mathcal{S}}^*))$ be a strategy profile, and let $(\overline{q}_1, \ldots, \overline{q}_{\mathcal{S}})$ be the induced reaction of the market. This strategy profile is a Nash equilibrium, if for any two suppliers, S_i, S_j, the pair of strategies $((q_i^*, p_i^*), (q_j^*, p_j^*))$ is a Nash equilibrium for a market with two suppliers (with evaluation function defined by Equation (7.20)) and demand $\tilde{d} = d - \sum_{k \notin \{i,j\}} \overline{q}_k$.

Hence using the above Proposition 7.4, we know that necessarily at equilibrium the prices proposed by the suppliers are equal, and the quantities q_i^* induce a reaction of the market such that $\overline{q}_i \in \widehat{Q}_i(p^*)$.

Let us first extend the previous definition of the set $\mathcal{I}$ by $\mathcal{I} = \emptyset$ if $\sum_{j=1}^{\mathcal{S}} Q_j < d$, and otherwise,

$$\begin{array}{l} \mathcal{I} \stackrel{\text{def}}{=} [\mathcal{I}^-, \mathcal{I}^+], \\ \text{where} \\ \mathcal{I}^- = \min\{p,\ \sum_{j=1}^{\mathcal{S}} \max(q \in \widehat{Q}_j(p)) \geq d\}, \\ \mathcal{I}^+ = \max\{p,\ \sum_{j=1}^{\mathcal{S}} \min(q \in \widehat{Q}_j(p)) \leq d\}. \end{array} \tag{7.24}$$

We have the following

THEOREM 7.1 *Suppose we have $\mathcal{S}$ suppliers on a market with maximal price $p_{\max}$ and demand d.*

1. *If $p_{\max} < \min\{p \in \mathcal{I}\}$-**Excess demand case**, any strategy profile $((q_1^*, p_{\max}), \ldots, (q_{\mathcal{S}}^*, p_{\max}))$, with $q_j^* \in \widehat{Q}_j(p_{\max})$, $j = 1, \ldots, \mathcal{S}$, is a Nash equilibrium. In that case we have $\sum_{j=1}^{\mathcal{S}} q_j^* < d$.*

2. *If $p_{\max} = \min\{p \in \mathcal{I}\}$, any strategy profile $((q_1^*, p_{\max}), \ldots, (q_{\mathcal{S}}^*, p_{\max}))$ where $q_j^* \in \widehat{Q}_j(p_{\max})$, $j = 1, \ldots, \mathcal{S}$ is a Nash equilibrium. In that case we may have $\sum_{j=1}^{\mathcal{S}} q_j^* \geq d$ or $\sum_{j=1}^{\mathcal{S}} q_j < d$.*

3. *If $p_{\max} > \min\{p \in \mathcal{I}\}$- **Excess supply case**, any strategy profile $((q_1^*, p^*), \ldots, (q_{\mathcal{S}}^*, p^*))$, such that $p^* \in \mathcal{I}$, $p^* \leq p_{\max}$ and which induces a reaction $(\overline{q}_1, \ldots, \overline{q}_{\mathcal{S}})$, $\overline{q}_j \geq q_j^*$, $j = 1, \ldots, \mathcal{S}$, such that*

   (a) $\sum_{j=1}^{\mathcal{S}} \overline{q}_j = d$,

   (b) $\overline{q}_j \in \widehat{Q}_j(p^*)$, $j = 1, \ldots, \mathcal{S}$,

   (c) for any $j = 1, \ldots, \mathcal{S}$

$$(d - \sum_{k \neq j} q_k^*) p_{\max} - C_j(d - \sum_{k \neq j} q_k^*) \leq \overline{q}_j p^* - C_j(\overline{q}_j), \tag{7.25}$$

   is a Nash equilibrium.

The previous results show that a Nash equilibrium always exists for the case where the profit is used by the suppliers as an evaluation function. For convenience we have supposed the existence of a maximal price $p_{\max}$. On real markets we observe that, usually this maximal price is infinity, since most markets do not impose a maximal price on electricity. Hence the interesting case is the case where $p_{\max} > \min\{p \in \mathcal{I}\}$. The case with $p_{\max} \leq \min\{p \in \mathcal{I}\}$, can be interpreted as a monopolistic situation. The demand is so large compared with the maximal price that each supplier can behave as if it is alone on the market.

When $p_{\max}$ is large enough, Proposition 7.4 and Theorem 7.1 exhibit some conditions for a strategy profile to be a Nash equilibrium. We can make several remarks.

REMARK 7.7 Note that conditions (7.25) are satisfied for $q_j^* = d$. Hence we can conclude that, provided that the market reacts in such a way that $\overline{q}_j \in \widehat{Q}_j$, the strategy profile $((d, p^*), \ldots, (d, p^*))$ is a Nash equilibrium. Nevertheless, this equilibrium is not realistic. As a matter of fact, to implement this equilibrium, the suppliers have to propose to the market a quantity that is higher than the optimal quantity, and which possibly may lead to a negative profit (when $p_{\max}$ is very large). The second aspect that may appear unrealistic is the fact that the suppliers give up their power of decision. As a matter of fact, they announce high quantities, so that (7.25) is satisfied, and let the market choose the appropriate $\overline{q}_j$.

Example. We consider the market, with demand $d = 10$ and maximal price $p_{\max} = +\infty$, with the five suppliers already described page 147. In order to be able to compare the equilibria for both criteria, market share and profit, we suppose that the marginal cost is equal to the minimal price function displayed in the table page 148, i.e., $C_j' = \mathcal{L}_j$.

The following table displays the quantities $o(p) \stackrel{\text{def}}{=} \sum_{j=1}^{5} \min\{q \in \widehat{Q}_j(p)\}$ and $O(p) \stackrel{\text{def}}{=} \sum_{j=1}^{5} \max\{q \in \widehat{Q}_j(p)\}$.

p	$\in [0, 10[$	$= 10$	$\in]10, 15[$	$= 15$	$\in]15, 20[$	$= 20$	$\in]20, 23[$
$o(p)$	0	0	2	2	9	9	15
$O(p)$	0	2	2	9	9	15	15

From the above table we deduce that $\mathcal{I} = \{20\}$, hence the only possible equilibrium price is $p^* = 20$. As a matter of fact, we have $O(20) = 15 > 10 = d$, and for any $p < 20$, $O(p) < 10 = d$, and $o(20) = 9 < 10 = d$, and for any $p > 20$, $o(p) > 10 = d$. Hence $p^* = 20 \in \mathcal{I}$ as defined by Equation (7.24).

Now concerning the quantities, the equilibrium depend upon the additional rule of the market. We suppose that the market chooses $\overline{q}_i \in \widehat{Q}_i$, $\forall i \in \{1,\dots,5\}$, and then to give preference to S_1, then to S_2, etc.

The equilibrium is $q_1^* \geq 3$, $q_2^* \geq 0$, $q_3^* \geq 2$, $q_4^* \geq 4$, $q_2^* \geq 0$, $q_5^* \geq 0$.

The fact that the market wants to choose quantities $\overline{q}_i \subset \widehat{Q}_i(20)$ implies that $\overline{q}_1 \in \widehat{Q}_1(20) = 3$, $\overline{q}_2 \in \widehat{Q}_2(20) = [0,5]$, $\overline{q}_3 \in \widehat{Q}_3(20) = 2$, $\overline{q}_4 \in \widehat{Q}_4(20) = [4,5]$, $\overline{q}_5 = 0$, and the preference for S_2 compared to S_5 implies that $\overline{q}_1 = 3$, $\overline{q}_2 = 1$, $\overline{q}_3 = 2$, $\overline{q}_4 = 4$, $\overline{q}_5 = 0$.

If the preference would have been S_5 then S_4 then S_3 etc. the equilibrium would have been the same, but we would have had

$$\overline{q}_1 = 3, \quad \overline{q}_2 = 0, \quad \overline{q}_3 = 2, \quad \overline{q}_4 = 5, \quad \overline{q}_5 = 0.$$

5. Conclusion

We have shown in the previous sections that for both criteria, market share and profit maximization, it is possible to find a Nash equilibrium for a number $\mathcal{S}$ of suppliers. It is noticeable that for both cases the equilibrium price involved is the same (i.e., p^* given by Equation (7.14) for market share maximization and $p^* \in \mathcal{I}$ defined by Equation (7.24) for profit maximization), only the quantities proposed by the suppliers differ.

Nevertheless as already discussed in Remark 7.7, for profit maximization, the equilibrium strategies involved are not realistic in the interesting cases ($p_{\max}$ large). This may suggest that on these unregulated markets where suppliers are interested in instantaneous profit maximization, an equilibrium never occurs. Prices may becomes arbitrarily high and anticipation of the market behavior, and particularly market price, basically impossible. An extended discussion on this topic can be found in Bossy et al. (2004).

This paper contains some modeling aspects that could be considered in more detail in future works. A first extension would be to consider more general suppliers. As a matter of fact, in the current paper, the evaluation functions chosen are more suitable for providers. Indeed, for profit maximization, we assumed that we had a production cost only for that part of electricity which is actually sold. This would fit the case where suppliers are producers. They produce only the electricity sold. The evaluation function chosen does not fit the case of traders who may have already bought some electricity and try to sell at best price the maximal electricity they have. The extension of our results in that case should not be difficult.

We supposed that every supplier perfectly knows the evaluation function of the other suppliers, and in particular their marginal costs. In general this is not true. Hence some imperfect information version of the model should probably be investigated.

An other extension worthwhile would be to consider the multi market case, since the suppliers have the possibility to sell their electricity on several markets. This aspect has been briefly discussed in Bossy et al. (2004). It leads to a much more complex model, which in particular involves a two level game where at both levels the agents strive to set a Nash equilibrium.

Acknowledgments. This work was partially performed while Geert Jan Olsder was on leave at CNRS/i3s/UNSA, Sophia Antipolis, France during the academic year 02/03.

References

Başar, T. and Olsder, G.J. (1999). *Dynamic Noncooperative Game Theory.* SIAM, Philadelphia.

Bossy, M., Maïzi, N., Olsder, G.J., Pourtallier, O., and Tanré, E. (2004). Using game theory for the electricity market. *Research report INRIA RR-5274.*

Hämäläinen, R.P. (1996). Agent-based modeling of the electricity distribution system. *Proceedings of the 15th IAESTED International Conference*, pages 344–346, Innsbruck, Austria.

Kempert, C. and Tol, R. (2000). The Liberalisation of the German Electricity Market: Modeling an Oligopolistic Structure by a Computational Game Theoretic Modeling Tool. Working paper, Department of Economics I, Oldenburg University, Oldenburg, Germany.

Luce, D. and Raifa, H. (1957). *Games and Decisions.* Wiley, New York.

Madlener, R. and Kaufmann, M. (2002). Power Exchange Spot Market Trading in Europe: Theoretical Considerations and Empirical Evidence. Contract no. ENSK-CT-2000-00094 of the 5th Framework Programme of the European Community.

Madrigal, M. and Quintana, V.H. (2001). Existence and determination of competitive equilibrium in unit commitment power pool auctions: Price setting and scheduling alternatives. *IEEE Transactions on Power Systems*, 16:380–388.

Murto, P. (2003). On Investment, Uncertainty and Strategic Interaction with Applications to Energy Markets. Ph.D. Thesis, Helsinki University of Technology.

Office of Electricity Regulation (OFFER, now OFGEM). (1998). Pool Price, A consultation by OFFER. Technical Report, Birmingham, U.K.

Ruusunen, J. (1994). Barter contracts in intertemporal energy exchange. *European Journal of Operational research*, 75:589–599.

Stoft, S. (2002). *Power System Economics-Disigning Markets for Electricity*. IEEE Press, John Wiley & Sons, Inc. Publication.

Supatchiat, C., Zhang, R.Q., and Birge, J.R. (2001). Equilibrium values in a competitive power exchange market. *Computational Economics*, 17:93–121.

Chapter 8

EFFICIENCY OF BERTRAND AND COURNOT: A TWO STAGE GAME

Michèle Breton
Abdalla Turki

Abstract We consider a differentiated duopoly where firms invest in research and development (R&D) to reduce their production cost. The objective of this study is to derive and compare Bertrand and Cournot equilibria, and then examine the robustness of the literature's results, especially those of Qiu (1997). We find that The main results of this study are as follows: (a) Bertrand competition is more efficient if R&D productivity is low, industry spillovers are weak, or products are very different. (b) Cournot competition is more efficient if R&D productivity is high and R&D spillovers and products' degree of substitutability are not very small. (c) Cournot competition may lead to higher outputs, higher consumer surpluses and lower prices, provided that R&D productivity is very high and spillovers and degree of substitutability of firms' products are moderate to high. (d) Cournot competition results in higher R&D investments compared to Bertrand's. These results show that the relative efficiencies of Bertrand and Cournot equilibria are sensitive to the suggested specifications, and hence far from being robust.

1. Introduction

This paper compares the relative efficiency of Cournot and Bertrand equilibria in a differentiated duopoly. In a Cournot game, players compete by choosing their outputs while in a Bertrand game, the strategies are the prices of these products. In a seminal paper, Singh and Vives (1984) show that (i) Bertrand competition is always more efficient than Cournot competition, (ii) Bertrand prices (*quantities*) are lower (*higher*) than Cournot prices (*quantities)* if the goods are substitutes (*complements*), and (iii) it is a dominant strategy for a firm to choose quantity (*price*) as its strategic variable provided that the goods are substitutes (*complements*).

These findings attracted the economists' attention and gave rise to two main streams in the literature. The first one extends the above model in different ways. However, it treats the firms' costs of production as constants and assumes that firms face the same demand and cost structure in both types of competition (Vives (1985), Okuguchi (1987), Dastidar (1997), Hackner (2000), Amir and Jin (2001), and Tanaka (2001)).

The second stream of research aims at examining the robustness of the findings in Singh and Vives (1984) in a duopoly where firms' strategies include investments in research and development (R&D), in addition to prices or quantities (see for instance, Delbono and Denicolo (1990), Motta (1993), Qiu (1997) and Symeonidis (2003)). A general result is that the findings in Singh and Vives (1984) may not always hold. For instance, in a two stage model of a duopoly producing substitute goods and engaging in process R&D (to reduce their unit production cost), Qiu (1997) finds that (i) although Cournot competition induces more R&D effort than Bertrand competition the latter still results in lower prices and higher quantities, (ii) Bertrand competition is more efficient if either R&D productivity is low, or spillovers are weak, or products are very different, and (iii) Cournot competition is more efficient if either R&D productivity is high, spillovers are strong, and products are close substitutes. Similar results are obtained by Symeonidis (2003) for the case where the duopoly engage in product R&D.

This paper belongs to this second stream, extending the model of Qiu (1997) in three ways. First, costs of production are assumed to be quadratic in the production levels, rather than linear (decreasing returns to scale). Second, we incorporate the specification of the R&D cost function suggested by Amir (2000), making it depend on the spillover level to correct for the eventual biases introduced by postulating additive spillovers on R&D outputs rather than inputs. Finally, the firms' benefits from each other in reducing their costs of production are assumed to depend not only on the R&D spillovers but also on the degree of substitutability of their products.

We show in this setting that Cournot competition may lead to higher outputs, higher consumer surpluses and lower prices when R&D productivity is high. We also show that when R&D productivity is high, Cournot competition is more efficient. Finally, we show that our results are robust, whether the investment cost is independent or not of the spillover level.

The rest of this paper is organized as follows. In Section 2, we outline the model. In Section 3 and 4, we respectively characterize Cournot and

Bertrand equilibria. In Section 5, we compare the results and we briefly conclude in Section 6.

2. The Model

Consider an industry formed of two firms producing differentiated but substitutable goods. Each firm independently undertakes cost-reducing R&D investments, and chooses the price p_i or the quantity q_i of its product, so as to maximize its profits. If the firms choose quantity (*price*) levels, then it is said that they engage in a Cournot (*Bertrand*) game.

Following Singh and Vives (1984), the representative consumer's preferences are described by the following utility function:

$$U(q_1, q_2) = a(q_1 + q_2) - \frac{1}{2}(q_1^2 + q_2^2) - \eta q_1 q_2, \tag{8.1}$$

where a is positive constant and $0 \leq \eta < 1$. The parameter η represents the degree of product differentiation; products become closer to substitutes as this parameter approaches 1. The resulting market inverse demands are linear and given by

$$p_i = a - q_i - \eta q_j, \quad i, j = 1, 2, i \neq j. \tag{8.2}$$

Denote by x_i firm i's R&D investment. The unit production cost is assumed to depend on both firms' investments in R&D as well as on the quantity produced and has the following form

$$C_i(q_i, x_i, x_j) = (c + \frac{r}{2}q_i - x_i - \eta\beta x_j),\ i, j = 1, 2, i \neq j, \tag{8.3}$$

where $0 < c < a$, $r \geq 0$ and $0 \leq \beta \leq 1$. The parameter β is the industry degree of R&D spillover and $\eta\beta$ represents the effective R&D spillover. We assume that the unit cost is strictly positive.

The cost specification in (8.3) differs from the one proposed by Qiu (1997) in two aspects. First, the benefit in cost reduction from rival's R&D depends on the effective spillover rate $\eta\beta$ and not only on β. An explanation of our assumption lies in the fact that the products must be related, i.e., $\eta \neq 0$, for spillovers to be beneficial to firms (another way to achieve this is to restrict the spillover rate in Qui's model to values that are smaller than η). Second, firm i's production cost function is assumed to be quadratic in its level of output rather than linear, which allows to model decreasing returns to scale.

The investment cost incurred by player i is assumed quadratic in R&D effort, i.e.,

$$F_i(x_i) = \frac{\nu + \delta\eta\beta}{2}x_i^2, i = 1, 2, \tag{8.4}$$

where $\delta \geq 0$. For $\delta > 0$, the cost is increasing in the effective spillover level $\eta\beta$. Thus higher R&D effective spillover leads to higher R&D costs for each firm. For $\delta = 0$, the cost is independent of the spillover, as in Qui's model. When $\delta > 0$, the cost function is steeper in R&D and is similar to the one proposed by Amir (2000)[1].

The total profit of firm i, to be maximized, is given by

$$\pi_i = (p_i - C_i(q_i, x_i, x_j))\, q_i - F_i(x_i). \tag{8.5}$$

We shall in the sequel compare consumer surplus (CS) and total welfare (TW) under Bertrand and Cournot modes of play. Recall that consumer surplus is defined as consumer's utility minus total expenditures evaluated at equilibrium, i.e.,

$$CS = U(q_1^*, q_2^*) - p_1^* q_1^* - p_2^* q_2^*. \tag{8.6}$$

Total welfare is defined as the sum of consumer surplus and industry's profit, i.e.,

$$TW = CS + \pi_1^* + \pi_2^*, \tag{8.7}$$

where the superscript $*$ refers to (Bertrand or Cournot) equilibrium values.

REMARK 8.1 *Our model is symmetric, i.e., all parameters involved are the same for both players. This assumption allows us to compare Bertrand and Cournot equilibria in a setting where any difference would be due to the choice of the strategic variables and nothing else. We shall confine our interest to symmetric equilibria in both Cournot and Bertrand games.*

3. Cournot Equilibrium

In the non-cooperative two stage Cournot game, firms select their R&D investments and output levels, independently and sequentially. To

[1]Notice that following d'Aspremont and Jacquemin (1988, 1990), the models used in Qui (1997) postulate that the possible R&D spillovers take place in the final R&D outcomes. However, Kamien et al. (1992), among others, presume that such spillovers take place in the R&D dollars (spending). Amir (2000) makes an extensive comparison between these two models (i.e. d'Aspremont and Jacquemin (1988, 1990), and Kamien et al. (1992)), and assesses their validity. He concludes that the latter's predictions and results are more valid and robust. Furthermore, in order to make the above two models equivalent, Amir (2000) suggests a certain specification of the R&D cost function of the first above-mentioned models. As he shows, such specification does make the R&D cost functions steeper, and hence do correct for the biases introduced by postulating additive spillovers on R&D outputs rather than inputs.

characterize Cournot equilibrium, it will be convenient to use the following notation:

$$\begin{aligned}
A &= a - c \\
V &= \nu + \delta\eta\beta \\
B &= \eta\beta + 1 \\
R_1 &= r + 2 \\
G_1 &= r + 2 + \eta \\
D_1 &= r + 2 - \eta \\
E_1 &= r + 2 - \eta^2\beta \\
\Delta_1 &= D_1 G_1^2 V - B E_1 R_1.
\end{aligned}$$

We assume that the parameters satisfy the conditions:

$$D_1^2 G_1^2 V - E_1^2 R_1 > 0 \tag{8.8}$$

$$D_1 G_1 V \left(G_1 - \frac{(a-c)(1+\eta)}{a} \right) - B E_1 R_1 > 0. \tag{8.9}$$

The following proposition characterizes Cournot equilibrium.

PROPOSITION 8.1 *Assuming (8.8-8.9), in the unique symmetric subgame perfect Cournot equilibrium, output and R&D investment strategies are given by*

$$q^C = \frac{A D_1 G_1 V}{\Delta_1}, \tag{8.10}$$

$$x^C = \frac{A E_1 R_1}{\Delta_1}. \tag{8.11}$$

Proof. Firm i's profit function is given by:

$$\pi_i = \left(a - q_i - \eta q_j - (c + \frac{r}{2} q_i - x_i - \eta\beta x_j)\right) q_i - \frac{\nu + \delta\eta\beta}{2} x_i^2 ,$$
$$i, j = 1, 2, i \neq j. \qquad \text{(cournot profit function)}$$

Cournot sub-game perfect equilibria are derived by maximizing the above profit function sequentially as follows: Given any first-stage R&D outcome (x_i, x_j), firms choose output to maximize their respective market profits. Individual profit functions are strictly concave in q_i and, assuming an interior solution, the first order conditions yield the unique Nash-Cournot equilibrium output:

$$q_i^* = \frac{A D_1 - \eta x_j (1 - \beta R_1) + x_i E_1}{D_1 G_1}, i, j = 1, 2, i \neq j.$$

In the first stage firms choose R&D levels to maximize their respective profits taking the equilibrium output into account. After substituting for the equilibrium output levels q_i^*, individual profit functions are concave in x_i if (8.8) is satisfied. The first order conditions yield the unique symmetric R&D equilibrium given in (8.11), which is interior if $D_1G_1^2V - BE_1R_1 = \Delta_1 > 0$, which is implied by (8.9). Substituting for x^C in q_i^* yields the Cournot equilibrium output (8.10). □

The equilibrium Cournot price and profit are given by

$$p^C = a - \frac{AD_1G_1V(1+\eta)}{\Delta_1},$$

$$\pi^C = \frac{A^2R_1V\left(D_1^2G_1^2V - E_1^2R_1\right)}{2\Delta_1^2}$$

which are respectively positive if (8.9) and (8.8) are satisfied. Inserting Cournot equilibrium values in (8.6) and (8.7) provides the following consumer surplus and total welfare

$$CS^C = \frac{(AD_1G_1V)^2(1+\eta)}{\Delta_1^2}, \tag{8.12}$$

$$TW^C = A^2V\frac{D_1^2G_1^2V(1+\eta+R_1) - E_1^2R_1^2}{\Delta_1^2}. \tag{8.13}$$

The following proposition compares the results under the two specifications of the investment cost function in R&D, i.e., for $\delta > 0$ and $\delta = 0$.

PROPOSITION 8.2 *Cournot's output, investment in R&D, consumer surplus and total welfare are lower and price is higher when the cost function is steeper in R&D investments (i.e., $\delta > 0$).*

Proof. Compute $\frac{d}{d\delta}\left(x^C\right) = -\frac{AD_1E_1G_1^2R_1\eta\beta}{\left(D_1G_1^2(\nu+\delta\eta\beta)-BE_1R_1\right)^2} < 0$, indicating that equilibrium R&D investment is decreasing in δ. Denote $x^{C\delta}$ the Cournot equilibrium investment corresponding to a given value of δ. Straightforward computations lead to

$$x^{C\delta} - x^{C0} = -\frac{AD_1E_1G_1^2R_1\delta\eta\beta}{\left(D_1G_1^2(\nu+\delta\eta\beta)-BE_1R_1\right)\left(D_1G_1^2\nu-BE_1R_1\right)}.$$

Notice that:

$$q^C = \frac{A+Bx^C}{G_1},$$

$$\begin{aligned} p^C &= a - q^C (1+\eta), \\ CS^C &= \left(q^C\right)^2 (1+\eta), \end{aligned}$$

yielding the results for output, price and consumer surplus.

To assess the effect of δ on profits, compute

$$\frac{d}{d\delta}\left(\pi^C\right) = -\frac{1}{2}A^2 E_1 R_1^2 \eta\beta \frac{D_1 G_1^2 V (2BD_1 - E_1) - BE_1^2 R_1}{\left(D_1 G_1^2 V - BE_1 R_1\right)^3}.$$

Using

$$2BD_1 - E_1 > 0,$$

$$D_1 G_1^2 V (2BD_1 - E_1) - BE_1^2 R_1 > 2BE_1 R_1 (BD_1 - E_1),$$

we see that the sign of the derivative depends on the parameter's values, so that profit can increase with δ when ν is small and β is less than $\frac{1}{R_1}$ (see appendix for an illustration). However, the impact of δ on total welfare remains negative:

$$\begin{aligned} \frac{d}{d\delta}\left(TW^C\right) &= -A^2 E_1 R_1 \eta\beta \\ &\quad \frac{D_1 G_1^2 V (2BD_1 (1 + R_1 + \eta) - E_1 R_1) - BE_1^2 R_1^2}{\left(D_1 G_1^2 V - BE_1 R_1\right)^3} \\ &< -2A^2 BE_1^2 R_1^2 \eta\beta \frac{(D_1 B (1 + R_1 + \eta) - E_1 R_1)}{\left(D_1 G_1^2 V - BE_1 R_1\right)^3} < 0. \end{aligned}$$

□

4. Bertrand Equilibrium

In the non-cooperative two stage Bertrand game, firms select their R&D investments and prices, independently and sequentially. From (8.2), we define the demand functions

$$q_i = \frac{(1-\eta)a - p_i + \eta p_j}{(1-\eta^2)}, i, j = 1, 2, i \neq j. \tag{8.14}$$

To characterize Bertrand equilibrium, it will be convenient to use the following notation:

$$\begin{aligned} G_2 &= r + 2 + \eta - \eta^2 \\ D_2 &= r + 2 - \eta - \eta^2 \\ E_2 &= r + 2 - \eta^2\beta - \eta^2 \end{aligned}$$

$$\begin{aligned} R_2 &= r + 2 - 2\eta^2 \\ \Delta_2 &= D_2 G_2^2 V - B E_2 R_2. \end{aligned}$$

We assume that the parameters satisfy the conditions:

$$D_2^2 G_2^2 V - E_2^2 R_2 > 0 \tag{8.15}$$

$$D_2 G_2 V \left(G_2 - \frac{(a-c)(\eta+1)}{a} \right) - B E_2 R_2 > 0. \tag{8.16}$$

The following proposition characterizes Bertrand equilibrium.

PROPOSITION 8.3 *Assuming (8.15-8.16), in the unique symmetric subgame perfect Bertrand equilibrium, price and R&D investment strategies are given by*

$$p^B = \frac{D_2 G_2 V (a G_2 - A(\eta+1)) - a B E_2 R_2}{\Delta_2}, \tag{8.17}$$

$$x^B = \frac{A E_2 R_2}{\Delta_2} \tag{8.18}$$

Proof. The proof is similar to that of Proposition 1 and is omitted. □

The equilibrium Bertrand output and profit are given by

$$q^B = \frac{A D_2 G_2 V}{\Delta_2}. \tag{8.19}$$

$$\pi^B = \frac{A^2 R_2 V \left(D_2^2 G_2^2 V - E_2^2 R_2 \right)}{2 \Delta_2^2}$$

which are respectively positive if (8.16) and (8.15) are satisfied. Consumer surplus and total welfare are then given by:

$$CS^B = \frac{(A D_2 G_2 V)^2 (1+\eta)}{\Delta_2^2}, \tag{8.20}$$

$$TW^B = A^2 V \frac{D_2^2 G_2^2 V (1 + \eta + R_2) - R_2^2 E_2^2}{\Delta_2^2}. \tag{8.21}$$

The following proposition compares the Bertrand equilibrium under the two specifications of the investment cost function in R&D, i.e., for $\delta > 0$ and $\delta = 0$.

PROPOSITION 8.4 *Bertrand's output, investment in R&D, consumer surplus and total welfare are lower and price is higher when the cost function is steeper in R&D investments (i.e., $\delta > 0$).*

Proof. The impact of δ on output, investment, consumer surplus and total welfare is obtained as in the proof of Proposition 2. Again, the effect of δ on profit depends on the parameter's value, and can be positive when η is large and r is less than $\frac{1-\beta}{(2\beta+1)}$ (see appendix for an illustration). □

5. Comparison of Equilibria

In this section we compare Cournot outputs, prices, profits, R&D levels, consumer surplus and total welfare with their Bertrand counterparts. We assume that conditions (8.8), (8.9), (8.15) and (8.16) are satisfied by the parameters. These conditions are necessary and sufficient for the equilibrium prices, quantities and profits to be positive in both modes of play.

PROPOSITION 8.5 *Cournot R&D investments are always higher than Bertrand's.*

Proof. From (8.11) and (8.18) we have

$$x^C - x^B = AV\frac{D_2E_1G_2^2R_1 - D_1E_2G_1^2R_2}{\Delta_1\Delta_2},$$

where the denominator is positive under conditions (8.9) and (8.16). Substituting the parameter values in the numerator yields

$$\begin{aligned}
D_2E_1G_2^2R_1 - D_1E_2G_1^2R_2 \;=\; & \\
& \eta^3r^3\left(\eta+\eta\beta+1\right) \\
& +\eta^3r^2\left(\left(2-\eta^2\right)\left(3\eta\beta+1\right)+\left(2-\eta\right)\left(\eta+2\right)\left(\eta+1\right)+\eta^2\right) \\
& +\eta^3r\left(\left(12-11\eta^2-\eta^3+\eta^4\right)\left(\eta\beta+1\right)+\eta\left(2-\eta\right)\left(\eta+2\right)\left(\eta+1\right)\right) \\
& +2\eta^3\left(1-\eta\right)\left(2-\eta\right)\left(\eta+2\right)\left(\eta+1\right)\left(\eta\beta+1\right)
\end{aligned}$$

which is positive. □

This finding is similar to that of Qiu (1997) and can be explained in the same way. Qiu analyzes the factors that induce firms to undertake R&D under both Cournot and Bertrand settings, using general demand and R&D cost functions. He decomposes these factors into four main categories: (i) strategic effects, (ii) spillover effects, (iii) size effects and (iv) cost effects. He shows that while the last three factors do have similar impact on both Cournot and Bertrand, the strategic factor does induce more R&D under the Cournot and discourages it under Bertrand. Therefore he concludes that due to the strategic factor, R&D investment under Cournot will always be higher than that of Bertrand.

PROPOSITION 8.6 *Cournot output and consumer surplus could be higher than Bertrand's.*

Proof. Use (8.10) and (8.19) to get

$$q^C - q^B = AV\frac{(D_1G_1\Delta_2 - D_2G_2\Delta_1)}{\Delta_1\Delta_2}.$$

The numerator is the difference of two positive numbers, and is positive if

$$V < \frac{B\left(G_2D_2R_1E_1 - D_1G_1E_2R_2\right)}{D_2G_2D_1G_1\eta^2},$$

that is, when V is small while satisfying conditions (8.8), (8.9), (8.15) and (8.16), which are lower bounds on V. The intersection of these conditions is not empty (see appendix for examples). This shows that the output under Cournot could be higher than that of Bertrand when R&D productivity is high. Consequently, consumer surplus under Cournot could be higher than under Bertrand. □

These results are different from those of Qiu and Singh and Vives as well. Furthermore, they contrast the fundamental results of the oligopoly theory. As expected, if R&D productivity is low, then the traditional results, which state that output and consumer surplus (prices) under Bertrand are higher (lower) than Cournot's, still hold. Notice that Proposition 6 holds when $r = 0$ and $\delta = 0$.[2]

The difference in profits can be written:

$$\begin{aligned}\pi^C - \pi^B &= \frac{V^2A^2}{2B^2}\left(D_1G_1^3\frac{R_1B\left(\eta\beta-1\right)+D_1G_1V}{\Delta_1^2}\right.\\ &\qquad \left.-D_2G_2^3\frac{R_2B\left(\eta\beta-1\right)+G_2D_2V}{\Delta_2^2}\right)\\ &= \frac{V^2A^2}{2\Delta_1^2\Delta_2^2}\left(\alpha V^2+\gamma V+\lambda\right),\end{aligned}$$

where

$$\begin{aligned}\alpha &= D_2^2G_2^2D_1^2G_1^2\eta^3\left(R_1\left(2+\eta\right)+2\eta\right)>0\\ \gamma &= 2D_2R_2R_1G_1^2D_1G_2^2B\eta^3\left(\eta\beta-1\right)+D_1^2G_1^4E_2^2R_2^2-D_2^2G_2^4E_1^2R_1^2\\ \lambda &= \left(1-\eta\beta\right)R_2R_1B\left(D_2G_2^3E_1^2R_1-E_2^2G_1^3D_1R_2\right).\end{aligned}$$

The sign of the above expression is not obvious. Extensive numerical investigations failed to provide an instance where the difference is

[2] This apparent contradiction with Qiu's result is due to his using a restrictive condition on the parameter values which is sufficient but not necessary for the solutions to make sense.

negative, provided conditions (8.8), (8.9), (8.15) and (8.16) are satisfied. Since V can take arbitrarily large values and $\alpha > 0$, the difference is increasing in ν when sufficiently large. This result is similar to that of Qiu (1997) and Singh and Vives (1984) and indicates that firm should prefer Cournot equilibrium, and more so when productivity of R&D is low.

As a consequence, it is apparent that total welfare under Cournot can be higher than that under Bertrand (when for instance both consumer surplus and profit are higher). Examples of positive and negative differences in total welfare are provided in the appendix. With respect to the conclusions in Qiu, we find that efficiency of Cournot competition does not require that η and β both be very high, provided R&D productivity is high. Again, this is true for $r = 0$ and $\delta = 0$.

REMARK 8.2 *It can be easily verified that if $\eta = 0$, i.e., the products are not substitutes, then Bertrand and Cournot equilibria coincide. This means that each firm becomes a monopoly in its market and it does not matter if the players choose prices or quantities as strategic variables.*

6. Concluding Remarks

In this paper, we developed a two stage games for a differentiated duopoly. Each of the two firms produces one variety of substitute goods. It also engages in process R&D investment activities, which aims at reducing its production costs, incidentally reducing also the competitor's costs because of spillovers which are proportional to the degree of substitability of the products. The firms intend to maximize their total profit functions by choosing the optimal levels of either their outputs/or prices as well as R&D investments. We derived and compared the Bertrand and the Cournot equilibria, and then examined the robustness of the literature's results, especially those of Qiu (1997).

The main results of this study are as follows: (a) Bertrand competition is more efficient if R&D productivity is low and effective spillovers are weak. (b) Cournot competition may lead to higher outputs, higher consumer surpluses and lower prices, provided that R&D productivity is very high. (c) Cournot competition results in higher R&D investments compared to Bertrand's. (d) A steeper investment cost specification lowers output, investment and consumer surplus in both kinds of competition but does not change qualitatively the results about their comparative efficiencies. (e) These results are robust to a convex production cost specification.

Appendix:

	1	2	3	4	5	6
a	400	400	400	400	900	300
c	300	300	300	300	800	200
η	0.8	0.8	0.9	0.9	0.9	0.9
β	0.1	0.1	0.2	0.2	0.9	0.7
r	1	1	0.3	0.3	0	1
v	0.4	0.4	1	1	0.6	10
δ	0	1	0	1	0	0
x^C	276	154	60	46	268	2
q^C	105	70	54	49	202	27
p^C	62	124	297	307	17	249
π^C	1232	1682	1510	1453	19129	1036
CS^C	19724	8810	5573	4542	77192	1348
TW^C	22188	12174	8593	7448	115450	3419
x^B	201	123	40	32	40	2
q^B	100	74	63	59	82	33
p^B	220	117	280	288	244	237
π^B	581	1039	444	493	811	750
CS^B	18108	9768	7533	6553	12805	2113
TW^B	19271	11846	8421	7539	14428	3613
$q^C - q^B$	4	-4	-9	-10	119	-7
$\pi^C - \pi^B$	651	643	1066	960	18318	286
$CS^C - CS^B$	1616	-958	-1960	-2010	64387	-766
$TW^C - TW^B$	2917	328	172	-90	101022	-194

Columns 1 and 2 show an example where increasing δ increases Cournot profit.
Columns 3 and 4 show an example where increasing δ increases Bertrand profit.
Columns 1 and 5 show examples where Cournot quantities and consumer surplus are larger than Bertrand's.
Columns 1, 2, 3 and 5 show examples where total welfare is larger under Cournot, columns 4 and 6 where it is smaller.

References

Amir, R. (2000). Modelling imperfectly appropriable R&D via spillovers. *International Journal of Industrial Organization*, 18:1013–1032.

Amir, R. and Jin, J.Y. (2001). Cournot and Bertrand equilibria compared: substitutability, complementarity, and concavity. *International Journal of Industrial Organization*, 19:303–317.

d'Aspremont, C. and Jacquemin, A. (1988). Cooperative and non cooperative R&D in duopoly with spillovers. *American Economic Review*, 78:1133–1137.

d'Aspremont, C., and Jacquemin, A. (1990). Cooperative and Noncooperative R&D in Duopoly with Spillovers: Erratum. *American Economic Review*, 80:641–642.

Dastidar, K.G. (1997). Comparing Cournot and Bertrand in a homogeneous product market. *Journal of Economic Theory*, 75:205–212.

Delbono, F. and Denicolo, V. (1990). R&D in a symmetric and homogeneous oligopoly. *International Journal of Industrial Organization*, 8:297–313.

Hackner, J. (2000). A note on price and quantity competition in differentiated oligopolies. *Journal of Economic Theory*, 93:233–239.

Kamien, M., Muller, E., and Zang, I. (1992). Research Joint Ventures and R&D Cartels. *American Economic Review*, 82:1293–1306.

Motta, M. (1993). Endogenous quality choice: price versus quality competition. *Journal of Industrial Economics*, 41:113–131.

Okuguchi, K. (1987). Equilibrium prices in Bertrand and Cournot oligopolies. *Journal of Economic Theory*, 42:128–139.

Qiu, L. (1997). On the dynamic efficiency of Bertrand and Cournot equilibria. *Journal of Economic Theory*, 75:213–229.

Singh, N. and Vives, X. (1984). Price and quantity competition in a differentiated duopoly. *Rand Journal of Economics*, 15:546–554.

Symeonidis, G. (2003). Comparing Cournot and Bertrand equilibria in a differentiated duopoly with product R&D. *International Journal of Industrial Organization*, 21:39–55.

Tanaka, Y. (2001). Profitability of price and quantity strategies in an oligopoly. *Journal of Mathematical Economics*, 35:409–418.

Vives, X. (1985). On the efficiency of Bertrand and Cournot equilibria with product differentiation. *Journal Economic Theory*, 36:166–175.

Chapter 9

CHEAP TALK, GULLIBILITY, AND WELFARE IN AN ENVIRONMENTAL TAXATION GAME

Herbert Dawid
Christophe Deissenberg
Pavel Ševčik

Abstract We consider a simple dynamic model of environmental taxation that exhibits time inconsistency. There are two categories of firms, Believers, who take the tax announcements made by the Regulator to face value, and Non-Believers, who perfectly anticipate the Regulator's decisions, albeit at a cost. The proportion of Believers and Non-Believers changes over time depending on the relative profits of both groups. We show that the Regulator can use misleading tax announcements to steer the economy to an equilibrium that is Pareto superior to the solutions usually suggested in the literature. Depending upon the initial proportion of Believers, the Regulator may prefer a fast or a low speed of reaction of the firms to differences in Believers/Non-Believers profits.

1. Introduction

The use of taxes as a regulatory instrument in environmental economics is a classic topic. In a nutshell, the need for regulation usually arises because producing causes detrimental emissions. Due to the lack of a proper market, the firms do not internalize the impact of these emissions on the utility of other agents. Thus, they take their decisions on the basis of prices that do not reflect the true social costs of their production. Taxes can be used to modify the prices confronting the firms so that the socially desirable decisions are taken.

The problem has been exhaustively investigated in static settings where there is no room for strategic interaction between the Regulator and the firms. Consider, however, the following situation: (a) The emission taxes have a dual effect, they incite the firms to reduce production *and* to undertake investments in abatement technology. This is typically the case when the emissions are increasing in the output and decreasing in the abatement technology; (b) Emission reduction is socially desirable, the reduction of production is not; and (c) The investments are irreversible. In that case, the Regulator must find an optimal compromise between implementing high taxes to motivate high investments, and keeping the taxes low to encourage production. The fact that the investments are irreversible introduces a strategic element in the problem. If the firms are naive and believe his announcements, the Regulator can insure high production *and* important investments by first declaring high taxes and reducing them once the corresponding investments have been realized. More sophisticated firms, however, recognize that the initially high taxes will not be implemented, and are reluctant to invest in the first place. In other words, one is confronted with a typical time inconsistency problem, which has been extensively treated in the monetary policy literature following Kydland and Prescott (1977) and Barro and Gordon (1983). In environmental economics, the time inconsistency problem has received yet only limited attention, although it frequently occurs. See among others Gersbach and Glazer (1999) for a number of examples and for an interesting model, Abrego and Perroni (1999), Batabyal (1996a), Batabyal (1996b), Dijkstra (2002), Marsiliani and Renström (2000), Petrakis and Xepapadeas (2003).

The time inconsistency is directly related to the fact that the situation described above defines a Stackelberg game between the Regulator (the leader) and the firms (the followers). As noted in the seminal work of Simaan and Cruz (1973a,b), inconsistency arises because the Stackelberg equilibrium is not defined by mutual best responses. It implies that the follower uses a best response in reaction the leader's action, but not that the leader's action is itself a best response to the follower's. This opens the door to a re-optimization by the leader once the follower has played. Thus, a Regulator who announces that he will implement the Stackelberg solution is not credible. An usual conclusion is that, in the absence of additional mechanisms, the economy is doomed to converge towards the less desirable Nash solution.

A number of options to insure credible solutions have been considered in the literature – credible binding commitments by the Stackelberg leader, reputation building, use of trigger strategies by the followers, etc. See McCallum (1997) for a review in a monetary policy context. Schematically, these solutions aim at assuring the time consistency of Stackelberg solution with either the Regulator or the firms as a leader. Usually, these solutions are not efficient and can be Pareto-improved.

In this paper, we suggest a new solution to the time inconsistency problem in environmental policy. We show that non-binding tax announcements can increase the payoff not only of the Regulator, but also of all firms, if these include any number of naive *Believers* who take the announcements at face value. Moreover, if firms tend to adopt the behavior of the most successful ones, a stable equilibrium may exist where a positive fraction of firms are Believers. This equilibrium Pareto-dominates the one where all firms anticipate perfectly the Regulator's action. To attain the superior equilibrium, the Regulator builds reputation and leadership by making announcements and implementing taxes in a way that generates good results for the Believers, rather than by pre-committing to his announcements.

This Pareto-superior equilibrium does not always exist. Depending upon the model parameters (most crucially: upon the speed with which the firms that follow different strategies react to differences in their respective profits, i.e., upon the *flexibility* of the firms) it may be rational for the Regulator to steer the Pareto-inferior fully rational equilibrium. This paper, thus, stresses the importance of the flexibility in explaining the policies followed by a Regulator, the welfare level realized, and the persistence or decay of private confidence in the Regulator's announcements.

The potential usefulness of employing misleading announcements to Pareto-improve upon standard game-theoretic equilibrium solutions was suggested for the case of general linear-quadratic dynamic games in Vallée et al. (1999) and developed by the same authors in subsequent papers. An early application to environmental economics is Vallée (1998). The Believers/Non-Believers dichotomy was introduced in Deissenberg and Alvarez Gonzalez (2002), who study the credibility problem in monetary economics in a discrete-time framework with reinforcement learning. A similar monetary policy problem has been investigated in Dawid and Deissenberg (2004) in a continuous-time setting akin to the one used in the present work.

The paper is organized as follows. In Section 2, we present the model of environmental taxation, introduce the imitation-type dynamics that determine the evolution of the number of Believers in the economy, and derive the optimal reaction functions of the firms. In Section 3, we discuss the solution of the static problem one obtains by assuming a constant proportion of Believers. Section 4 is devoted to the analysis of the dynamic problem and to the presentation of the main results. Section 5 concludes.

2. The Model

We consider an economy consisting of a Regulator R and of a continuum of atomistic, profit-maximizing firms i with identical production technology. Time τ is continuous. To keep notation simple, we do not index the variables with either i or τ, unless useful for a better understanding.

In a nutshell, the situation we consider is the following. The Regulator can tax the firms in order to incite them to reduce their emissions. Taxes, however, have a negative impact on the employment. Thus, R has to choose them in order to achieve an optimal compromise between emissions reduction and employment. The following sequence of events occurs in every τ:

- R makes a non-binding announcement $t^a \geq 0$ about the tax level he will implement. The tax level is defined as the amount each firm has to pay per unit of its emissions.

- Given t^a, the firms form expectations t^e about the actual level of the environmental tax. As will be described in more detail later, there are two different ways for an individual firm to form its expectations. Each firm i decides about its level of emission reduction v_i based on its expectation t_i^e and makes the necessary investments.

- R chooses the actual level of tax $t \geq 0$.

- Each firm i produces a quantity x_i generating emissions $x_i - v_i$.

- The individual firms revise the way they form their expectations (that is, revise their beliefs) depending on the profits they have realized.

The Firms

Each firm produces the same homogenous good using a linear production technology: The production of x units of output requires x units of labor and generates x units of environmentally damaging emissions. The production costs are given by:

$$c(x) = wx + c_x x^2, \tag{9.1}$$

where x is the output, $w > 0$ the fixed wage rate, and $c_x > 0$ a parameter. For simplicity's sake, the demand is assumed infinitely elastic at the given price $\tilde{p} > w$. Let $p := \tilde{p} - w > 0$.

At each point of time, each firm can spend an additional amount of money γ in order to reduce its current emissions. The investment

$$\gamma(v) = c_v v^2, \tag{9.2}$$

with $c_v > 0$ a given parameter, is needed in order to reduce the firm's current emissions by $v \in [0, x]$. The investment in one period has no impact on the emissions in future periods. Rather than expenditures in emission-reducing capital, γ can therefore be interpreted as the additional costs resulting of a temporary switch to a cleaner resource – say, of a switch from coal to natural gas.

Depending on the way they form their expectations t^e, we consider two types of firms, Believers B and Non-Believers NB. The fraction of Believers in the population is denoted by $\pi \in [0, 1]$. Believers consider the Regulator's announcement to be truthful and set $t^e = t^a$. Non-Believers disregard the announcement and anticipate perfectly the actual tax level, $t^e = t$. Making perfect anticipations at any point of time, however, is costly. Thus, Non-Believers occur costs of $\delta > 0$ per unit of time.

The firms are profit-maximizers. As will become apparent in the following, one can assume without loss of substance that they are myopic, that is, maximize in every τ their current profit.

The Regulator R

The Regulator's goal is to maximize over an infinite horizon the cumulated discounted value of an objective function with the employment, the emissions, and the tax revenue as arguments. In order to realize this objective, it has two instruments at his disposal, the announced instantaneous tax level t^a, and the actually realized level t.

The objective function is given by:

$$\begin{aligned}\Phi(t^a, t) &= \int_0^\infty e^{-\rho\tau}\phi(t^a, t)d\tau \qquad (9.3)\\ &:= \int_0^\infty e^{-\rho\tau}\Big[k(\pi x^b + (1-\pi)x^{nb}) - \kappa(\pi(x^b - v^b)\\ &+(1-\pi)(x^{nb} - v^{nb})) + t(\pi(x^b - v^b) + (1-\pi)(x^{nb} - v^{nb}))\Big]\, d\tau,\end{aligned}$$

where x^b, x^{nb}, v^b, v^{nb} denote the optimal production respectively investment chosen by the Believers B and the Non-Believers NB, and where k and κ are strictly positive weights placed by R on the average employment respectively on the average emissions (remember that output and employment are in a one-to-one relationship in this economy). The strictly positive parameter ρ is a social discount factor.

Belief Dynamics

The firms' beliefs (B or NB) change according to a imitation-type dynamics, see Dawid (1999), Hofbauer and Sigmund (1998). The firms meet randomly two-by-two, each pairing being equiprobable. At each encounter the firm with the lower current profit adopts the belief of the other firm with a probability proportional to the current difference between the individual profits. This gives rise to the dynamics:

$$\dot{\pi} = \beta\pi(1-\pi)(g^b - g^{nb}), \tag{9.4}$$

where g^b and g^{nb} denote the profits of a Believer and of a Non-Believer:

$$g^b = px^b - c_x(x^b)^2 - t(x^b - v^b) - c_v(v^b)^2,$$
$$g^{nb} = px^{nb} - c_x(x^{nb})^2 - t(x^{nb} - v^{nb}) - c_v(v^{nb})^2 - \delta.$$

Notice that $\pi(1-\pi)$ reaches its maximum for $\pi = \frac{1}{2}$ (the value of π for which the probability of encounter between firms with different profits is maximized), and tends towards 0 for $\pi \to 0$ and $\pi \to 1$ (for extreme values of π, almost all firms have the same profits). The parameter $\beta \geq 0$, that depends on the adoption probability of the other's strategy may be interpreted as a measure of the willingness to change strategies, that is, of the flexibility of the firms.

Equation (9.4) implies that by choosing the value of (t^a, t) at time τ, the Regulator not only influences its instantaneous objective but also the future proportion of Bs in the economy. This, in turn, has an impact on the values of its objective function. Hence, although there are no explicit dynamics for the economic variables v and x, the R faces a non-trivial inter-temporal optimization problem.

Optimal Decisions of the Firms

Since the firms are atomistic, each single producer is too small to influence the dynamics of π. Thus, the single firm does not take into account any inter-temporal effect and, independently of its true planing horizon, *de facto* maximizes its current profit in every τ. Each firm chooses its investment v after it has learned t^a but *before* t has been made public. However, it fixes its production level x *after* v and t are known. The firms being price takers, the optimal production decision is:

$$x = \frac{p-t}{2c_x}. \tag{9.5}$$

The thus defined optimal x does not depend upon t^a, neither directly nor indirectly. As a consequence, both Bs and NBs choose the same production level (9.5) as a function of the realized tax t alone, $x^b = x^{nb} = x$.

The profit of a firm given that an investment v has been realized is:

$$g(v;t) = \frac{(p-t)^2}{4c_x} + tv.$$

When the firms determine their investment v, the actual tax rate is not known. Therefore, they solve:

$$\max_v [g(v;t^e) - c_v v^2],$$

with $t^e = t$ if the firm is a NB and $t^e = t^a$ if the firm is a B. The interior solution to this problem is:

$$v^b = \frac{t^a}{2c_v}, \quad v^{nb} = \frac{t}{2c_v}. \tag{9.6}$$

The net emissions $x - v$ of any firm will remain non-negative after the investment, i.e., $v \in [0, x]$ will hold, if:

$$p \geq \frac{c_v + c_x}{c_v} \max[t, t^a]. \tag{9.7}$$

Given above expressions (9.5) for x and (9.6) for v, it is straightforward to see that the belief dynamics can be written as:

$$\dot{\pi} = \beta\pi(1-\pi)\left(\frac{-(t^a-t)^2}{4c_v} + \delta\right). \tag{9.8}$$

The two effects that govern the evolution of π become now apparent. Large deviations of the realized tax level t from t^a induce a decrease in the stock of believers, whereas the stock of believers tends to grow if the cost δ necessary to form rational expectations is high.

Using (9.5) and (9.6), the instantaneous objective function ϕ of the Regulator becomes:

$$\phi(t^a, t) = (k - \kappa + t)\frac{p-t}{2c_x} + \frac{\kappa - t}{2c_v}(\pi t^a + (1-\pi)t). \tag{9.9}$$

3. The static problem

In the model, there is only one source of dynamics, the beliefs updating (9.4). Before investigating the dynamic problem, it is instructive to cursorily consider the solution [$] of the static case in which R maximizes the integrand in (9.3) for a given value of π.

From (9.9), one recognizes easily that at the optimum $t^{a\$}$ will either take its highest possible value or be zero, depending on wether $\kappa - t^{\$}$ is positive or negative. The case $\kappa - t^{\$} < 0$ corresponds to the uninteresting situation where the regulator values tax income more than emissions reduction and thus tries to increase the volume of emissions. We therefore restrict our analysis to the environmentally friendly case $t^{\$} < \kappa$. Note that (9.7) provides a natural upper bound $\bar{t}^a$ for t, namely:

$$\bar{t}^a = \frac{pc_v}{c_v + c_x}. \tag{9.10}$$

Assuming that the optimal announcement $t^{a\$}$ takes the upper value $\bar{t}^a$ just defined (the choice of another bound is inconsequential for the qualitative results), the optimal tax $t^{\$}$ is:

$$t^{\$} = \frac{1}{2}(\kappa + \bar{t}^a - \frac{c_v k}{c_v + c_x - c_x \pi}). \tag{9.11}$$

Note that $t^{\$}$ is decreasing in π. (As π increases, the announcement t^a becomes a more powerful instrument, making a recourse to t less necessary). The requirement $\kappa > t^{\$}$ is fulfilled for all π iff:

$$c_v(k + \kappa - p) + c_x\kappa > 0. \tag{9.12}$$

In the reminder of this paper, we assume that (9.12) holds.

Turning to the firms profits, one recognizes the difference $g^{nb} - g^b$ between the NB's and B's profits is increasing in $|t^{\$} - t^{a\$}|$:

$$g^{nb} - g^b = \frac{(t^{\$} - t^{a\$})^2}{4c_v} - \delta. \tag{9.13}$$

For $\delta = 0$, the profit of the NBs is always higher that the profit of the Bs whenever $t^a \neq t$, reflecting the fact the latter make a systematic error about the true value of t. The profit of the Bs, however, can exceed the one of the NBs if the learning cost δ is high enough. Since $t^{a\$}$ is constant and $t^{\$}$ decreasing in π, and since $t^{\$} < t^{a\$}$, the difference $t^{a\$} - t^{\$}$ increases in π. Therefore, the difference between the profits of the NBs and Bs, (9.13), is increasing in π.

Further analytical insights are exceedingly cumbersome to present due to the complexity of the functions involved. We therefore illustrate a remarkable, generic feature of the solution with the help of Figure 9.1[1].

[1]The parameter values underlying the figure are $c_v = 5$, $c_x = 3$, $\delta = 0$, $p = 6$, $k = 4$, $\kappa = 3$. The figure is qualitatively robust with respect to changes in these values.

Not only the Regulator's utility ϕ increases with π, so do also the profits of the Bs and NBs. For $\pi = 0$, the outcome is the Nash solution of a game between the NBs and the Regulator. This outcome is not efficient, leaving room for Pareto-improvement. As π increases, the Bs enjoy the benefits of a decreasing t, while their investment remains fixed at $v(\bar{t}^a)$. Likewise, the NBs benefit from the decrease in taxation. The lowering of the taxation is made rational by the existence of the Believers, who are led by the R's announcements to invest more than they would otherwise, and to subsequently emit less. Accordingly, the marginal tax income goes down as π increases and therefore R is induced to reduce taxation if the proportion of believers goes up.

The only motive that could lead R to reduce the spread between t and t^a, and in particular to choose $t^a < \bar{t}^a$, lies in the impact of the tax announcements on the beliefs dynamics. Ceteris paribus, R prefers a high proportion of Bs to a low one, since it has one more instrument (that is, t^a) to influence the Bs than to control the NBs. A high spread $t^a - t$, however, implies that the profits of the Bs will be low compared to those of the NBs. This, by (9.4), reduces the value of $\dot{\pi}$ and leads over time to a lower proportion of Bs, diminishing the instantaneous utility of R. Therefore, in the dynamic problem, R will have to find an optimal compromise between choosing a high value of t^a, which allows a low value of t and insures the Regulator a high instantaneous utility, and choosing a lower one, leading to a more favorable proportion of Bs in the future.

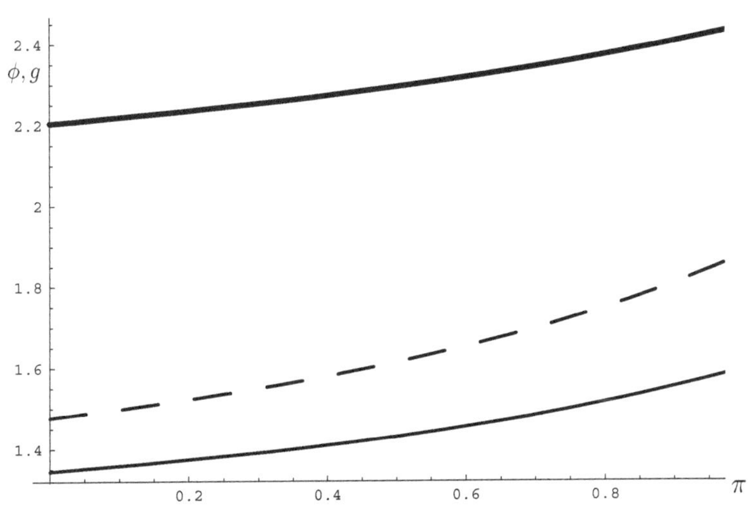

Figure 9.1. Profits of the Believers (solid line), of the Non-Believers (dashed line), and Regulator's utility (bold-line) as a function of π.

4. Dynamic analysis

4.1 Characterization of the optimal paths

As pointed out earlier, the Regulator faces a dynamic optimization problem because of the effect of his current action on the future stock of believers. This problem is given by:

$$\max_{0\leq t^a(\tau),t(\tau)} \Phi(t^a,t) \quad \text{subject to (9.8).}$$

The Hamiltonian of the problem is given by:

$$H(t^a,t,\pi,\lambda) = (k-\kappa+t)\frac{p-t}{2c_x} + \frac{\kappa-t}{2c_v}(\pi t^a + (1-\pi)t) + \lambda\beta\pi(1-\pi)\Big(-\frac{1}{4c_v}(t^a-t)^2+\delta\Big),$$

where λ denotes the co-state variable. The Hamiltonian is concave in (t,t^a) iff:

$$2\lambda\beta(1-\pi)(c_v+c_x)-\pi c_x > 0. \tag{9.14}$$

The optimal controls * are then given by:

$$t^*(\pi,\lambda) = \frac{\lambda\beta(1-\pi)(c_v(\hat{p}-k+\kappa)+c_x\kappa)-c_x\kappa\pi}{2\lambda\beta(1-\pi)(c_v+c_x)-c_x\pi} \tag{9.15}$$

$$t^{a*}(\pi,\lambda) = \frac{1}{2\lambda\beta(1-\pi)(c_v+c_x)-c_x\pi}\big[c_x\kappa(1-\pi) + \lambda\beta(1-\pi)(c_v(\hat{p}-k+\kappa)+c_x\kappa)-c_v(\hat{p}-k-\kappa)\big] \tag{9.16}$$

Otherwise, there are no interior optimal controls. In what follows we assume that (9.14) is satisfied along the optimal path. It can be easily checked that this is the case at the equilibrium discussed below.

The difference between the optimal announced and realized tax levels is given by:

$$t^{a*}-t^* = \frac{\kappa-t^*}{\lambda\beta(1-\pi)}. \tag{9.17}$$

Hence the optimal announced tax exceeds the realized one if and only if $t^*<\kappa$. As in the static case, we restrict the analysis to the environmentally friendly case $t^*<\kappa$ and assume that (9.12) holds.

According to Pontriagins' Maximum Principle, an optimal solution $(\pi(\tau),\lambda(\tau))$ has to satisfy the state dynamics (9.8) plus:

$$\dot{\lambda} = \rho\lambda - \frac{\partial H(t^{a*}(\pi,\lambda),t^*(\pi,\lambda),\pi,\lambda)}{\partial\pi}, \tag{9.18}$$

$$\lim_{\tau\to\infty} e^{-\rho\tau}\lambda(\tau)=0. \tag{9.19}$$

In our case the co-state dynamics are given by:

$$\dot{\lambda} = \rho\lambda - \frac{\kappa - t}{2c_v}(t^a - t) - \lambda\beta(1 - 2\pi)\left(-\frac{1}{4c_v}(t^a - t)^2 + \delta\right). \tag{9.20}$$

In order to analyze the long run behavior of the system we now provide a characterization of the steady states for different values of the public flexibility parameter β. Due to the structure of the state dynamics there are always trivial steady states for $\pi = 0$ and $\pi = 1$. For $\pi = 0$ the announcement is irrelevant. Its optimal value is therefore indeterminate. The optimal tax level is $t = \kappa$. For $\pi = 1$, the concavity condition (9.14) is violated and the optimal controls are like in the static case. In the following, we restrict our attention to $\pi < 1$. In Proposition 1, we first show under which conditions there exists an interior steady state where Believers coexist with Non-Believers. Furthermore, we discuss the stability of the steady states.

PROPOSITION 9.1 *Steady states and their stability:*

(i) For $0 < \beta \leq \frac{\rho}{2\delta}$ there exists no interior steady state. The steady state at $\pi = 0$ is stable.

(ii) For $\beta > \frac{\rho}{2\delta}$ there exists a unique stable interior steady state π^+ with:

$$\pi^+ = 1 - \frac{\rho}{2\beta\delta}. \tag{9.21}$$

The steady state at $\pi = 0$ is unstable.

Proof. We prove the existence and the stability of the interior steady state. The claims about the boundary steady state at $\pi = 0$ follow directly.

Equation (9.8) implies that, in order for $\dot{\pi} = 0$, to hold one must have:

$$(t^{a*} - t^*)^2 = 4c_v\delta. \tag{9.22}$$

That is, taking into account (9.17):

$$t^* = \kappa - 2\lambda^+\beta(1 - \pi^+)\sqrt{c_v\delta}. \tag{9.23}$$

Equation (9.22) implies that $\dot{\lambda} = 0$ is satisfied iff:

$$\rho\lambda - \frac{\kappa - t^*}{2c_v}(t^{a*} - t^*) = 0. \tag{9.24}$$

Using (9.17) for $t^{a*} - t^*$ in (9.24) we obtain the condition:

$$t^* = \kappa - \sqrt{2\rho\beta(1 - \pi^+)c_v}. \tag{9.25}$$

Combining (9.23) and (9.25) shows that

$$2\lambda^+\beta(1-\pi^+)\sqrt{c_v\delta} = \sqrt{2\rho\beta(1-\pi^+)c_v}$$

must hold at an interior steady state. This condition is equivalent to (9.21). For $\beta \leq \frac{\rho}{2\delta}$, (9.21) becomes smaller or equal zero. Therefore, an interior steady state is only possible for $\beta > \frac{\rho}{2\delta}$.

Using (9.21, 9.15, 9.23), one obtains for the value of the co-state at the steady state:

$$\lambda^+ = \frac{\delta\beta(c_x(2\sqrt{c_v\delta}+\kappa) + c_v(k+\kappa-p)) - \rho c_x\sqrt{c_v\delta}}{2\rho\beta\sqrt{c_v\delta}(c_v+c_x)}. \tag{9.26}$$

To determine the stability of the interior steady state we investigate the Jacobian matrix J of the canonical system (9.8, 9.20) at the steady state. The eigenvalues of J are given by:

$$e_{1,2} = \frac{tr(J) \pm \sqrt{tr(J)^2 - 4det(J)}}{2}.$$

Therefore, the steady state is saddle point stable if and only if the determinant of J is negative. Inserting (9.15, 9.16) into the canonical system (9.8, 9.20), taking the derivatives with respect to (π, λ) and then inserting (9.21, 9.26) gives the matrix J. Tedious calculations show that its determinant is given by:

$$det(j) = C\left(-\sqrt{\delta\beta}(c_v(k+\kappa-p) + c_x\kappa) - c_x\sqrt{c_v}(2\beta\delta - \rho)\right),$$

where C is a positive constant. The first of the two terms in the bracket is negative due to the assumption (9.12), the second is negative whenever $\beta > \frac{\rho}{2\delta}$. Hence, $det(J) < 0$ whenever an interior steady state exists, implying that the interior steady state is always stable. For $\beta = \frac{\rho}{2\delta}$ this stable steady state collides with the unstable one at $\pi = 0$. The steady state at $\pi = 0$ therefore becomes stable for $\beta \leq \frac{\rho}{2\delta}$. □

Since there is always only one stable steady state and since cycles are impossible in a control problem with a one-dimensional state, we can conclude from Proposition 9.1 that the stable steady state is always a global attractor. Convergence to the unique steady state is always monotone. The long run fraction of believers in the population is independent from the original level of trust. From (9.21) one recognizes that it decreases with ρ. An impatient Regulator will not attempt to build up a large proportion of Bs since the time and efforts needed now for an additional increase of π weighs heavily compared to the future benefits. By contrast, π^+ is increasing in β and δ. A high flexibility β of the firms means that the cumulated loss of potential utility occurred by the Regulator *en route* to π^+ will be small and easily compensated by the gains in the vicinity of and at π^+. Reinforcing this, the Regulator does not have to make Bs much better off than NBs in order to insure a fast reaction. As a result, for β large, the equilibrium π^+ is characterized by a large proportion of Bs and provide high profits respectively utility to all players. A high learning cost δ means that the Regulator can make the Bs better off than the NBs at low or no cost, implying again a high value of π at the steady state.

Note that it is never optimal for the Regulator to follow a policy that would ultimately insure that all firms are Believers, $\pi^+ = 1$. There are two concurrent reasons for that. On the one hand, as π increases, the Regulator has to deviate more and more from the statically optimal solution $t^{a\$}(\pi)$, $t^{\$}(\pi)$ to make believing more profitable than non-believing. On the other hand, the beliefs dynamics slow down. Thus, the discounted benefits from increasing π decrease.

For $\rho = 0.8$, $c_v = 5$, $c_x = 3$, $\delta = 0.15$, $p = 6$, $k = 4$, $\kappa = 3$ e.g., the profits and utility at the steady state $\pi^+ = 0.733333$ are $g^b = g^{nb} = 1.43785$, $\phi = 2.33043$. This steady state Pareto-dominates the fully rational equilibrium with $\pi = 0$, where $g^{nb} = 1.32708$, $\phi = 2.30844$. It also dominates the equilibrium attained when the belief dynamics (9.8) holds but the Regulator maximizes in each period his instantaneous utility ϕ instead of Φ. At this last equilibrium, $\pi = 0.21036$, $g^b = g^{nb} = 1.375$, $\phi = 2.23689$. Note that the last two equilibria cannot be compared, since the latter provides a higher profit to the firms but a lower utility to the Regulator.

This ranking of equilibria is robust with respect to parameter variations. A clear message emerges. As we contended at the beginning of this paper the suggested solution Pareto-improves on the static Nash equilibrium. This solution implies both a beliefs dynamics among the firms and a farsighted Regulator. A farsighted Regulator without beliefs dynamics is pointless. Beliefs dynamics with a myopic Regulator lead to a more modest Pareto-improvement. But it is the combination of beliefs dynamics and farsightedness that Pareto-dominates all other solutions.

4.2 The influence of public flexibility

An interesting question is whether the Regulator would prefer a population that reacts quickly to profit differences, shifting from Believing to Non-Believing or vice versa in a short time, or if it would prefer a less reactive population. In other words, would the Regulator prefer a high or a low value of β?

From Proposition 9.1 we know that the long run fraction of Believers is given by:

$$\pi^+ = \max\left[0, 1 - \frac{\rho}{2\beta\delta}\right].$$

A minimum level $\beta > \frac{\rho}{2\delta}$ of flexibility is necessary for the system to converge towards an interior steady state with a positive fraction of Bs. For β greater than $\frac{\rho}{2\delta}$, the fraction of Bs at the steady state increases with β, converging towards $\pi = 1$ as β goes to infinity. One might think that, since the Regulator always prefers a high proportion of Bs at equilibrium, he would also prefer a high value of β. Stated in a more formal manner, one might expect that the value function of the Regulator, $V^R(\pi_0)$, increases with β regardless of π_0. This, however, is not the case. The dependence of $V^R(\pi_0)$ on β is non-monotone and depends crucially on π_0. An analytical characterization of $V^R(\pi_0)$ being impossible, we use a numerical example to illustrate that point. The results are very robust with respect to parameter variations.

Figure 9.2 shows the steady state value $\pi^+ = \pi^+(\beta)$ of π for $\beta \in [1, 30]$. Figure 9.3 compares[2] $V^R(0.2)$ and $V^R(0.8)$ for the same values of β. The other parameter values are as before $\rho = 0.8$, $c_v = 5$, $c_x = 3$, $\delta = 0.15$, $p = 6$, $k = 4$, $\kappa = 3$ in both cases.

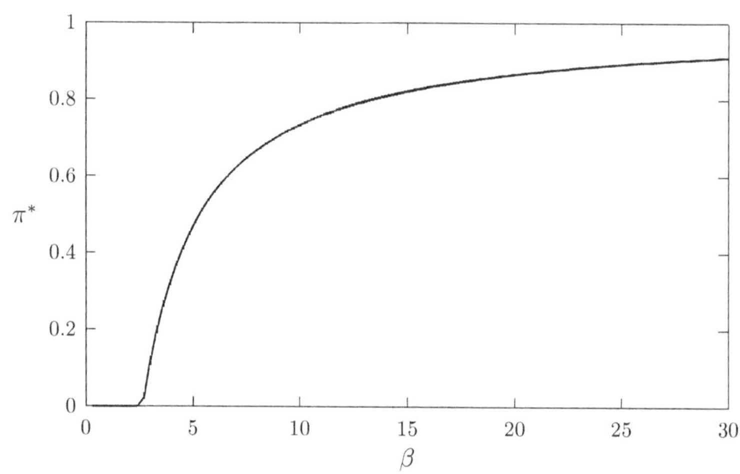

Figure 9.2. The proportion π of Believers at the steady state for $\beta \in [1, 30]$.

[2]The numerical calculations underlying this figure were carried out using a powerful proprietary computer program for the treatment of optimal control problems graciously provided by Lars Grüne, whose support is most gratefully acknowledged. See Grüne and Semmler (2002).

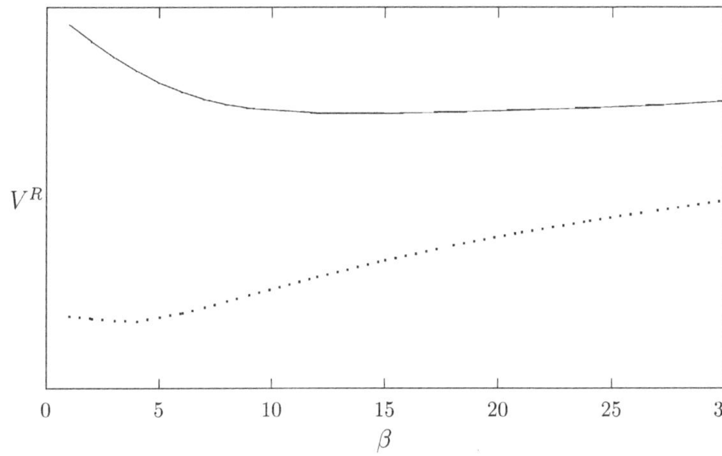

Figure 9.3. The value function of the Regulator for $\pi = 0.2$ (dotted line) and $\pi = 0.8$ (solid line) for $\beta \in [1, 30]$.

Figure 9.3 reveals that one always has $V^R(0.8) > V^R(0.2)$, reflecting the general result that $V^R(\pi_0)$ is increasing in π_0 for any β. Both value functions are not monotone in β but U-shaped. Combining Figure 9.2 and Figure 9.3 shows that the minimum of $V^R(\pi_0)$ is always attained for the value of β at which the steady state value $\pi^+(\beta)$ coincides with the initial fraction of Believers π_0. This result is quite intuitive. If $\pi^+(\beta) < \pi_0$, it is optimal for the Regulator to reduce π over time. The Regulator does it by announcing a tax t^a much greater than the tax t he will implement, making the Bs worse off than the NBs, but also increasing his own instantaneous benefits ϕ. Thus, R prefers that the convergence towards the steady state be as slow as possible. That is, V^R is decreasing in β. On the other hand, if $\pi^+ > \pi_0$, it is optimal for R to increase π over time. To do so, he must follow a policy that makes the Bs better off than the NBs, but this is costly in terms of his instantaneous objective function ϕ. It is therefore better for R if the firms react fast to the profit difference. The value function V^R increases with β. Summarizing, the Regulator prefers (depending on π_0) to be confronted with *either* very flexible *or* very inflexible firms. In-between values of β provide him smaller discounted benefit streams.

Whether a low or a high flexibility is preferable for R depends on the initial fraction π_0 of Believers. Our numerical analysis suggests that a Regulator facing a small π_0 prefer large values of β, whereas he prefers a low value of β when π_0 is large. This result may follow from the specific functional form used in the model rather than reflect any fundamental property of the solution.

If $\beta = 0$, the proportion of Bs remains fixed over time at π_0 – any initial value of $\pi_0 \in (0,1)$ corresponds to a stable equilibrium. The Pareto-improving character of the inner equilibrium then disappears. Given $\beta = 0$, condition (9.14) is violated. The Regulator can announce any tax level without having to fear negative long term consequences. Thus, it is in his interest to exploit the gullibility of the Bs to the maximum. To obtain a meaningful, Pareto-improving solution, some flexibility is necessary that assures that the firms are not kept captive of beliefs that penalize them. Only then will the Regulator be led to take into account the Bs interest.

5. Conclusions

The starting point of this paper is a situation frequently encountered in environmental economics (and similarly in other economic contexts as well): If all firms are perfectly rational Non-Believers who make perfect predictions of the Regulator's actions and discard the Regulators announcements as cheap talk, standard optimizing behavior leads to a Pareto-inferior outcome, although there are no conflicts of interest between the different firms and although the objectives of the firms and of the Regulator largely concur. We show that, in a static world, the existence of a positive fraction of Believers who take the Regulator's announcement at face value Pareto-improves the economic outcome. This property crucially hinges on the fact that the firms are atomistic and thus do not anticipate the collective impact of their individual decisions.

The static model is extended by assuming that the proportion of Believers and Non-Believers changes over time depending on the difference in the profits made by the two types of firms. The Regulator is assumed to recognize his ability to influence the evolution of the proportion of Believers by his choice of announced and realized taxes, and to be interested not only in his instantaneous but also in his future utility. It is shown that a rational Regulator will never steer the economy towards a Pareto-optimal equilibrium. However, his optimal policy may lead to a stable steady state with a strictly positive proportion of Believers that is Pareto-superior to the equilibrium where all agents perfectly anticipate his actions. Prerequisites therefore are a sufficiently patient Regulator and firms that occur sufficiently high costs for building perfect anticipations of the government actions and/or are suitably flexible, i.e., switch adequately fast between Believing and Non-Believing. The conjunction of beliefs dynamics for the firms *and* of a farsighted Regulator allows for a larger Pareto-improvement than either only beliefs dynamics or only farsightedness. Depending upon the initial proportion of Believers, the

Regulator is better off if the firms are very flexible or very inflexible. Intermediate values of the flexibility parameter are never optimal for the Regulator.

References

Abrego, L. and Perroni, C. (1999). Investment Subsidies and Time-Consistent Environmental Policy. Discussion paper, University of Warwick, U.K.

Barro, R. and D. Gordon. (1983). Rules, discretion and reputation in a model of monetary policy. *Journal of Monetary Economics*, 12: 101–122.

Batabyal, A. (1996a). Consistency and optimality in a dynamic game of pollution control I: Competition. *Environmental and Resource Economics*, 8:205–220.

Batabyal, A. (1996b). Consistency and optimality in a dynamic game of pollution control II: Monopoly. *Environmental and Resource Economics*, 8:315–330.

Biglaiser, G., Horowitz, J., and Quiggin, J. (1995). Dynamic pollution regulation. *Journal of Regulatory Economics*, 8:33–44.

Dawid, H. (1999). On the dynamics of word of mouth learning with and without anticipations. *Annals of Operations Research*, 89:273–295.

Dawid, H. and Deissenberg, C. (2004). On the efficiency-effects of private (dis-)trust in the government. Forthcoming in *Journal of Economic Behavior and Organization.*

Deissenberg, C. and Alvarez Gonzalez, F. (2002). Pareto-improving cheating in a monetary policy game. *Journal of Economic Dynamics and Control*, 26:1457–1479.

Dijkstra, B. (2002). Time Consistency and Investment Incentives in Environmental Policy. *Discussion paper 02/12*, School of Economics, University of Nottingham, U.K.

Gersbach, H., and Glazer, A. (1999). Markets and regulatory hold-up problems. *Journal of Environmental Economics and Management*, 37: 151–64.

Grüne, L., and Semmler, W. (2002). Using Dynamic Programming with Adaptive Grid Scheme for Optimal Control Problems in Economics. Working paper, Center for Empirical Macroeconomics, University of Bielefeld, Germany, 2002. Forthcoming in *Journal of Economic Dynamics and Control.*

Hofbauer, J. and Sigmund, K. (1998). *Evolutionary Games and Population Dynamics.* Cambridge University Press.

Kydland, F. and Prescott, E. (1977). Rules rather than discretion: The inconsistency of optimal plans. *Journal of Political Economy*, 85: 473–491.

Marsiliani, L. and Renström, T. (2000). Time inconsistency in environmental policy: Tax earmarking as a commitment solution. *Economic Journal*, 110:C123–C138.

McCallum, B. (1997). Critical issues concerning central bank independence. *Journal of Monetary Economics*, 39:99–112.

Petrakis, E. and Xepapadeas, A. (2003). Location decisions of a polluting firm and the time consistency of environmental policy. *Resource and Energy Economics*, 25:197–214.

Simaan, M. and Cruz, J.B. (1973a). On the Stackelberg strategy in nonzero-sum games. *Journal of Optimization Theory and Applications*, 11:533–555.

Simaan, M. and Cruz, J.B. (1973b). Additional aspects of the Stackelberg strategy in nonzero-sum games. *Journal of Optimization Theory and Applications*, 11:613–626.

Vallée, T., Deissenberg, C., and Başar, T. (1999). Optimal open loop cheating in dynamic reversed linear-quadratic Stackelberg games. *Annals of Operations Research*, 88:247–266.

Vallée, T. (1998). Comparison of Different Stackelberg Solutions in a Deterministic Dynamic Pollution Control Problem. Discussion Paper, LEN-C3E, University of Nantes, France.

Chapter 10

A TWO-TIMESCALE STOCHASTIC GAME FRAMEWORK FOR CLIMATE CHANGE POLICY ASSESSMENT

Alain Haurie

Abstract In this paper we show how a multi-timescale hierarchical non-cooperative game paradigm can contribute to the development of integrated assessment models of climate change policies. We exploit the well recognized fact that the climate and economic subsystems evolve at very different time scales. We formulate the international negotiation at the level of climate control as a piecewise deterministic stochastic game played in the "slow" time scale, whereas the economic adjustments in the different nations take place in a "faster" time scale. We show how the negotiations on emissions abatement can be represented in the slow time scale whereas the economic adjustments are represented in the fast time scale as solutions of general economic equilibrium models. We provide some indications on the integration of different classes of models that could be made, using an hierarchical game theoretic structure.

1. Introduction

The design and assessment of climate change policies, both at national and international levels, is a major challenge of the 21st century. The problem of climate change concerns all the developed and developing countries that together influence the same climate system through their emissions of greenhouse gases (GHG) associated with their economic development, and which will be differently affected by the induced climate change. Therefore the problem is naturally posed in a multi-agent dynamic decision analysis setting. The Kyoto and Marrakech protocol and the eventual continuation of the international negotiations in this domain illustrate this point. To summarize the issues let us quote Edwards et al., 2005 who conclude as follows their introductory paper of a book[1] on *the coupling of climate and economic dynamics*:

> *- Knowledge of the dynamics of the carbon cycle and the forcing by greenhouse gases permits us to predict global climate change due to anthropogenic influences on a time scale of a century (albeit with uncertainty).*
> *- Stabilizing the temperature change to an acceptable level calls for a drastic worldwide reduction of the GHG emissions level (to around a quarter of the 1990 level) over the next 50 years.*
> *- Climate inertia implies that many of those who will benefit (suffer) most from our mitigation actions (lack of mitigation) are not yet born.*
> *- The climate change impacts may be large and unequally distributed over the planet, with a heavier toll for some DCs.*
> *- The rapid rise of GHG emissions has accompanied economic development since the beginning of the industrial era; new ways of bypassing the* Kuznets curve *phenomenon have to be found for permitting DCs to enter into a necessary global emissions reduction scheme.*
> *- The energy system sustaining the world economy has to be profoundly modified; there are possible technological solutions but their implementations necessitate a drastic reorganization of the infrastructures with considerable economic and geopolitical consequences.*
> *- The policies to implement at international level must take explicitly into account the intergenerational and interregional equity issues.*
> *- The magnitude of the changes that will be necessary impose the implementation of market-based instruments to limit the welfare losses of the different parties (groups of consumers) involved.*
> *The global anthropogenic climate change problem is now relatively well identified... viable policy options will have to be designed as* equilibrium solutions to dynamic games played by different groups of nations. *The paradigm of dynamic games is particularly well suited to represent the conflict of a set of economic agents (here the nations) involved jointly in the control of a complex dynamical system (the climate), over a very long time horizon, with distributed and highly unequal costs and benefits...*

Integrated assessment models (IAMs) are the main tools for analyzing the interactions between climatic change and socio-economic development (see the recent survey by Toth, 2005). In a first category of IAMs one finds the models based on a paradigm of optimal economic growth à la Ramsey, 1928, to describe the world economy, associated with a simplified representation of the climate dynamics, in the form of a set of differential equations. The main models in this category are DICE94 (Nordhaus, 1994), DICE99 and RICE99 (Nordhaus and Boyer, 2000), MERGE (Manne et al., 1995), and more recently ICLIPS (Toth et al., 2003 and Toth, 2005). In a second category of IAMs one finds a represen-

[1]We refer the reader to this book for a more detailed presentation of the climate change policy issues.

tation of the world economy in the form of a computable general equilibrium model (CGEM), whereas the climate dynamics is studied through a (sometimes simplified) land-and-ocean-resolving (LO) model of the atmosphere, coupled to a 3D ocean general circulation model (GCM). This is the case of the MIT Integrated Global System Model (IGSM) which is presented by Prinn et al., 1999 as *"designed for simulating the global environmental changes that may arise as a result of anthropogenic causes, the uncertainties associated with the projected changes, and the effect of proposed policies on such changes"*.

Early applications of dynamic game paradigms to the analysis of GHG induced climate change issues are reported in the book edited by Carraro and Filar, 1995. Haurie, 1995, and Haurie and Zaccour, 1995, propose a general differential game formalism to design emission taxes in an imperfect competition context. Kaitala and Pohjola, 1995, address the issue of designing transfer payment that would make international agreements on greenhouse warming sustainable. More recent dynamic game models dealing with this issue are those by Petrosjan and Zaccour, 2003, and Germain et al., 2003. Carraro and Siniscalco, 1992, Carraro and Siniscalco, 1993, Carraro and Siniscalco, 1996, Buchner et al., 2005, have studied the dynamic game structure of Kyoto and post-Kyoto negotiations with a particular attention given to the "issue-linking" process, where agreement on the environmental agenda is linked with other possible international trade agreement (R&D sharing example). These game theoretic models have used very simple qualitative models or adaptations of the DICE or RICE models to represent the climate-economy interactions.

Another class of game theoretic models of climate change negotiations has been developed on the basis of IAMs incorporating a CGEM description of the world economy. Kemfert, 2005, uses such an IAM (WIAGEM which combines a CGEM with a simplified climate description) to analyze a game of climate policy cooperation between developed and developing nations. Haurie and Viguier, 2003b, Bernard et al., 2002, Haurie and Viguier, 2003a, Viguier et al., 2004, use a two-level game structure to analyze different climate change policy issues. At a lower level the World or European economy is described as a CGEM, at an upper level a negotiation game is defined where strategies correspond to strategic negotiation decisions taken by countries or group of countries in the Kyoto-Marrakech agreement implementation.

In this paper we propose a general framework based on a multi-timescale stochastic game theoretic paradigm to build IAMs for global climate change policies. The particular feature that we shall try to represent in our modeling exercise is the difference in timescales between the interacting economic and climate systems. In Haurie, 2003, and Haurie, 2002, some considerations have already been given to that issue. In the present paper we propose to use the formalism of hierarchical control and singular perturbation theory to take into account these features (we shall use in particular the formalism developed by Filar et al., 2001).

The paper is organized as follows: In Section 2 we propose a general modeling framework for the interactions between economic development and climate change. In particular we show that the combined economy-climate dynamical system is characterized by two timescales; in Section 3 we formulate the long term game of economy and climate control which we call *game of sustainable development.* In Section 4 we exploit the multi timescale structure of the controlled system to define a reduced order game, involving only the slow varying climate related variables. In Section 5 we propose a research agenda for developing IAMs based on the formalism proposed in this paper.

2. Climate and economic dynamics

In this section we propose a general control-theoretic modeling framework for the representation of the interactions between the climate and economic systems.

2.1 The linking of climate and economic dynamics

2.1.1 The variables of the climate-economy system.

We view the climate and the economy as two dynamical systems that are coupled. We represent the state of the economy at time t by an hybrid vector $\mathbf{e}(t) = (\zeta(t), \mathbf{x}(t))$ where $\zeta(t) \in \mathcal{I}$ is a discrete variable that represents a particular mode of the world economy (describing for example major technological advances, or major geopolitical reorganizations) whereas the variable $\mathbf{x}(t)$ represents physical capital stocks, production output, household consumption levels, etc. in different countries. We also represent the climate state as an hybrid vector $\mathbf{c}(t) = (\kappa(t), \mathbf{y}(t, \cdot))$, where $\kappa(t) \in \mathrm{L}$ is a discrete variable describing different possible climate modes, like e.g. those that may result from threshold events like the disruption of the thermohaline circulation or the melting of the ice-shields in Greenland or Antarctic, and $\mathbf{y}(t, \cdot) = (y(t, \omega)) : \omega \in \Omega)$ is a spatially distributed variable where the set Ω represents all the different locations on Earth where climate matters. The climate state variable $\mathbf{y}$ represents typically the average surface temperatures, the average precipitation levels, etc.

The world is divided into a set J of economic regions that we shall call *nations*. The linking between climate and economics occurs because climate change may cause damages, measured through a vulnerability index[2] $\upsilon_j(t)$ of region $j \in M$, and also because emissions abatement policies may exert a toll on the economy. The climate policies of nations is summarized by the state variables $z_j(t), j \in M$, which represent the cap on GHG emissions that they impose on their respective economies at time t. A control variable $u_j(t)$ will represent the abatement effort of region $j \in M$ which acts on the cap trajectory $z(\cdot)$ and through it on the evolution of the climate variables $\mathbf{c}(\cdot)$. We summarize the set of variables and indices in Table 10.1.

Table 10.1. List of variables in the climate economy model

Variable name	meaning
t	time index (fast)
$\tau = \frac{t}{\varepsilon}$	"stretched out" time index
ε	timescales ratio
$\omega \in \Omega$	spatial index
$j \in M = \{1, \ldots, m\}$	economic region (nation, player) index
$L_j(t)\, j \in M$	population in region $j \in M$ at time t
$\zeta(t) \in \mathcal{I}$	economic/geopolitical mode
$\kappa(t) \in \text{Ł}$	climate mode
$\xi(t) = (\zeta(t), \kappa(t)) \in \mathcal{I} \times \text{Ł}$	pooled mode indicator
$x_j(t) \in X_j$	economic state variable for Nation j
$\mathbf{x}(t) \in \mathbf{X}$	economic state variable (all countries)
$\mathbf{y}(t) = (y(t, \omega) \in Y(\omega))$	climate state variable
$z_j(t) \in Z_j$	cap on GHG emissions for Nation j
$\mathbf{z}(\mathbf{t}) \in \mathbf{Z}$	world global cap on GHG emissions
$\mathbf{s}(t) = (\mathbf{z}(t), \mathbf{y}(t))$	slow paced economy-climate state variables
$u_j(t) \in U_j$	GHG abatement effort of Nation j
$\upsilon_j(t) \in \Upsilon_j$	vulnerability indicator for Nation j

2.1.2 The dynamics. Our understanding of the carbon cycle and the temperature forcing due to the accumulation of CO_2 in the atmosphere is that the dynamics of anthropogenic climate change has a slow pace compared to the speed of adjustment in the "fast" economic systems. We shall therefore propose a modeling framework characterized by two timescales with a ratio ε between the fast and the slow one.

[2]This index is a functional of the climate state $\mathbf{c}(t)$.

Economic and climate mode dynamics. We assume that the economic and climate modes may evolve according to controlled jump processes described by the following transition rates

$$\begin{aligned} \mathrm{P}[\zeta(t+\delta)=\ell|\zeta(t)=k \text{ and } \mathbf{x}(t)] &= p_{k\ell}(\mathbf{x}(t))\delta+o(\delta) &(10.1)\\ \lim_{\delta\to 0}\frac{o(\delta)}{\delta} &= 0. &(10.2)\end{aligned}$$

and

$$\begin{aligned} \mathrm{P}[\kappa(t+\delta)=\iota|\kappa(t)=i \text{ and } \mathbf{y}(t)] &= q_{i\iota}(\mathbf{y}(t))\delta+o(\delta) &(10.3)\\ \lim_{\delta\to 0}\frac{o(\delta)}{\delta} &= 0. &(10.4)\end{aligned}$$

Combining the two jump processes we have to consider the pooled mode $(k,i)\in\mathcal{I}\times\mathrm{L}$ and the possible transitions $(k,i)\to(\ell,i)$ with rate $p_{k\ell}(\mathbf{x}(t))$ or transitions $(k,i)\to(i,\iota)$ with rate $q_{i\iota}(\mathbf{y}(t))$.

Climate dynamics. The climate state variable is spatially distributed (it is typically the average surface temperatures and the precipitation levels on different points of the planet). The climate state evolves according to complex earth system dynamics with a forcing term which is directly linked with the global GHG emissions levels $\mathbf{z}(t)$. We may thus represent the climate state dynamics as a distributed parameter system obeying a (generalized) differential equation which is indexed over the mode (climate regime) $\kappa(t)$

$$\begin{aligned} \dot{y}(t,\cdot) &= g^{\xi(t)}(\mathbf{z}(t),y(t,\cdot)) &(10.5)\\ y(0,\cdot) &= y^o(\cdot). &(10.6)\end{aligned}$$

Economic dynamics. The world economy can be described as a set of interconnected dynamical systems. The state dynamics of each region $j\in M$ is depending on the state of the other regions due to international trade and technology transfers. Each region is also characterized by the current cap on GHG emissions $z_j(t)$ and by its abatement effort $u_j(t)$. The nations $j\in M$ occupy respective territories $\Omega_j\subset\Omega$. The economic performance of Nation j will be affected by its vulnerability to climate change. In the most simple way one can represent this indicator as a vector $v_j(t)\in\mathbb{R}^p$ (e.g. the percentage loss of output for each of the p economic sectors) defined as a functional (e.g. an integral over Ω_j) of a distributed damage function $\tilde{d}_j^{\kappa(t)}(\omega,\tilde{y}(t,\omega))$ associated with climate mode $\xi(t)$ and distributed climate state $y(t,\cdot)$.

$$v_j(t)=\int_{\Omega_j}\tilde{d}_j^{\xi(t)}(\omega,\tilde{y}(t,\omega))\,d\omega. \qquad (10.7)$$

Now the dynamics of the nation-j economy is described by the differential equation

$$\dot{x}_j(t) = \frac{1}{\varepsilon} f_j^{\zeta(t)}(t, \mathbf{x}(t), v_j(t), z_j(t)) \quad j \in M \tag{10.8}$$

$$x_j(0) = x_j^o \tag{10.9}$$

$$\mathbf{x}(t) = (x_j(t))_{j \in M}. \tag{10.10}$$

We have indexed the velocities $f_j^{\zeta(t)}(t, \mathbf{x}(t), v_j(t), z_j(t))$ over the discrete economic/geopolitical state $\zeta(t)$ since these different possible modes have an influence on the evolution of the economic variables[3]. The factor $\frac{1}{\varepsilon}$ where ε is small expresses the fact that the economic adjustments take place in a much faster timescale than the ones for climate or modal states. Notice also that the feedback of climate on the economies is essentially represented by the influence of the vulnerability variables on the economic dynamics of different countries.

The GHG emission cap variable $z_j(t)$ for nation $j \in M$ is a controlled variable which evolves according to the dynamics

$$\dot{z}_j(t) = h_j^{\zeta(t)}(x_j(t), z_j(t), u_j(t)) \tag{10.11}$$

$$z_j(0) = z_j^o \tag{10.12}$$

where $u_j(t)$ is the reduction effort. This formulation should permit the analyst to represent the R&D and other adaptation actions that have been taken by a government in order to implement a GHG emissions reduction. Again we have indexed the velocity $h_j^{\zeta(t)}(x_j(t), z_j(t), u_j(t))$ over the economic/geopolitical modes $\zeta(t) \in \mathcal{I}$. The absence of factor $\frac{1}{\varepsilon}$ in front of velocities in Eqs. (10.11) indicates that we assume a slow pace in the emissions cap adjustments.

2.2 A two timescale system

We summarize below the equations characterizing the climate-economy system.

$$\dot{x}_j(t) = \frac{1}{\varepsilon} f_j^{\zeta(t)}(t, \mathbf{x}(t), z_j(t), v_j(t)) \quad j \in M$$

$$x_j(0) = x_j^o \quad j \in M$$

[3]We could have also indexed these state equations over the pooled modal state indicator $(\zeta(t), \kappa(t))$, assuming that the climate regime might also have an influence on the economic dynamics. For the sake of simplifying the model structure we do not consider the climate regime influence on economic dynamics.

$$
\begin{aligned}
\mathbf{x}(t) &= (x_j(t))_{j\in M} \\
\dot{z}_j(t) &= h^{\zeta(t)}(x_j(t), z_j(t), u_j(t)) \\
\mathbf{z}(t) &= \sum_{j\in M} z_j(t) \\
z_j(0) &= z_j^o \quad j \in M \\
\dot{y}(t,\cdot) &= g^{\kappa(t)}(\mathbf{z}(t), y(t,\cdot)) \\
y(0,\cdot) &= y^o(\cdot) \\
v_j(t) &= \int_{\Omega_j} \tilde{d}_j^{\kappa(t)}(\omega, \tilde{y}(t,\omega))\, d\omega \\
\mathrm{P}[\zeta(t+\delta) = \ell | \zeta(t) = k \text{ and } \mathbf{x}(t)] &= p_{k\ell}(\mathbf{x}(t))\delta + o(\delta) \quad k,\ell \in \mathcal{I} \\
\lim_{\delta\to 0} \frac{o(\delta)}{\delta} &= 0 \quad k,\ell \in \mathcal{I} \\
\mathrm{P}[\kappa(t+\delta) = \iota | \kappa(t) = i \text{ and } \mathbf{y}(t)] &= q_{i\iota}(\mathbf{y}(t))\delta + o(\delta) \quad k,\ell \in \text{Ł} \\
\lim_{\delta\to 0} \frac{o(\delta)}{\delta} &= 0 \quad i,\iota \in \text{Ł}.
\end{aligned}
$$

In the parlance of control systems we have constructed a two-timescale piecewise deterministic setting. The time index t refers to the "climate" and slow paced socio-economic evolutions. The parameter ε is the timescale ratio between the slow and fast timescales. The stretched out timescale is defined as $\tau = t/\varepsilon$. It will be used when we study the fast adjustments of the economy given climate, socio-economic and abatement conditions.

3. The game of sustainable development

In this section we describe the dynamic game that nations[4] play when they negotiate to define the GHG emissions cap in international agreements. The negotiation process is represented by the control variables $u_j(t)$, $j \in M$. We first define the strategy space then the payoffs that will be considered by the nations and we propose a characterization of the equilibrium solutions.

3.1 Strategies

We assume that the nations use piecewise open-loop (POL) strategies. This means that the controls $u_j(\cdot)$ are functions of time that can be adapted after each jump of the $\xi(\cdot)$ process. A strategy for Nation j is therefore a mapping from $(t, k, i, \mathbf{x}, \mathbf{z}, \mathbf{y}) \in [0,\infty) \times \mathcal{I} \times \text{Ł} \times \mathbf{X} \times \mathbf{Z} \times \mathbf{Y}$

[4] We identify the players as being countries (possibly group of countries in some applications).

into the class $\mathcal{U}_j^t$ of control functions $u_j(\tau) : \tau > t$. We denote $\underline{\gamma}$ the vector strategy of the m nations.

According to these strategies the nations (players) decide about the pace at which they will decrease their total GHG emissions over time. These decision can be revised if a modal change occurs (new climatic regime or new geopolitical configuration).

3.2 Payoffs

In this model, the nations control the system by modifying their GHG emission caps. As a consequence the climate change is altered, the damages are controlled and the economies are adapting in the fast timescale through the market dynamics. Climate and economic state have also an influence on the probability of switching to a different modal state (geopolitical or climate regime). The payoff to nations is based on a discounted sum of their welfare[5].

So, we associate with an initial state $(\xi^o; \mathbf{x}^o, \mathbf{z}^o, \mathbf{y}^o)$ at time $t^o = 0$ and a strategy m-tuple $\underline{\gamma}$ a payoff for each nation $j \in M$ defined as follows

$$\begin{aligned} J_j^{\xi^o}(\underline{\gamma}; t^o, \mathbf{x}^o, \mathbf{z}^o, \mathbf{y}^o) \;=\; & \mathrm{E}_{\underline{\gamma}}\Bigg[\int_{t^o}^{\infty} e^{-\rho_j(t-t^o)} W_j^{\xi(t)} \\ & (L_j(t), x_j(t), z_j(t), v_j(t), u_j(t))\, dt \\ & |\xi(0) = \mu, \mathbf{x}(0) = \mathbf{x}^o, \mathbf{z}(0) = \mathbf{z}^o, \mathbf{y}(0) = \mathbf{y}^o\Bigg], \\ & j = 1, \ldots, m \end{aligned} \tag{10.13}$$

where $W_j^{(k,i)}(L_j, x_j, z_j, v_j, u_j)$ represents the welfare of nation j at a given time when the population level is L_j, the cap level is z_j, the vulnerability is v_j and the cap reduction effort is u_j, while the economy/geopolitical-climate mode is (k, i). The parameter ρ_j is the discount rate used in nation j.

3.3 Equilibrium and DP equations

It is natural to assume that the nations will play a Nash equilibrium, i.e. a strategy m-tuple $\underline{\gamma}^*$ such that for each nation j, the strategy γ_j^* is the best reply to the strategy choice of the other nations. The equilibrium strategy m-tuple should therefore satisfy, for all $j \in M$ and

[5] The simplest expression of it could be $L(t)\mathcal{U}(C(t)/L(t))$ where $L(t)$ is the population size, $C(t)$ is the total consumption by households and $\mathcal{U}(\cdot)$ is the utility of per-capita consumption.

all initial conditions[6] $(t, k, i, \mathbf{x}^o, \mathbf{s}^o)$

$$V_j^*(k, i; t^o, \mathbf{x}^o, \mathbf{s}^o) = J_j^{(k,i)}(\underline{\gamma}^*; t^o, \mathbf{x}^o, \mathbf{s}^o) \geq J_j^{(k,i)}([\underline{\gamma}^{*-j}, \gamma_j]; t^o, \mathbf{x}^o, \mathbf{s}^o), \tag{10.14}$$

where the expression $[\underline{\gamma}^{*-j}, \gamma_j]$ stands for the strategy m-tuple obtained when Nation j modifies unilaterally its strategy choice.

The climate-economy system model has a piecewise deterministic structure. We say that we have defined a *piecewise deterministic differential game* (PDDG). A dynamic programming (DP) functional equation will characterize the optimal payoff function. It is obtained by applying the Bellman optimality principle when one considers the time T of the first jump of the ξ-process after the initial time t^o. This yields

$$\begin{aligned} V_j^*(k, i; t^o, \mathbf{x}^o, \mathbf{s}^o) = \text{equil.}\Big(\mathrm{E}\Big[\int_{t^o}^{T} & e^{-\rho_j(t-t^o)} W_j^{(k,i)} \\ & (L_j(t), x_j(t), z_j(t), v_j(t), u_j(t))\, dt \\ & + e^{-\rho_j(T-t^o)} V_j^*(\zeta(T), \kappa(T); T, \mathbf{x}(T), \mathbf{s}(T))\Big]\Big) \quad j \in M \end{aligned} \tag{10.15}$$

where the equilibrium is taken with respect to the trajectories defined by the solutions of the state equations

$$\begin{aligned} \dot{x}_j(t) &= \frac{1}{\varepsilon} f_j^k(t, \mathbf{x}(t), z_j(t), v_j(t)) \quad j \in M \\ x_j(0) &= x_j^o \quad j \in M \\ \mathbf{x}(t) &= (x_j(t))_{j\in M} \\ \dot{z}_j(t) &= h^k(x_j(t), z_j(t), u_j(t)) \\ \mathbf{z}(t) &= \sum_{j\in M} z_j(t) \\ z_j(0) &= z_j^o \quad j \in M \\ \dot{y}(t, \cdot) &= g^i(\mathbf{z}(t), y(t, \cdot)) \\ y(0, \cdot) &= y^o(\cdot) \\ v_j(t) &= \int_{\Omega_j} \tilde{d}_j^i(\omega, \tilde{y}(t, \omega))\, d\omega. \end{aligned}$$

The random time T of the next jump, with the new discrete state $(\zeta(T), \kappa(T))$ reached at this jump time are random events with probability obtained from the transition rates $p_{k\ell}(\mathbf{x}(t))$ and $q_{i\iota}(\mathbf{y}(t))$. We

[6] Here we use the notation $\mathbf{s} = (\mathbf{z}, \mathbf{y})$ for the slow varying continuous state variables.

can use this information to define a family of associated open-loop differential games.

3.4 A family of implicit OLE games

We recall here[7] that we can associate with a POL equilibrium in a PDDG a family of implicitly defined Open-Loop Equilibrium (OLE) problem for a class of deterministic differential games.

$$
\begin{aligned}
V_j^*(k, i; \mathbf{x}^o, \mathbf{s}^o) = \text{equil.}_{\underline{u}(\cdot)} \int_0^\infty & e^{-\rho_j t + \int_0^t \lambda^{(k,i)}(\mathbf{x}(s), \mathbf{y}(s))ds} \\
& \Big\{ L_j^{k,i}(x_j(t), z_j(t), v_j(t), u_j(t)) \\
& + \sum_{\ell \in \mathcal{I} - k} p_{k,\ell}(\mathbf{x}(s)) V_j^*((\ell, i); \mathbf{x}(t), \mathbf{s}(t)) \\
& + \sum_{\iota \in \text{Ł} - i} q_{i,\iota}(\mathbf{y}(s)) V_j^*((k, \iota); \mathbf{x}(t), \mathbf{s}(t)) \Big\} dt \quad j \in M
\end{aligned}
\tag{10.16}
$$

where the equilibrium is taken with respect to the trajectories defined by the solutions of the state equations

$$
\begin{aligned}
\dot{x}_j(t) &= \frac{1}{\varepsilon} f_j^k(t, \mathbf{x}(t), z_j(t), v_j(t)) \quad j \in M \\
x_j(0) &= x_j^o \quad j \in M \\
\mathbf{x}(t) &= (x_j(t))_{j \in M} \\
\dot{z}_j(t) &= h^k(x_j(t), z_j(t), u_j(t)) \\
\mathbf{z}(t) &= \sum_{j \in M} z_j(t) \\
z_j(0) &= z_j^o \quad j \in M \\
\dot{y}(t, \cdot) &= g^i(\mathbf{z}(t), y(t, \cdot)) \\
y(0, \cdot) &= y^o(\cdot) \\
v_j(t) &= \int_{\Omega_j} \tilde{d}_j^{\kappa(t)}(\omega, \tilde{y}(t, \omega))\, d\omega
\end{aligned}
$$

and where we have introduced the notation

$$
\lambda^{(k,i)}(\mathbf{x}(s), \mathbf{y}(s)) = \sum_{\ell \in \mathcal{I} - k} p_{k\ell}(\mathbf{x}(s)) + \sum_{\iota \in \text{Ł} - i} q_{i\iota}(\mathbf{y}(s)). \tag{10.17}
$$

[7]See Haurie, 1989, and Haurie and Roche, 1994, for details.

3.5 Economic interpretation

These auxiliary OLE problems offer an interesting economic interpretation. The nations that are competing in their abatement policies have to take into account the combined dynamic effect on welfare of the climate change damages and abatement policy costs; this is represented by the term

$$L_j^{(k,i)}(x_j(t), z_j(t), v_j(t), u_j(t))$$

in the reward expression (10.16). But they have also to trade-off the risks of modifying the geopolitical mode or the climate regime; the valuation of these risks at each time t is given by the terms

$$\sum_{\ell \in \mathcal{I}-k} p_{k,\ell}(\mathbf{x}(s)) V_j^*((\ell, i); \mathbf{x}(t), \mathbf{s}(t)) + \sum_{\iota \in \mathrm{L}-i} q_{i,\iota}(\mathbf{y}(s)) V_j^*((k, \iota); \mathbf{x}(t), \mathbf{s}(t))$$

in the integrand of (10.16). Furthermore the associated deterministic problem involves an endogenous discount term $e^{-\rho_j t + \int_0^t \lambda^{(k,i)}(\mathbf{x}(s), \mathbf{y}(s)) ds}$ which is related to pure time preference (discount rate ρ_j) and controlled probabilities of jumps (pooled jump rate $\lambda^{(k,i)}(\mathbf{x}(s), \mathbf{y}(s))$). In solving these dynamic games the analysts will face several difficulties. The first one, which is related to the DP approach is to obtain a correct evaluation of the value functions $V_j^*((k, i); \mathbf{x}, \mathbf{s})$; this implies a computationally demanding fixed point calculation in a high dimensional space. A second difficulty will be to find the solution of the associated OLEs. These are problems with very large state space, in particular in the representation of the coupled economies of the different nations $j \in M$. A possible way to circumvent this difficulty is to exploit the hierarchical structure of this dynamic game that is induced by the difference in timescale between the evolution of the economy related state variables and those linked with climate.

4. The hierarchical game and its limit equilibrium problem

In this section we propose to define an approximate dynamic game which is much simpler to analyze and to solve numerically in IAMs. This approximation is proposed by extending formally, using[8] "analogy reasoning" some results obtained in the field of control systems (one-player games) under the generic name of *singular perturbation theory.*

[8]Analogy reasoning is a sin in mathematics. It is used here to delineate the results (theorems to prove) that are needed to justify the proposed approach.

We will take our inspiration mostly from Filar et al., 2001. The reader will find in this paper a complete list of references on the theory of singularly perturbed control systems.

4.1 The singularly perturbed dynamic game

We describe here the possible extension of the fundamental technique used in singular perturbation theory for control systems, which leads to an averaging of the fast part of the system and a "lifting up" of the control problem to the upper-level slow paced system.

4.1.1 The local economic equilibrium problem. If at a given time $\bar{t}$, the nations have adopted GHG emissions caps represented by $\mathbf{z}^{\bar{t}}$, the state of climate $\mathbf{y}^{\bar{t}}$ is generating damages $v_j^{\bar{t}}$, $j \in M$, we call local economic equilibrium problem the solution $\bar{\mathbf{x}}^{\bar{t}}$ of the set of algebraic equations

$$0 = f_j^{\mu}(\bar{t}, \bar{\mathbf{x}}^{\bar{t}}, z_j^{\bar{t}}, v_j^{\bar{t}}) \quad j \in M. \tag{10.18}$$

We shall now use a modification of the time scale, called the *stretched out timescale*. It is obtained when we pose $\tau = \frac{t}{\varepsilon}$. Then we denote $\tilde{x}(\tau) = x(\tau\varepsilon)$ the economic trajectory when represented in this stretched out time.

ASSUMPTION 10.1 *We assume that the following holds for fixed values of time $\bar{t}$, and slow paced state and control variables $(z_j^{\bar{t}}, v_j^{\bar{t}}, u_j^{\bar{t}})$, $j \in M$.*

$$\begin{aligned} & L_j^{k,i}(\bar{x}_j^{\bar{t}}, z_j^{\bar{t}}, v_j^{\bar{t}}, u_j^{\bar{t}}) + \sum_{\ell \in \mathcal{I}-k} p_{k,\ell}(\bar{\mathbf{x}}^{\bar{t}}) v_j((\ell, i); \mathbf{s}^{\bar{t}}) \\ = & \lim_{\theta \to \infty} \frac{1}{\theta} \int_0^{\theta} \{L_j^{k,i}(\tilde{x}_j(t), \tilde{z}_j(t), \tilde{v}_j(t), u_j(t)) \\ & + \sum_{\ell \in \mathcal{I}-k} p_{k,\ell}(\tilde{\mathbf{x}}(s)) v_j((\ell, i); , \mathbf{s}^{\bar{t}})\}\, dt \quad j \in M \end{aligned} \tag{10.19}$$

s.t.

$$\dot{\tilde{x}}_j(\tau) = f_j^k(\bar{t}, \tilde{\mathbf{x}}(\tau), z_j^{\bar{t}}) \quad j \in M \tag{10.20}$$

$$\tilde{x}_j(0) = x_j^o \quad j \in M \tag{10.21}$$

$$\tilde{x}_j(\theta) = x_j^f \quad j \in M. \tag{10.22}$$

The problem (10.19)-(10.22) consists in averaging the part of the instantaneous reward in the associated OLE game that depends on the fast economic variable $\mathbf{x}$. This averaging is made over the $\tilde{\mathbf{x}}(\cdot)$ trajectory, when the timescale has been stretched out and when a potential function $v_j((\ell, \iota); , \mathbf{s}^{\bar{t}})$ is used instead of the true equilibrium value function

$V_j^*((k,\iota);\mathbf{x},\mathbf{s})$. *The condition says that this averaging problem has a value which is given by the reward associated with the local economic equilibrium* $\bar{x}_j^{\bar{t}}$, $j \in M$ *corresponding to the solution of*

$$0 = f_j^k(\bar{t}, \bar{\mathbf{x}}^{\bar{t}}, z_j^{\bar{t}}, v_j^{\bar{t}}) \quad j \in M. \tag{10.23}$$

Clearly this assumption holds if the economic equilibrium is a stable point for the dynamical system (10.20) in the stretched out timescale.

4.1.2 The limit equilibrium problem. In control systems an assumption similar to Assumption 10.1 permits one to "lift" the optimal control problem to the "upper level" system which is uniquely concerned with the slow varying variables. The basic idea consists in splitting the time axis $[0,\infty)$ into a succession of small intervals of length $\Delta(\varepsilon)$ which will tend to 0 together with the timescale ratio ε in such a way that $\frac{\Delta(\varepsilon)}{\varepsilon} \to \infty$. Then using the averaging property (10.19)-(10.22) when $\varepsilon \to 0$ one defines an approximate control problem called *the limit control problem* which is uniquely defined at the level of the slow paced variable.

We propose here the analogous *limit equilibrium problem* for the multi-agent system that we are studying. We look for equilibrium (potential) value functions $v_j^*((k,i);,\mathbf{s}), j \in M$ (where we use the notation $\mathbf{s} = (\mathbf{z},\mathbf{y})$) that satisfy the family of associated OLE defined as follows

$$\begin{aligned}
v_j^*((k,i);,\mathbf{s}^o) &= \text{equil.}\, L_j^{k,i}(\bar{x}_j(t), z_j(t), v_j(t), u_j(t)\\
&\quad + \sum_{\ell \in \mathcal{I}-k} p_{k,\ell}(\bar{\mathbf{x}}(t)) v_j^*((\ell,i);\mathbf{s}(t))\\
&\quad + \sum_{\iota \in \mathbf{L}-i} q_{i,\iota}(\mathbf{y}(t)) v_j^*((k,\iota);\mathbf{s}(t)), \quad j \in M\\
\text{s.t.}&\\
0 &= f_j^k(\bar{t}, \bar{\mathbf{x}}(t), z_j(t), v_j(t)), \quad j \in M\\
\dot{z}_j(t) &= h^k(x_j(t), z_j(t), u_j(t)), \quad j \in M\\
\mathbf{z}(t) &= \sum_{j \in M} z_j(t), \quad j \in M\\
z_j(0) &= z_j^o, \quad j \in M\\
\dot{y}(t,\cdot) &= g^i(\mathbf{z}(t), y(t,\cdot))\\
y(0,\cdot) &= y^o(\cdot)\\
v_j(t) &= \int_{\Omega_j} \tilde{d}_j^i(\omega, \tilde{y}(t,\omega))\, d\omega, \quad j \in M
\end{aligned}$$

In this problem, the economic variables $\bar{\mathbf{x}}(t)$ are not state variables anymore; they are not appearing in the arguments of the equilibrium value functions $v_j^*((k,i);,\mathbf{s}), j \in M$. They serve now as auxiliary variables in the definition of the nations rewards when they have selected an emission cap policy. The reduction in state space dimension is therefore very important and it can be envisioned to solve numerically these equations in simulations of IAMs.

5. A research agenda

The reduction of complexity obtained in the limit equilibrium problem is potentially very important. An attempt to solve numerically the resulting dynamic game could be considered, although a high dimensional state variable, the climate descriptor $\mathbf{y}(t)$ remains in this limit problem. More research is needed before such an attempt could succeed. We give below a few hints about the topics that need further clarification.

5.1 Comparison of economic and climate timescales

The pace of anthropogenic climate change is still a matter of controversies. A better understanding of the influence of GHG emissions on climate change should emerge from the development of better intermediate complexity models. Recent experiments by Drouet et al., 2005a, and Drouet et al., 2005b, on the coupling of economic growth models with climate models tend to clarify the difference in adjustment speeds between the two systems.

5.2 Approximations of equilibrium in a two-timescale game

The study of control systems with two timescales has been developed under the generic name of "singular perturbation" theory. A rigorous extension of the control results to a game-theoretic equilibrium solution environment still remains to be done.

5.3 Viability approach

Aubin et al., 2005 propose an approach more encompassing than game theory to study the dynamics of climate-economy systems. The concept of viability could be introduced in the piecewise deterministic formalism proposed here instead of the more "teleonomic" equilibrium solution concept.

6. Conclusion

In this paper we have proposed to use a formalism directly inspired from the system control and dynamic game literature to model the climate-economy interplay that characterizes the climate policy negotiations. The Kyoto protocol is the first example of international agreement on GHG emissions abatement. It should be followed by other complex negotiations between nations with long term economic and geopolitical consequences at stake. The framework of stochastic piecewise deterministic games with two-timescales offers an interesting paradigm for the construction of IAMs dealing with long term international climate policy. The examples, given in the introduction, of the first experiments with the use of hierarchical dynamic games to study real life policies in the realm of the Kyoto protocol tend to show that the approach could lead to interesting policy evaluation tools. It is remarkable that economic growth models as well as climate models are very close to the general system control paradigm. In a proposed research agenda we have indicated the type of developments that are needed for making this approach operational for climate policy assessment.

Acknowledgments. This research has been supported by the Swiss NSF under the NCCR-Climate program. I thank in particular Laurent Viguier for the enlightening collaboration on the economic modeling of climate change and Biancamaria D'Onofrio and Francesco Moresino for fruitful exchanges on singularly perturbed stochastic games.

References

Aubin, J.-P., Bernardo, T., and Saint-Pierre, P. (2005). A viability approach to global climate change issues. In: Haurie, A. and Viguier, L. (eds), *The Coupling of Climate and Economic Dynamics*, pp. 113–140, Springer.

Bernard, A., Haurie, A., Vielle, M., and Viguier, L. (2002). A Two-level Dynamic Game of Carbon Emissions Trading Between Russia, China, and Annex B Countries. Working Paper 11, NCCR-WP4, Geneva. To appear in *Journal of Economic Dynamics and Control.*

Buchner, B., Carraro, C., Cersosimo, I., and Marchiori, C. (2005). Linking climate and economic dynamics. In: Haurie, A. and Viguier, L. (eds), *Back to Kyoto? US participation and the linkage between R&D and climate cooperation, Advances in Global Change Research*, Springer.

Carraro, C. and Filar, J.A. (1995). *Control and Game-Theoretic Models of the Environment*, volume 2 of *Annals of the International Society of Dynamic Games*. Birhaüser, Boston.

Carraro, C. and Siniscalco, D. (1992). The international dimension of environmental policy. *European Economic Review*, 36:379–87.

Carraro, C. and Siniscalco, D. (1993). Strategies for the international protection of the environment. *Journal of Public Economics*, 52: 309–28.

Carraro, C. and Siniscalco, D. (1996). R&D Cooperation and the Stability of International Environmental Agreements. In: C. Carraro and D. Siniscalco (eds), *International Environmental Negotiations*, Kluwer Academic Publishers.

Drouet, L., Beltran, C., Edwards, N.R., Haurie, A.B., Vial, J.-P., and Zachary, D.S. (2005a). An oracle method to couple climate and economic dynamics. In: Haurie, A. and Viguier, L. (eds), *The Coupling of Climate and Economic Dynamics*, Springer.

Drouet, L., Edwards, N.R., and Haurie (2005b). Coupling climate and economic models in a cost-benefit framework: A convex optimization approach. *Environmental Modeling and Assessment.*

Edwards, N.R., Greppin, H., Haurie, A.B., and Viguier, L. (2005). Linking climate and economic dynamics. In: Haurie, A. and Viguier, L. (eds), *The Coupling of Climate and Economic Dynamics, Advances in Global Change Research*, Springer.

Filar, J., Gaitsgory, V., and Haurie, A. (2001). Control of singularly perturbed hybrid stochastic systems. *IEEE Transactions on Automatic Control*, 46(2):179–190.

Germain, M., Toint, P., Tulkens, H., and de Zeeuw, A. (2003). Transfers to sustain dynamic core-theoretic cooperation in international stock pollutant control. *Journal of Economic Dynamics and Control*, 28: 79–99.

Haurie, A. (1989). Piecewise deterministic differential games. In: Bernhard, P. and Başar, T. (eds), *Differential Games and Applications*, Lecture Notes in Control and Information, volume 119, pp. 114–127, Springer Verlag.

Haurie, A. (1995). Environmental coordinatio in dynamic oligopolistic markets. *Group Decision anf Negotiation*, 4:46–67.

Haurie, A. (2002). Turnpikes in multi-discount rate environments and GCC policy evaluation. In: Zaccour, G. (ed.), *Optimal Control and Differential Games: essays in Honor of Steffen Jørgensen*, volume 5 of *Advances in Computational Management Science*, Kluwer Academic Publishers.

Haurie, A. (2003). Integrated assessment modeling for global climate change: an infinite horizon optimization viewpoint. *Environmental Modeling and Assessment*, 8(3):117–132.

Haurie, A. and Roche, M. (1994). Turnpikes and computation of piecewise open-loop equilibria in stochastic differential games. *Journal of Economic Dynamics and Control*, 18:317–344.

Haurie, A. and Viguier, L. (2003a). A stochastic dynamic game of carbon emissions trading. *Environmental Modeling & Assessment*, 8(3): 239–248.

Haurie, A. and Viguier, L. (2003b). A stochastic game of carbon emissions trading. *Environmental Modeling and Assessment*, 8(3):239–248.

Haurie, A. and Zaccour, G. (1995). Differential game models of global environmental management. In: Carraro, C. and Filar, J. (eds), *Control and Game-Theoretic Models of the Environment*, volume 2 of *Annals of the International Society of Dynamic Games*, pp. 3–23, Birkhäuser.

Kaitala, V. and Pohjola, M. (1995). Sustainable international agreements on greenhouse warming - a game theory study. In: Carraro, C. and Filar, J. (eds), *Control and Game-Theoretic Models of the Environment*, volume 2 of *Annals of the International Society of Dynamic Games*, pp. 67–87, Birkhäuser.

Kemfert, C. (2005). Linking climate and economic dynamics. In: Haurie, A. and Viguier, L. (eds), *Climate policy cooperation games between developed and developing nations: A quantitative applied analysis, Advances in Global Change Research*, Springer.

Manne, A.S., Mendelsohn, R., and Richels, R.G. (1995). Merge: A model for evaluating regional and global effects of ghg reduction policies. *Energy Policy*, 23:17–34.

Nordhaus, W. D. (1994). *Managing the Global Commons: The Economics of Climate Change*. MIT Press, Cambridge, MA.

Nordhaus, W. D. and Boyer, J. (2000). *Warming the World: Economic Models of Global Warming*. MIT Press, Cambridge, MA.

Petrosjan, L. and Zaccour, G. (2003). Time-consistent shapley value allocation of pollution cost reduction. *Journal of Economic Dynamics and Control*, 27:381–398.

Prinn, R., Jacoby, H., Sokolov, A., Wang, C., Xiao, X., Yang, Z., Eckaus, R., P. Stone, D. Ellerman, Melillo, J., Fitzmaurice, J., Kicklighter, D., Holian, G., , and May, Y. Liu (1999). Integrated global system model for climate policy assessment: Feedbacks and sensitivity studies. *Climatic Change*, 41(3/4): 469–546.

Ramsey, F. P. (1928). A mathematical theory of saving. *Economic Journal*, 38(152):543–59.

Toth, F.L. (2005). Coupling climate and economic dynamics: recent achievements and unresolved problems. In: Haurie, A. and Viguier, L. (eds), *Coupling Climate and Economic Dynamics*, Springer.

Toth, F.L., Bruckner, T., Füssel, H.-M., Leimbach, M., and Petschel-Held, G. (2003). Integrated assessment of long-term climate policies: Part 1 - model presentation. *Climatic Change*, 56:37–56.

Viguier, L., Vielle, M., Haurie, A., and Bernard, A. (2004). A two-level computable equilibrium model to assess the strategic allocation of emission allowances within the european union. To appear in *Computers & Operations Research.*

Chapter 11

A DIFFERENTIAL GAME OF ADVERTISING FOR NATIONAL AND STORE BRANDS

Salma Karray
Georges Zaccour

Abstract We consider a differential game model for a marketing channel formed by one manufacturer and one retailer. The latter sells the manufacturer's product and may also introduce a private label at a lower price than the manufacturer's brand. The aim of this paper is twofold. We first assess in a dynamic context the impact of a private label introduction on the players' payoffs. If this is beneficial for the retailer to propose his brand to consumers and detrimental to the manufacturer, we wish then to investigate if a cooperative advertising program could help the manufacturer to mitigate the negative impact of the private label.

1. Introduction

Private labels (or store brand) are taking increasing shares in the retail market in Europe and North America. National manufacturers are threatened by such private labels that can cannibalize their market shares and steal their consumers, but they can also benefit from the store traffic generated by their presence. In any event, the store brand introduction in a product category affects both retailers and manufacturers marketing decisions and profits. This impact has been studied using static game models with prices as sole decision variables. Mills (1995, 1999) and Narasimhan and Wilcox (1998) showed that for a bilateral monopoly, the presence of a private label gives a bigger bargaining power to the retailer and increases her profit, while the manufacturer gets lower profit. Adding competition at the manufacturing level, Raju et al. (1995) identified favorable factors to the introduction of a private label for the retailer. They showed in a static context that price competition between the store and the national brands, and between national brands has considerable impact on the profitability of the private label introduction.

Although price competition is important to understand the competitive interactions between national and private labels, the retailer's promotional decisions do also affect the sales of both product (Dhar and Hoch 1997). Many retailers do indeed accompany the introduction of a private label by heavy store promotions and invest more funds to promote their own brand than to promote the national ones in some product categories (Chintagunta et al. 2002).

In this paper, we present a dynamic model for a marketing channel formed by one manufacturer and one retailer. The latter sells the manufacturer's product (the national brand) and may also introduce a private brand which would be offered to consumers at a lower price than the manufacturer's brand. The aim of this paper is twofold. We first assess in a dynamic context the impact of a private label introduction on the players' profits. If we find the same results obtained from static models, i.e., that it is beneficial for the retailer to propose his brand to consumers and detrimental to the manufacturer, we wish then to investigate if a cooperative advertising program could help the manufacturer to mitigate, at least partially, the negative impact of the private label.

A *cooperative advertising* (or promotion) program is a cost sharing mechanism where a manufacturer pays part of the cost incurred by a retailer to promote the manufacturer's brand. One of the first attempts to study cooperative advertising, using a (static) game model, is Berger (1972). He studied a case where the manufacturer gives an advertising allowance to his retailer as a fixed discount per item purchased and showed that the use of quantitative analysis is a powerful tool to maximize the profits in the channel. Dant and Berger (1996) used a Stackelberg game to demonstrate that advertising allowance increases retailer's level of local advertising and total channel profits. Bergen and John (1997) examined a static game where they considered two channel structures: A manufacturer with two competing retailers and two manufacturers with two competing retailers. They showed that the participation of the manufacturers in the advertising expenses of their dealers increases with the degree of competition between these dealers, with advertising spillover and with consumer's willingness to pay. Kim and Staelin (1999) also explored the two-manufacturers, two-retailers channel, where the cooperative strategy is based on advertising allowances.

Studies of cooperative advertising as a coordinating mechanism in a dynamic context are of recent vintages (see, e.g., Jørgensen et al. (2000, 2001), Jørgensen and Zaccour (2003), Jørgensen et al. (2003)).

Jørgensen et al. (2000) examine a case where both channel members make both long and short term advertising efforts, to stimulate current sales and build up goodwill. The authors suggest a cooperative advertising program that can take different forms, i.e., a full-support program where the manufacturer contributes to both types of the retailer's advertising expenditures (long and short term) or a partial-support program where the manufacturer supports only one of the two types of retailer advertising. The authors show that all three cooperative advertising programs are Pareto-improving (profit-wise) and that both players prefer the full support program. The conclusion is thus that a coop advertising program is a coordinating mechanism in also a dynamic setting. Due to the special structure of the game, long term advertising strategies are constant over time. This is less realistic in a dynamic game with an infinite time horizon. A more intuitive strategy is obtained in Jørgensen et al. (2001). This paper reconsiders the issue of cooperative advertising in a two-member channel in which there is, however, only one type of advertising of each player. The manufacturer advertises in national media while the retailer promotes the brand locally. The sales response function is linear in promotion and concave in goodwill. The dynamics are a Nerlove-Arrow-type goodwill evolution equation, depending only on the manufacturer's national advertising. In this case, one obtains a nondegenerate Markovian advertising strategy, being linearly decreasing in goodwill.

In Jørgensen et al. (2000, 2001), it is an assumption that the retailer's promotion affects positively the brand image (goodwill stock). Jørgensen, et al. (2003) study the case where promotions damage the brand image and ask the question whether a cooperative advertising program is meaningful in such context. The answer is yes if the initial brand image is "weak" or if the initial brand image is at an "intermediate" level and retailer promotions are not "too" damaging to the brand image.

Jørgensen and Zaccour (2003) suggest an extension of the setup in Jørgensen et al. (2003). The idea now is that excessive promotions, and not instantaneous action, is harmful to the brand image.

To achieve our objective, we shall consider three scenarios or games:

1. Game N: the retailer carries only the National brand and no cooperative advertising program is available. The manufacturer and the retailers play a noncooperative game and a feedback Nash equilibrium is found.

2. Game S: the retailer offers a Store brand along with the manufacturer's product and there is no cooperative advertising program.

The mode of play is noncooperative and a feedback Nash equilibrium is the solution concept.

3. Game C: the retailer still offers both brands and the manufacturer proposes to the retailer a Cooperative advertising program. The game is played à la Stackelberg with the manufacturer as leader. As in the two other games, we adopt a feedback information structure.

Comparing players' payoffs of the first two games allows to measure the impact of the private label introduction by the retailer. Comparing the players' payoffs of the last two games permits to see if a cooperative advertising program reduces the harm of the private label for the manufacturer. A necessary condition for the coop plan to be attractive is that it also improves the retailer's profit, otherwise the will not accept to implement it.

The remaining of this paper is organized as follows: In Section 2 we introduce the differential game model and define rigorously the three above games. In Section 3 we derive the equilibria for the three games and compare the results in Section 4. In Section 5 we conclude.

2. Model

Let the marketing channel be formed of a manufacturer (player M) and a retailer (player R). The manufacturer controls the rate of national advertising for his brand $A(t), t \in [0, \infty)$. Denote by $G(t)$ the goodwill of the manufacturer's brand, which dynamics evolve à la Nerlove and Arrow (1962):

$$\dot{G}(t) = \lambda A(t) - \delta G(t), \quad G(0) = G_0 \geq 0, \tag{11.1}$$

where λ is a positive scaling parameter and $\delta > 0$ is the decay rate.

The retailer controls the promotion efforts for the national brand, denoted by $p_1(t)$, and for the store brand, denoted by $p_2(t)$.
We consider that promotions have an immediate impact on sales and do not affect the goodwill of the brand. The demand functions for the national brand (Q_1) and for the store brand (Q_2) are as follows:

$$\begin{aligned} Q_1(p_1, p_2, G) &= \alpha p_1(t) - \beta p_2(t) + \theta G(t) - \mu G^2(t), & (11.2)\\ Q_2(p_1, p_2, G) &= \alpha p_2(t) - \psi p_1(t) - \gamma G(t), & (11.3) \end{aligned}$$

where $\alpha, \beta, \theta, \mu, \psi$ andγ are positive parameters.

Thus, the demand for each brand depends on the retailer's promotions for both brands and on the goodwill of the national brand. Both demands are linear in promotions.

We have assumed for simplicity that the sensitivity of demand to own promotion is the same for both brands considering that the retailer is using usually the same media and methods to promote both brands. However, the cross effect is different allowing for asymmetry in brand substitution. We assume that own brand promotion has a greater impact on sales, in absolute value, than competitive brand promotion, i.e., $\alpha > \beta$ and $\alpha > \psi$. This assumption mirrors the one usually made on prices in oligopoly theory. We further suppose that the marginal effect of promoting the national brand on the sales of the store brand is higher than the marginal effect of promoting the store brand on the sales of the national brand, i.e., $\psi > \beta$. This actually means that the manufacturer's brand enjoys a priori a stronger consumer preference than the retailer's one. Putting together these inequalities leads to the following assumption

$$\text{A1} : \alpha > \psi > \beta > 0.$$

Finally, the demand for the national brand is concave increasing in its goodwill (i.e., $\frac{\partial Q_1}{\partial G} = \theta - 2\mu G > 0, \forall G > 0$) and the demand for the store brand is decreasing in that goodwill.

Denote by $D(t), 0 \leq D(t) \leq 1$, the coop participation rate of the manufacturer in the retailer's promotion cost of the national brand. We assume as in, e.g., Jørgensen et al. (2000, 2003), that the players face quadratic advertising and promotion costs. The net cost incurred by the manufacturer and the retailer are as follows

$$\begin{aligned} C_M(A) &= \frac{1}{2}u_M A^2(t) + \frac{1}{2}u_R D(t)\, p_1^2(t), \\ C_R(p_1, p_2) &= \frac{1}{2}u_R\Big\{\big[1 - D(t)\big]p_1^2(t) + p_2^2(t)\Big\}, \end{aligned}$$

where $u_R, u_M > 0$.

Denote by m_0 the manufacturer's margin, by m_1 the retailer's margin on the national brand and by m_2 her margin on the store brand. Based on empirical observations, we suppose that the retailer has a higher margin on the private label than on the national brand, i.e., $m_2 > m_1$. Ailawadi and Harlam (2004) found indeed that for product categories where national brands are heavily advertised, the percent retail margins are significantly higher for store brands than for national brands.

We denote by r the common discount rate and we assume that each player maximizes her stream of discounted profit over an infinite horizon. Omitting the time argument when no ambiguity may arise, the optimization problems of players M and R in the different games are as follows:

- Game C: Both brands are offered and a coop program is available.

$$\max_{A,D} J_M^C = \int_0^{+\infty} e^{-rt}\Big[m_0\big(\alpha p_1 - \beta p_2 + \theta G - \mu G^2\big) - \frac{u_M}{2}A^2 - \frac{u_R}{2}Dp_1^2\Big]dt,$$

$$\max_{p_1,p_2} J_R^C = \int_0^{+\infty} e^{-rt}\Big[m_1\big(\alpha p_1 - \beta p_2 + \theta G - \mu G^2\big) + m_2\big(\alpha p_2 - \psi p_1 - \gamma G\big) - \frac{1}{2}u_R\Big[\big(1-D\big)p_1^2 + p_2^2\Big]\Big]dt.$$

- Game S: Both brands are available and there is no coop program.

$$\max_{A} J_M^S = \int_0^{+\infty} e^{-rt}\Big[m_0\big(\alpha p_1 - \beta p_2 + \theta G - \mu G^2\big) - \frac{u_M}{2}A^2\Big]dt,$$

$$\max_{p_1,p_2} J_R^S = \int_0^{+\infty} e^{-rt}\Big[m_1\big(\alpha p_1 - \beta p_2 + \theta G - \mu G^2\big) + m_2\big(\alpha p_2 - \psi p_1 - \gamma G\big) - \frac{u_R}{2}\Big(p_1^2 + p_2^2\Big)\Big]dt.$$

- Game N: Only manufacturer's brand is offered and there is no coop program.

$$\max_{A} J_M^N = \int_0^{+\infty} e^{-rt}\Big[m_0\big(\alpha p_1 + \theta G - \mu G^2\big) - \frac{u_M}{2}A^2\Big]dt,$$

$$\max_{p_1} J_R^N = \int_0^{+\infty} e^{-rt}\Big[m_1\big(\alpha p_1 + \theta G - \mu G^2\big) - \frac{u_R}{2}p_1^2\Big]dt.$$

3. Equilibria

We characterize in this section the equilibria of the three games. In all cases, we assume that the players adopt stationary Markovian strategies, which is rather standard in infinite-horizon differential games. The following proposition gives the result for Game N.

PROPOSITION 11.1 *When the retailer does not sell a store brand and the manufacturer does not provide any coop support to the retailer, stationary feedback Nash advertising and promotional strategies are given by*

$$p_1^N = \frac{\alpha m_1}{u_R},$$

$$A^N(G) = X + YG,$$

where

$$X = \frac{2m_0\theta\lambda}{\left(r + 2\sqrt{\Delta_1}\right)u_M}, \quad Y = \frac{r + 2\delta - 2\sqrt{\Delta_1}}{2\lambda},$$

$$\Delta_1 = \left(\delta + \frac{r}{2}\right)^2 + \frac{2\mu m_0 \lambda^2}{u_M}.$$

Proof. A sufficient condition for a stationary feedback Nash equilibrium is the following: Suppose there exists a unique and absolutely continuous solution $G(t)$ to the initial value problem and there exist bounded and continuously differentiable functions $V_i : \Re_+ \to \Re, i \in \{M, R\}$, such that the Hamilton-Jacobi-Bellman (HJB) equations are satisfied for all $G \geq 0$:

$$rV_M(G) = \max_A \Big\{ m_0\left(\alpha p_1 + \theta G - \mu G^2\right) - \frac{1}{2}u_M A^2 + V_M'(G)\left(\lambda A - \delta G\right) \mid A \geq 0 \Big\}, \tag{11.4}$$

$$rV_R(G) = \max_{p_1} \Big\{ m_1\left(\alpha p_1 + \theta G - \mu G^2\right) - \frac{1}{2}u_R p_1^2 + V_R'(G)\left(\lambda A - \delta G\right) \mid p_1 \geq 0 \Big\}. \tag{11.5}$$

The maximization of the right-hand-side of equations (11.4) and (11.5) yields the following advertising and promotional rates:

$$A(G) = \frac{\lambda}{u_M}V_m'(G), \quad p_1 = \frac{\alpha m_1}{u_R}.$$

Substituting the above in (11.4) and (11.5) leads to the following expressions

$$rV_M(G) = m_0\left(\frac{\alpha^2 m_1}{u_R} + \theta G - \mu G^2\right) + \frac{\lambda^2}{2u_M}\left[V_M'(G)\right]^2 - \delta G V_M'(G), \tag{11.6}$$

$$rV_R(G) = m_1\left(\frac{\alpha^2 m_1}{2u_R} + \theta G - \mu G^2\right) + V_R'(G)\left[\frac{\lambda^2}{u_M}V_M'(G) - \delta G\right]. \tag{11.7}$$

It is easy to show that the following quadratic value functions solve the HJB equations;

$$V_M(G) = a_1 + a_2 G + \frac{1}{2} a_3 G^2, \quad V_R(G) = b_1 + b_2 G + \frac{1}{2} b_3 G^2,$$

where $a_1, a_2, a_3, b_1, b_2, b_3$ are constants. Substitute $V_M(G)$, $V_R(G)$ and their derivatives into equations (11.6) and (11.7) to obtain:

$$r\left(a_1 + a_2 G + \frac{a_3}{2} G^2\right) = \frac{m_0 \alpha^2 m_1}{u_R} + \frac{\lambda^2 a_2^2}{2u_M}$$
$$+ \left(m_0 \theta - \delta a_2 + \frac{\lambda^2 a_2 a_3}{u_M}\right) G - \left(\mu m_0 + \delta a_3 - \frac{\lambda^2 a_3^2}{2u_M}\right) G^2,$$

$$r\left(b_1 + b_2 G + \frac{1}{2} b_3 G^2\right) = \frac{\alpha^2 m_1^2}{2u_R} + \frac{\lambda^2}{u_M} a_2 b_2$$
$$+ \left(m_1 \theta - \delta b_2 + \frac{\lambda^2}{u_M}(a_2 b_3 + a_3 b_2)\right) G - \left(m_1 \mu + \delta b_3 - \frac{\lambda^2}{u_M} b_3 a_3\right) G^2.$$

By identification, we obtain the following values for the coefficients of the value functions:

$$a_3 = \frac{\left(\delta + \frac{r}{2}\right) \pm \sqrt{\Delta_1}}{\lambda^2 / u_M}, \qquad b_3 = -\frac{m_1 \mu}{\frac{r}{2} + \delta - \frac{\lambda^2}{u_M} a_3}$$

$$a_2 = \frac{m_0 \theta}{r + \delta - \frac{\lambda^2}{u_M} a_3}, \qquad b_2 = \frac{m_1 \theta + \frac{\lambda^2}{u_M} b_3 a_2}{r + \delta - \frac{\lambda^2}{u_M} a_3}$$

$$a_1 = \frac{m_0 \alpha^2 m_1}{r u_R} + \frac{\lambda^2 a_2^2}{2 r u_M}, \qquad b_1 = \frac{\alpha^2 m_1^2}{2 r u_R} + \frac{\lambda^2 a_2 b_2}{r u_M}$$

where

$$\Delta_1 = \left(\delta + \frac{r}{2}\right)^2 + \frac{2\mu m_0 \lambda^2}{u_M}.$$

To obtain an asymptotically stable steady state, choose the negative solution for a_3. Note that the identified solution must satisfy the constraint $A(G) > 0$. Since $\frac{\lambda}{u_M} V_M'(G) = A(G)$, this assumption is true for $G \in \left[0, \bar{G}^N\right]$, where

$$\bar{G}^N = -\frac{a_2}{a_3}, \quad A(G) = \frac{\lambda}{u_M} V_M'(G) = \frac{2 m_0 \theta \lambda}{\left(r + 2\sqrt{\Delta_1}\right) u_M} + \frac{r + 2\delta - 2\sqrt{\Delta_1}}{2\lambda} G.$$

□

The above proposition shows that the retailer promotes always the manufacturer's brand at a positive constant rate and that the advertising strategy is decreasing in the goodwill. The next proposition characterizes the feedback Nash equilibrium in Game S.

PROPOSITION 11.2 *When the retailer does sell a store brand and the manufacturer does not provide any coop support to the retailer, assuming an interior solution, stationary feedback Nash advertising and promotional strategies are given by*

$$p_1^S = \frac{\alpha m_1 - \psi m_2}{u_R}, \quad p_2^S = \frac{\alpha m_2 - \beta m_1}{u_R}, \quad A^S(G) = A^N(G).$$

Proof. The proof proceeds exactly as the previous one and we therefore print only important steps. The HJB equations are given by:

$$rV_M(G) = \max_A \Big\{ m_0\big(\alpha p_1 - \beta p_2 + \theta G - \mu G^2\big) - \frac{u_M}{2}A^2 + V_M'(G)\big(\lambda A - \delta G\big) \mid A \geq 0 \Big\},$$

$$rV_R(G) = \max_{p_1,p_2} \Big\{ m_1\big(\alpha p_1 - \beta p_2 + \theta G - \mu G^2\big) + m_2\big(\alpha p_2 - \psi p_1 - \gamma G\big) - \frac{u_R}{2}\big(p_1^2 + p_2^2\big) + V_R'(G)\big(\lambda A - \delta G\big) \mid (p_1, p_2) \geq 0 \Big\}.$$

The maximization of the right-hand-side of the above equations yields the following advertising and promotional rates:

$$A(G) = \frac{\lambda}{u_M} V_m'(G), \quad p_1 = \frac{\alpha m_1 - \psi m_2}{u_R}, \quad p_2 = \frac{\alpha m_2 - \beta m_1}{u_R}.$$

We next insert the values of $A(G)$, p_1 and p_2 from above in the HJB equations and assume that the resulting equations are solved by the following quadratic functions:

$$V_M(G) = s_1 + s_2 G + \frac{1}{2} s_3 G^2, \quad V_R(G) = k_1 + k_2 G + \frac{1}{2} k_3 G^2,$$

where $k_1, k_2, k_3, s_1, s_2, s_3$ are constants. Following the same procedure as in the proof of the previous proposition, we obtain

$$s_3 = \frac{\left(\delta + \frac{r}{2}\right) \pm \sqrt{\Delta_2}}{\lambda^2 / u_M}, \qquad k_3 = -\frac{m_1 \mu}{\frac{r}{2} + \delta - \frac{\lambda^2}{u_M} s_3},$$

$$s_2 = \frac{m_0\theta}{r+\delta-\frac{\lambda^2}{u_M}s_3}, \qquad k_2 = \frac{m_1\theta - m_2\gamma + \frac{\lambda^2}{u_M}k_3 s_2}{r+\delta-\frac{\lambda^2}{u_M}s_3},$$

$$s_1 = \frac{m_0}{ru_R}\Big(\alpha\left(m_1\alpha - m_2\psi\right) - \beta\left(m_2\alpha - m_1\beta\right)\Big) + \frac{\lambda^2}{2ru_M}s_2^2,$$

$$k_1 = \frac{1}{2ru_R}\Big(\left(m_1\alpha - m_2\psi\right)^2 + \left(m_2\alpha - m_1\beta\right)^2\Big) + \frac{\lambda^2}{ru_M}k_2 s_2,$$

where

$$\Delta_2 = \Delta_1 = \left(\delta + \frac{r}{2}\right)^2 + \frac{2\mu m_0\lambda^2}{u_M}.$$

In order to obtain an asymptotically stable steady state, we choose for s_3 the negative solution. The assumption $A(G) > 0$ holds for $G \in \left[0, \bar{G}^S\right]$, where $\bar{G}^S = -\frac{s_2}{s_3}$. Note also that $s_3 = a_3$, $s_2 = a_2$ and $b_3 = k_3$. Thus $A^S(G) = A^N(G)$ and $\bar{G}^S = \bar{G}^N$. □

REMARK 11.1 Under A1 $(\alpha > \psi > \beta > 0)$ and the assumption that $m_2 > m_1$, the retailer will always promote his brand, i.e., $p_2^S = \frac{\alpha m_2 - \beta m_1}{u_R} > 0$. For $p_1^S = \frac{\alpha m_1 - \psi m_2}{u_R}$ to be positive and thus the solution to be interior, it is necessary that $(\alpha m_1 - \psi m_2) > 0$. This means that the retailer will promote the national brand if the marginal revenue from doing so exceeds the marginal loss on the store brand. This condition has thus an important impact on the results and we shall come back to it in the conclusion.

In the last game, the manufacturer offers a coop promotion program to her retailer and acts as leader in a Stackelberg game. The results are summarized in the following proposition.

PROPOSITION 11.3 *When the retailer does sell a store brand and the manufacturer provides a coop support to the retailer, assuming an interior solution, stationary feedback Stackelberg advertising and promotional strategies are given by*

$$p_1^C = \frac{2\alpha m_0 + (\alpha m_1 - \psi m_2)}{2u_R}, \quad p_2^C = \frac{\alpha m_2 - \beta m_1}{u_R},$$

$$A^C(G) = A^S(G), \quad D = \frac{2\alpha m_0 - (\alpha m_1 - \psi m_2)}{2\alpha m_0 + (\alpha m_1 - \psi m_2)}.$$

Proof. We first obtain the reaction functions of the follower (retailer) to the leader's announcement of an advertising strategy and a coop support rate. The later HJB equation is the following

$$rV_R(G) = \max_{p_1,p_2}\Big\{ m_1\big(\alpha p_1 - \beta p_2 + \theta G - \mu G^2\big) + m_2\big(\alpha p_2 - \psi p_1 - \gamma G\big) \tag{11.8}$$

$$- \frac{u_R}{2}\big((1-D)\,p_1^2 + p_2^2\big) + V_R'(G)\,(\lambda A - \delta G) \mid (p_1,p_2) \geq 0\Big\}.$$

Maximization of the right-hand-side of (11.8) yields

$$p_1 = \frac{\alpha m_1 - \psi m_2}{u_R\,(1-D)}, \quad p_2 = \frac{\alpha m_2 - \beta m_1}{u_R}. \tag{11.9}$$

The manufacturer's HJB equation is:

$$rV_M(G) = \max_{A,D}\Big\{ m_0\big(\alpha p_1 - \beta p_2 + \theta G - \mu G^2\big) - \frac{u_M}{2}A^2 - \frac{1}{2}u_R D p_1^2$$

$$+ V_M'(G)\,\big(\lambda A - \delta G\big) \mid A \geq 0, \quad 0 \leq D \leq 1\Big\}.$$

Substituting for promotion rates from (11.9) into manufacturer's HJB equation yields

$$rV_M(G) = \max_{A,D}\Big\{ m_0\big(\alpha\frac{\alpha m_1 - \psi m_2}{u_R\,(1-D)} - \beta\frac{\alpha m_2 - \beta m_1}{u_R} + \theta G - \mu G^2\big)$$

$$- \frac{u_M}{2}A^2 - \frac{u_R}{2}D\Big(\frac{\alpha m_1 - \psi m_2}{u_R\,(1-D)}\Big)^2 + V_M'(G)\,(\lambda A - \delta G)\Big\}$$

Maximizing the right-hand-side leads to

$$A(G) = \frac{\lambda}{u_M}V_M'(G), \quad D = \frac{2\alpha m_0 - (\alpha m_1 - \psi m_2)}{2\alpha m_0 + (\alpha m_1 - \psi m_2)}. \tag{11.10}$$

Using (11.9) and (11.10) provides the retailer's promotional strategies

$$p_1 = \frac{2\alpha m_0 + (\alpha m_1 - \psi m_2)}{2u_R}, \quad p_2 = \frac{\alpha m_2 - \beta m_1}{u_R}.$$

Following a similar procedure to the one in the proof of Proposition 11.1, it is easy to check that following quadratic value functions provide unique solutions for the HJB equations,

$$V_M(G) = n_1 + n_2 G + \frac{1}{2}n_3 G^2, \quad V_R(G) = l_1 + l_2 G + \frac{1}{2}l_3 G^2,$$

where $n_1, n_2, n_3, l_1, l_2, l_3$ are constants given by:

$$n_3 = \frac{\left(\delta + \frac{r}{2}\right) \pm \sqrt{\Delta_3}}{\lambda^2/u_M}, \qquad l_3 = -\frac{m_1\mu}{\frac{r}{2} + \delta - \frac{\lambda^2}{u_M}n_3},$$

$$n_2 = \frac{m_0\theta}{r+\delta-\frac{\lambda^2}{u_M}n_3}, \qquad l_2 = \frac{m_1\theta - m_2\gamma + \frac{\lambda^2}{u_M}l_3 n_2}{r+\delta-\frac{\lambda^2}{u_M}n_3},$$

$$n_1 = \frac{m_0}{ru_R}\Big(\alpha\big(\alpha m_0 + \frac{1}{2}(m_1\alpha - m_2\psi)\big) - \beta\big(m_2\alpha - m_1\beta\big)\Big)$$
$$+ \frac{\lambda^2}{2ru_M}n_2^2 - \frac{1}{2ru_R}\Big(\alpha^2 {m_0}^2 - \frac{1}{4}(m_1\alpha - m_2\psi)^2\Big),$$

$$l_1 = \frac{(m_1\alpha - m_2\psi)}{2ru_R}\Big(\alpha m_0 + \frac{1}{2}(m_1\alpha - m_2\psi)\Big)$$
$$+ \frac{(m_2\alpha - m_1\beta)^2}{2ru_R} + \frac{\lambda^2 l_2 n_2}{ru_M},$$

where $\Delta_3 = \Delta_2 = \Delta_1 = \left(\delta + \frac{r}{2}\right)^2 + 2\mu m_0 \frac{\lambda^2}{u_M}$.

To obtain an asymptotically stable steady state, we choose the negative solution for n_3. Note that $n_3 = s_3 = a_3$, $n_2 = s_2 = a_2$, $l_3 = k_3 = b_3$ and $l_2 = k_2$. Thus $A^C(G) = A^S(G) = A^N(G)$. □

REMARK 11.2 As in Game S, the retailer will always promote her brand at a positive constant rate. The condition for promoting the manufacturer's brand is $(2\alpha m_0 + \alpha m_1 - \psi m_2) > 0$ (the numerator of p_1^C has to be positive). The condition for an interior solution in Game S was that $(\alpha m_1 - \psi m_2) > 0$. Thus if p_1^S is positive, then p_1^C is also positive.

REMARK 11.3 The support rate is constrained to be between 0 and 1. It is easy to verify that if $p_1^C > 0$, then a necessary condition for $D < 1$ is that $(\alpha m_1 - \psi m_2) > 0$, i.e., $p_1^S > 0$. Assuming $p_1^C > 0$, otherwise there is no reason for the manufacturer to provide a support, the necessary condition for having $D > 0$ is $(2\alpha m_0 - \alpha m_1 + \psi m_2) > 0$.

4. Comparison

In making the comparisons, we assume that the solutions in the three games are interior. The following table collects the equilibrium strategies and value functions obtained in the three games.

In terms of strategies, it is readily seen that the manufacturer's advertising strategy $(A(G))$ is the same in all three games. This is probably a by-product of the structure of the model. Indeed, advertising does not affect sales directly but do it through the goodwill. Although the later has an impact on the sales of the store brand, this does not affect the profits earned by the manufacturer. The retailer adopts the same promotional strategy for the private label in the games where such brand is available, i.e., whether a coop program is offered or not. This is also due to the simple structure of our model.

Table 11.1. Summary of Results

	Game N	Game S	Game C
p_1	$\frac{\alpha m_1}{u_R}$	$\frac{\alpha m_1 - \psi m_2}{u_R}$	$\frac{2\alpha m_0 + (\alpha m_1 - \psi m_2)}{2u_R}$
p_2		$\frac{\alpha m_2 - \beta m_1}{u_R}$	$\frac{\alpha m_2 - \beta m_1}{u_R}$
$A(G)$	$A^N(G)$	$A^N(G)$	$A^N(G)$
D			$\frac{2\alpha m_0 - (\alpha m_1 - \psi m_2)}{2\alpha m_0 + (\alpha m_1 - \psi m_2)}$
$V_M(G)$	$a_1 + a_2 G + \frac{a_3}{2} G^2$	$s_1 + a_2 G + \frac{a_3}{2} G^2$	$n_1 + a_2 G + \frac{a_3}{2} G^2$
$V_R(G)$	$b_1 + b_2 G + \frac{b_3}{2} G^2$	$k_1 + k_2 G + \frac{b_3}{2} G^2$	$l_1 + k_2 G + \frac{b_3}{2} G^2$

The remaining and most interesting item is how the retailer promotes the manufacturer's brand in the different games. The introduction of the store brand leads to a reduction in the promotional effort of the manufacturer's brand ($p_1^N - p_1^S = \frac{\psi m_2}{u_R} > 0$). The coop program can however reverse the course of action and increases the promotional effort for the manufacturer's brand $\left(p_1^C - p_1^S = \frac{2\alpha m_0 - \alpha m_1 + \psi m_2}{2u_R} > 0\right)$. This result is expected and has also been obtained in the literature cited in the introduction. What is not clear cut is whether the level of promotion could reach back the one in the game without the store brand. Indeed, $\left(p_1^N - p_1^C\right)$ is positive if the condition that $(\alpha m_1 + \psi m_2 > 2\alpha m_0)$ is satisfied.

We now compare the players' payoffs in the different games and thus answer the questions raised in this paper.

Proposition 11.4 *The store brand introduction is harmful for the manufacturer for all values of the parameters.*

Proof. From the results of Propositions 11.1 and 11.2, we have:

$$V_M^S(G_0) - V_M^N(G_0) = s_1 - a_1 = -\frac{m_0}{ru_R}\left[m_2\psi\alpha + \beta\left(m_2\alpha - m_1\beta\right)\right] < 0.$$

□

For the retailer, we cannot state a clear-cut result. Compute,

$$\begin{aligned} V_R^S(G_0) - V_R^N(G_0) &= k_1 - b_1 + (k_2 - b_2)\,G_0 \\ &= \frac{1}{2ru_R}\left[(m_1\alpha - m_2\psi)^2 + (m_2\alpha - m_1\beta)^2 - \alpha^2 m_1^2\right] \\ &\quad + \frac{4\lambda^2 m_0 m_2 \theta\gamma}{ru_M\left(r + \sqrt{\Delta_2}\right)^2} + \frac{2m_2\gamma}{r\left(r + \sqrt{\Delta_2}\right)}G_0. \end{aligned}$$

Thus for the retailer to benefit from the introduction of a store brand, the following condition must be satisfied

$$V_R^S(G_0) - V_R^N(G_0) > 0 \Leftrightarrow G_0 > \frac{(r+\sqrt{\Delta_2})}{4m_2\gamma u_R}\alpha^2 m_1^2 - \frac{2\lambda^2 m_0\theta}{u_M(r+\sqrt{\Delta_2})} - \frac{(r+\sqrt{\Delta_2})}{4m_2\gamma u_R}\left[(m_1\alpha - m_2\psi)^2 + (m_2\alpha - m_1\beta)^2\right].$$

The above inequality says that the retailer will benefit from the introduction of a store brand unless the initial goodwill of the national one is "too low". One conjecture is that in such case the two brands would be too close and no benefit is generated for the retailer from the product variety. The result that the introduction of a private label is not always in the best interest of a retailer has also been obtained by Raju et al. (1995) who considered price competition between two national brands and a private label.

Turning now to the question whether a coop advertising program can mitigate, at least partially, the losses for the manufacturer, we have the following result.

PROPOSITION 11.5 *The cooperative advertising program is profit Pareto-improving for both players.*

Proof. Recall that $k_2 = l_2$, $k_3 = l_3 = n_3$ and $n_2 = s_2$. Thus for the manufacturer, we have

$$V_M^3(G_0) - V_M^2(G_0) = n_1 - s_1 = \frac{1}{8ru_R}\left[2\alpha m_0 - (\alpha m_1 - \psi m_2)\right]^2 > 0.$$

For the retailer

$$V_R^C(G_0) - V_R^S(G_0) = l_1 - k_1 = \frac{1}{4ru_R}(m_1\alpha - m_2\psi)(2\alpha m_0 - m_1\alpha + m_2\psi)$$

which is positive. Indeed, $(m_1\alpha - m_2\psi) = u_r p_1^S$ which is positive by the assumption of interior solution and $(2\alpha m_0 - m_1\alpha + m_2\psi)$ which is also positive (it is the numerator of D). □

The above proposition shows that the answer to our question is indeed yes and, importantly, the retailer would be willing to accept a coop program when suggested by the manufacturer.

5. Concluding Remarks

The results so far obtained rely heavily on the assumption that the solution of Game S is interior. Indeed, we have assumed that the retailer will promote the manufacturer's brand in that game. A natural question is that what would happen if it were not the case? Recall that we required that

$$p_1^S = \frac{\alpha m_1 - \psi m_2}{u_R} > 0 \Leftrightarrow \alpha m_1 > \psi m_2.$$

If $\alpha m_1 > \psi m_2$ is not satisfied, then $p_1^S = 0$ and the players' payoffs should be adjusted accordingly. The crucial point however is that in such event, the constraint on the participation rate in Game C would be impossible to satisfy. Indeed, recall that

$$D = \frac{2\alpha m_0 - (\alpha m_1 - \psi m_2)}{2\alpha m_0 + \alpha m_1 - \psi m_2},$$

and compute

$$1 - D = \frac{2\left(\alpha m_1 - \psi m_2\right)}{2\alpha m_0 + \alpha m_1 - \psi m_2}.$$

Hence, under the condition that $(\alpha m_1 - \psi m_2 < 0)$, the retailer does not invest in any promotions for the national brand after introducing the private label $\left(p_1^S = 0\right)$. In this case, the cooperative advertising program can be implemented only if the retailer does promote the national brand and the manufacturer offers the cooperative advertising program i.e., a positive coop participation rate, which is possible only if $(2\alpha m_0 + \alpha m_1 - \psi m_2) > 0$.

Now, suppose that we are in a situation where the following conditions are true

$$\alpha m_1 - \psi m_2 < 0 \quad \text{and} \quad 2\alpha m_0 + \alpha m_1 - \psi m_2 > 0 \tag{11.11}$$

In this case, the retailer does promote the manufacturer's product $(p_1^C > 0)$, however we obtain $D > 1$. This means that the manufacturer has to pay more than the actual cost to get her brand promoted by the retailer in Game C and the constraint $D < 1$ has to be removed.

For $p_1^S = 0$ and when the conditions in (11.11) are satisfied, it is easy to show that the effect of the cooperative advertising program on the profits of retailer and the manufacturer are given by

$$V_R^C\left(G_0\right) - V_R^S\left(G_0\right) = \frac{(\alpha m_1 - \psi m_2)}{4u_r}\left(2\alpha m_0 + m_1\alpha - m_2\psi\right) < 0$$

$$V_M^C\left(G_0\right) - V_M^S\left(G_0\right) = \frac{1}{8u_r}\left(2m_0\alpha + \alpha m_1 - \psi m_2\right)^2 > 0$$

In this case, even if the manufacturer is willing to pay the retailer more then the costs incurred by advertising the national brand, the retailer will not implement the cooperative program.

To wrap up, the message is that the implementability of a coop promotion program depends on the type of competition one assumes between the two brands and the revenues generated from their sales to the retailer. The model we used here is rather simple and some extensions are desirable such as, e.g., letting the margins or prices be endogenous.

References

Ailawadi, K.L. and Harlam, B.A. (2004). An empirical analysis of the determinants of retail margins: The role of store-brand share. *Journal of Marketing*, 68(1):147–165.

Bergen, M. and John, G. (1997). Understanding cooperative advertising participation rates in conventional channels. *Journal of Marketing Research*, 34(3):357–369.

Berger, P.D. (1972). Vertical cooperative advertising ventures. *Journal of Marketing Research*, 9(3):309–312.

Chintagunta, P.K., Bonfrer , A., and Song, I. (2002). Investigating the effects of store brand introduction on retailer demand and pricing behavior. *Management Science*, 48(10):1242–1267.

Dant, R.P. and Berger, P.D. (1996). Modeling cooperative advertising decisions in franchising. *Journal of the Operational Research Society*, 47(9):1120–1136.

Dhar, S.K. and S.J. Hoch (1997). Why store brands penetration varies by retailer. *Marketing Science*, 16(3):208–227.

Jørgensen, S., Sigué, S.P., and Zaccour, G. (2000). Dynamic cooperative advertising in a channel. *Journal of Retailing*, 76(1):71–92.

Jørgensen, S., Taboubi, S., and Zaccour, G. (2001). Cooperative advertising in a marketing channel. *Journal of Optimization Theory and Applications*, 110(1):145–158.

Jørgensen, S., Taboubi, S., and Zaccour, G. (2003). Retail promotions with negative brand image effects: Is cooperation possible? *European Journal of Operational Research*, 150(2):395–405.

Jørgensen, S. and Zaccour, G. (2003). A differential game of retailer promotions. *Automatica*, 39(7):1145–1155.

Kim, S.Y. and Staelin, R. (1999). Manufacturer allowances and retailer pass-through rates in a competitive environment. *Marketing Science*, 18(1):59–76.

Mills, D.E. (1995). Why retailers sell private labels. *Journal of Economics & Management Strategy*, 4(3):509–528.

Mills, D.E. (1999). Private labels and manufacturer counterstrategies. *European Review of Agricultural Economics*, 26:125–145.

Narasimhan, C. and Wilcox, R.T. (1998). Private labels and the channel relationship: A cross-category analysis. *Journal of Business*, 71(4): 573–600.

Nerlove, M. and Arrow, K.J. (1962). Optimal advertising policy under dynamic conditions. *Economica*, 39:129–142.

Raju, J.S., Sethuraman, R., and Dhar, S.K. (1995). The introduction and performance of store brands. *Management Science*, 41(6):957–978.

Chapter 12

INCENTIVE STRATEGIES FOR SHELF-SPACE ALLOCATION IN DUOPOLIES

Guiomar Martín-Herrán
Sihem Taboubi

Abstract We examine the issue of shelf-space allocation in a marketing channel where two manufacturers compete for a limited shelf-space at a retail store. The retailer controls the shelf-space to be allocated to brands while the manufacturers make advertising decisions to build their brand image and to increase final demand (pull strategy). Manufacturers also offer an incentive designed to induce the retailer to allocate more shelf-space to their brands (push strategy). The incentive takes the form of a shelf dependent display allowance. The problem is formulated as a Stackelberg differential game played over an infinite horizon, with manufacturers as leaders. Stationary feedback equilibria are computed, and numerical simulations are carried out in order to illustrate how channel members should allocate their marketing efforts.

1. Introduction

The increasing competition in almost all industries and the proliferation of retailers' private brands in the last decades induce a hard battle for shelf-space between manufacturers at retail stores. Indeed, according to a Food Marketing Institute Report (1999), about 100,000 grocery products are available nowadays on the market, and every year, thousands more new products are introduced. Comparing this number to the number of products that can be placed on the shelves of a typical supermarket (40,000 products) justifies the huge marketing efforts deployed by manufacturers to persuade their dealers to keep their brands on the shelves.

Manufacturers invest in promotional and advertising activities designed to final consumers (pull strategies) and spend in trade promotions designed for their dealers (push strategies). Trade promotions are incentives granted to retailers in return for promoting merchandise at their stores. When these incentives are designed to induce retailers to give a better display to the brand, they are called slotting (or display) allowances.

Shelf-space allocation is a decision that has to be taken by the retailer. However, this issue involves also the other channel members (i.e. manufacturers) as long as they can influence the shelf-space allocation decisions of retailers.

Studies on shelf-space allocation can be found in marketing and operational research literature. Some of these studies adopted a normative perspective where the authors investigated optimal shelf-space allocation decisions (e.g., Corstjens and Doyle (1981); Corstjens and Doyle (1983); Zufreyden (1986); Yang (2001)), while others examined the issue of shelf-space allocation in a descriptive manner by proving that shelf-space has a positive impact on sales (e.g., Curhan (1973); Drèze, Hoch and Purk (1994); Desmet and Renaudin (1998)).

All the studies mentioned above suppose that shelf-space allocation is a decision taken at the retail level, but they neglected the impact of this decision on the marketing decisions of manufacturers, and the impact of manufacturers' decisions on the shelf-space allocation policy of the retailers.

Studies that examined the marketing decisions in channels by taking into account the dynamic long term interactions between channel members adopted differential games as a framework, are, among others, Chintagunta and Jain (1992); Jørgensen, Sigué and Zaccour (2000); Jørgensen, Sigué and Zaccour (2001); Jørgensen, Taboubi and Zaccour (2001); Jørgensen and Zaccour (1999); Taboubi and Zaccour (2002). An exhaustive survey of this literature is presented in Jørgensen and Zaccour (2004).

In the marketing channels literature, studies that suggested the use of incentive strategies as a coordinating mechanisms of the channel used mainly a static framework (e.g., Jeuland and Shugan (1983); Bergen and John (1997)). More recently, Jørgensen and Zaccour (2003) extended this work to a dynamic setting where a manufacturer and a retailer implement two-sided incentives designed to induce each other to choose the coordinating pricing and advertising levels.

To our best knowledge, the only studies that investigated the shelf-space allocation issue by considering the whole marketing channel are those of Jeuland and Shugan (1983) and Wang and Gerchack (2001). Both studies examined the shelf-space allocation decisions as a way to reach channel cooperation (i.e., total channel profit maximization). Wang and Gerchack (2001) design an incentive that takes the form of an inventory-holding subsidy and that leads the retailer to select the coordinating inventory level (i.e., the inventory level that the manufacturer would have allocated to the brand in case of total channel profit maximization).

An important shortcoming of both studies is that the shelf-space allocation decision is considered as a single variable related to the brand of a unique manufacturer (manufacturers of competing brands are passive players in the game). Furthermore, both studies used a static setting, thus ignored the long-term effects of some marketing variables (e.g., advertising and promotional efforts) and the intertemporal interactions that take place between channel members.

Recent works by Martín-Herrán and Taboubi (2004) and Martín-Herrán, Taboubi and Zaccour (2004) examine the issue of shelf-space allocation by taking into account the interactions in the marketing channel and these interactions are not confined to a one-shot game. Both studies assumed that the competing manufacturers can influence the shelf-space allocation decisions of the unique retailer through advertising targeting their consumers (pull strategies) which builds the brand image and increase sales. In both studies, manufacturers' influence on retailer's shelf-space allocation is indirect. Manufacturers do not use push mechanisms to influence directly retailer's decisions.

In the present study we examine the case of a channel where a retailer sells the brands of two competing manufacturers. Retailer's shelf-space allocation decisions can be influenced directly through the use of incentive strategies by both manufacturers (push strategies), or indirectly through advertising (pull strategies).

By considering a dynamic model, we take into account the carry-over effects of manufacturers' advertising investments that build their brand images and the long-term interactions between the partners in the marketing channel.

The paper is organized as follows. Section 2 introduces the model. Section 3 gives the analytical solutions to the shelf-space, advertising, and incentive strategy problem of channel members. Section 4 and 5 present some numerical results to illustrate the findings, and Section 6 concludes.

2. The model

The model is designed to solve the problem of shelf-space allocation, advertising investments, and incentive decisions in a competitive marketing channel. The network is composed of a unique retailer selling the brands of two competing manufacturers.

The retailer has a limited shelf-space in her store. She must decide on how to allocate this limited resource between both brands.

Let $S_i(t)$ denote the shelf-space to be allocated to brand i at time t and consider that the total shelf-space available at the retail store is a constant, normalized to 1. Hence, the relationship between the shelf-spaces given to both brands can be written as

$$S_2(t) = 1 - S_1(t).$$

We assume that shelf-space costs are linear and the unit cost equal for both brands[1]. Without loss of generality, we consider that these costs of shelf-space allocation are equal to zero.

The manufacturers compete on the shelf-space available at the retail store. They control their advertising strategies in national media $(A_i(t))$ in order to increase their goodwill stocks $(G_i(t))$. We consider that the advertising costs are quadratic:

$$C(A_i(t)) = \frac{1}{2}u_i A_i^2(t), \quad i = 1, 2,$$

where u_i is a positive parameter.

Furthermore, in order to increase the final demand for their brands at the retail store, each manufacturer can offer an incentive with the aim that the retailer assigns a greater shelf-space to his brand.

The display allowance takes the form of a shelf dependent incentive to the retailer. We suppose that the incentives given by manufacturers 1 and 2 are, respectively

$$I_1(S_1) = \omega_1 S_1, \quad I_2(S_1) = \omega_2(1 - S_1).$$

The manufacturers control the incentive coefficient functions $\omega_1(t)$ and $\omega_2(t)$, which have to take positive values. The incentive $I_i(S_i)$ is a linear side-payment which rewards the retailer, with the objective that she allocates more shelf-space to brand i.

The retailer faces a demand function for brand i, $D_i = D_i(S_i, G_i)$, that takes the following form:

$$D_i(t) = S_i(t)\left[a_i G_i(t) - \frac{1}{2}b_i S_i(t)\right], \quad i = 1, 2, \tag{12.1}$$

where a_i, b_i are positive parameters and a_i captures the cross effect of shelf-space and goodwill on sales. The interaction between $G_i(t)$ and

[1]Shelf-space costs are costs of removing one item from the shelves and replacing it by another, and by putting price information on products.

$S_i(t)$ means that the goodwill effect on the sales of the brand is enhanced by his share of the shelf-space. Notice that the quadratic term $1/2b_iS_i^2(t)$ in the sales function of each brand captures the decreasing marginal effects of shelf-space on sales, which means that every additional unit of the brand on the shelf leads to lower additional sales than the previous one (see, for example, Bultez and Naert (1988)).

The demand function must have some features that, in turn, impose conditions on shelf-space and the model parameters:

$$D_i(t) \geqslant 0, \quad \partial D_i(t)/\partial G_i(t) \geqslant 0, \quad \partial D_i(t)/\partial S_i(t) \geqslant 0, \quad i = 1, 2,$$

and the resulting constraints on shelf-space are:

$$1 - \frac{a_2}{b_2}G_2(t) \leq S_1(t) \leq \frac{a_1}{b_1}G_1(t). \tag{12.2}$$

The goodwill for brand i is a stock that captures the long-term effects of the advertising of manufacturer i. It evolves according to the Nerlove and Arrow (1962) dynamics:

$$\frac{dG_i(t)}{dt} = \alpha_i A_i(t) - \delta G_i(t), \quad G_i(0) = G_{i0} > 0, \quad i = 1, 2, \tag{12.3}$$

where α_i is a positive parameter that captures the efficiency of the advertising investments of manufacturer i, and δ is a decay rate that reflects the depreciation of the goodwill stock, because of oblivion, product obsolence or competitive advertising.

The game is played over an infinite horizon and firms have a constant and equal discount rate ρ. To focus on the shelf-space allocation and incentive problems, we consider a situation that involves brands in the same product category with relatively similar prices. Hence, the retailer and both manufacturers have constant retail and wholesale margins for brand i, being fixed at the beginning of the game, and denoted by π_{R_i} and π_{M_i}[2].

By choosing the amount of shelf-space to allocate to brand 1, the retailer aims at maximizing her profit flow derived from selling the products of the two brands and the side-payments received from the manufacturers:

$$J_R = \int_0^\infty \exp(-\rho t) \sum_{i=1}^{2} (\pi_{R_i} D_i(t) + \omega_i(t) S_i(t))\, dt. \tag{12.4}$$

[2]This assumption was used in Chintagunta and Jain (1992), Jørgensen, Sigué and Zaccour (2000), and Jørgensen, Taboubi and Zaccour (2001).

Manufacturer i controls his advertising investment, $A_i(t)$, and the incentive coefficient function $\omega_i(t)$. His aim is to maximize his profit flow, taking into account the cost of implementing this strategy:

$$J_{M_i} = \int_0^\infty \exp(-\rho t)\left(\pi_{M_i} D_i(t) - \frac{1}{2} u_i A_i^2(t) - \omega_i S_i(t)\right) dt. \qquad (12.5)$$

To recapitulate, we have defined by (12.3), (12.4) and (12.5) a differential game that takes place between two competing manufacturers selling their brands through a common retailer. The game has five control variables $S_1(t)$, $A_1(t)$, $A_2(t)$, $w_1(t)$, $w_2(t)$ (one for the retailer and two for each manufacturer) and two state variables $G_1(t)$, $G_2(t)$. The controls are constrained by $0 < S_1(t) < 1$, $A_1(t) \geqslant 0$, $A_2(t) \geqslant 0$, $w_1(t) \geqslant 0$, $w_2(t) \geqslant 0$, and the conditions given in (12.2)[3]. The state constraints $G_i(t) \geqslant 0, i = 1, 2$, are automatically satisfied.

3. Stackelberg game

The differential game played between the different channel members is a Stackelberg game where the retailer is the follower and the manufacturers are the leaders. The sequence of the game is the following: the manufacturers as leaders announce, simultaneously, their advertising and incentive strategies. The retailer reacts to this information by choosing the shelf-space level that maximizes her objective functional (12.4). The manufacturers play a non-cooperative game à la Nash. We employ the hypothesis that both manufacturers only observe the evolution of their own goodwill, not that of their competitor[4].

Since the game is played over an infinite time horizon and is autonomous, we suppose that strategies depend on the current level of the state variable only.

The following proposition characterizes the retailer's reaction function.

PROPOSITION 12.1 *If $S_1 > 0$, the retailer's reaction function for shelf-space allocation is given by*[5]*:*

$$S_1(G_1, G_2) = \frac{\omega_1 - \omega_2 + a_1 \pi_{R1} G_1 - a_2 \pi_{R2} G_2 + b_2 \pi_{R2}}{b_1 \pi_{R1} + b_2 \pi_{R2}}. \qquad (12.6)$$

[3]We do not take into account these conditions in the problem resolution. However, we check their fulfillment "a posteriori".

[4]This assumption is mainly set for model's tractability, see Roberts and Samuelson (1988), Jørgensen, Taboubi and Zaccour (2003) and Taboubi and Zaccour (2002). In Martín-Herrán and Taboubi (2004), we prove that the qualitative results still hold whenever the hypothesis is removed.

[5]From now on, the time argument is often omitted when no confusion can arise.

Proof. The retailer's optimization problem is to choose the shelf-space level that maximizes (12.4) subject to the dynamics of the goodwill stocks given in (12.3). The shelf-space decision does not affect the differential equations in (12.3) and therefore, her optimal shelf-space decision is the solution of the following static optimization problem:

$$\max_{S_1} \left\{ \sum_{i=1}^{2} (\pi_{R_i} D_i + \omega_i S_i) \right\}.$$

The expression in (12.6) is the unique interior shelf-space allocation solution of the problem above. □

The proposition states that the shelf-space allocated to each brand is positively affected by its own goodwill and negatively affected by the goodwill stock of the competing brand. Shelf-space allocation depends also on the retail margins of both brands and the parameters of the demand functions.

Furthermore, the state-dependent (see Proposition 12.2 below) coefficient functions of both manufacturers' incentive strategies (ω_i) affect the shelf-space allocation decisions. Indeed, the term $\omega_1 - \omega_2$ in the numerator indicates that the shelf-space allocated to brand 1 is greater under the implementation of the incentive (than without it) if and only if $\omega_1 - \omega_2 > 0$. That is, manufacturer 1 attains his objective by giving the incentive to the retailer only if his incentive coefficient ω_1 is greater than the incentive coefficient ω_2 selected by the other manufacturer. Furthermore, when only one manufacturer offers an incentive, the shelf-space allocated to the other brand is reduced compared to the case where none gives an incentive.

3.1 Manufacturers' incentive strategies

Manufacturers play a Nash game and, as leaders in the Stackelberg game, they know the shelf-space that will be allocated to the brands by the retailer. Both manufacturers decide at the same time their advertising investments and values of the incentive coefficients ω_i. The manufacturers maximize their objective functionals, where the shelf-space has been replaced by its expression in (12.6), subject to the dynamics of their own brand goodwill.

The following proposition characterizes manufacturers' incentive state-dependent coefficient functions at the equilibrium.

PROPOSITION 12.2 *If $\omega_i > 0$, manufacturers' equilibrium incentive coefficients are*

$$\begin{aligned}\omega_i(G_i, G_j) &= -\frac{[b_i(\pi_{M_i} + \pi_{R_i}) + b_j\pi_{R_j}][b_i\pi_{R_i} + b_j(\pi_{M_j} + 2\pi_{R_j})]}{b_i(\pi_{M_i} + 3\pi_{R_i}) + b_j(\pi_{M_j} + 3\pi_{R_j})} \\ &\quad + \frac{(b_i\pi_{R_i} + b_j\pi_{R_j})(\pi_{M_i} - \pi_{R_i}) + b_j\pi_{M_i}(\pi_{M_j} + \pi_{R_j})}{b_i(\pi_{M_i} + 3\pi_{R_i}) + b_j(\pi_{M_j} + 3\pi_{R_j})} a_i G_i \\ &\quad + \frac{(\pi_{M_j} + \pi_{R_j})[b_i(\pi_{M_i} + \pi_{R_i}) + b_j\pi_{R_j}]}{b_i(\pi_{M_i} + 3\pi_{R_i}) + b_j(\pi_{M_j} + 3\pi_{R_j})} a_j G_j, \\ &\qquad i, j = 1, 2,\ i \neq j. \quad (12.7)\end{aligned}$$

Proof. Since the incentives do not affect the dynamics of the goodwill stocks, the manufacturers solve the static optimization problem:

$$\max_{\omega_i} \left[\pi_{M_i} D_i - \frac{1}{2} u_i A_i^2 - \omega_i S_i \right], \quad i = 1, 2,$$

where $S_2 = 1 - S_1$ and S_1 is given in (12.6).

Equating to zero the partial derivative of manufacturer i's objective function with respect to ω_i, we obtain a system of two equations for $\omega_i, i = 1, 2$. Solving this system gives the manufacturers' incentive coefficient functions at the equilibrium in (12.7). □

The proposition states that the incentive coefficients at the equilibrium depend on both channel members' goodwill stocks. The equilibrium value of ω_i is increasing in $G_j, j \neq i$. This means that each manufacturer increases his incentive coefficient when the goodwill stock of the competing brand increases, a behavior that can be explained by the fact that the retailer's shelf-space allocation rule indicates that the shelf-space given to a brand increases with the increase of his own goodwill stock. Hence, the manufacturer of the competing brand has to increase his incentive in order to try to increase his share of the shelf-space.

The incentive coefficient of a manufacturer could be increasing or decreasing in his goodwill stock, depending on the parameter values. The following corollary gives necessary and sufficient conditions ensuring a *negative* relationship between the manufacturer's incentive coefficient function and his own goodwill stock.

COROLLARY 12.1 *Necessary and sufficient conditions guaranteeing that ω_i is a decreasing function of G_i are given by:*

$$(b_i\pi_{R_i} + b_j\pi_{R_j})(\pi_{M_i} - \pi_{R_i}) + b_j\pi_{M_i}(\pi_{M_j} + \pi_{R_j}) < 0, \quad i, j = 1, 2,\ i \neq j. \quad (12.8)$$

In the case of symmetric demand functions, $a_1 = a_2 = a, b_1 = b_2 = b$, *and symmetric margins,* $\pi_{M1} = \pi_{M2} = \pi_M, \pi_{R1} = \pi_{R2} = \pi_R$, *the inequalities in (12.8) reduce to:*

$$\pi_R > (3 + \sqrt{17})/4\pi_M.$$

Proof. Inequalities (12.8) can be derived straightforwardly from the expressions in (12.7). The inequality applying in the symmetric case is obtained from (12.8). □

The inequality for the symmetric case indicates that the manufacturer of one brand will decrease his shelf-space incentive when the goodwill level of his brand increases if the retailer's margin is great enough, compared to the manufacturers' margins.

The next corollary establishes a necessary and sufficient condition guaranteeing that the implementation of the incentive mechanism allows manufacturer i to attain his objective of having a greater shelf-space allocation at the retail store.

COROLLARY 12.2 *The shelf-space allocated to brand i is greater when the incentive strategies are implemented than without, if and only if the following condition holds:*

$$\begin{aligned} b_i^2\pi_{R_i}^2 - b_j^2\pi_{R_j}^2 &+ b_i b_j(\pi_{M_j}\pi_{R_i} - \pi_{M_i}\pi_{R_j}) \\ &- [2b_i\pi_{R_i}^2 + b_j(\pi_{M_j}\pi_{R_i} - (\pi_{M_i} - 2\pi_{R_i})\pi_{R_j})]a_i G_i \\ &+ [2b_j\pi_{R_j}^2 - b_i(\pi_{M_j}\pi_{R_i} - (\pi_{M_i} + 2\pi_{R_i})\pi_{R_j})]a_j G_j > 0, \\ &\qquad i, j = 1, 2,\ i \neq j. \end{aligned} \tag{12.9}$$

In the symmetric case inequality (12.9) reduces to:

$$G_i - G_j < 0, \quad i, j = 1, 2,\ i \neq j.$$

Proof. From the expression of the retailer's reaction function in Proposition 12.1, we have that the shelf-space given to brand i is increased with the incentive whenever the difference $\omega_i - \omega_j$ is positive. This later inequality replacing the optimal expressions of ω_k in (12.7) can be rewritten as in inequality (12.9).

Exploiting the symmetry hypothesis, inequality (12.9) becomes:

$$-\frac{2a(G_i - G_j)\pi_R^2}{\pi_M + 3\pi_R} > 0, \quad i, j = 1, 2,\ i \neq j, \tag{12.10}$$

which is equivalent to $G_i - G_j < 0$. □

The result in (12.10) indicates that the shelf-space allocated by the retailer to brand i under the incentive policy is greater than without the incentive if and only if the goodwill of brand i is lower than that of his competitor. This result means that if the main objective of manufacturer i, when applying the incentive program, is to attain a greater shelf-space allocation for his brand, then he attains this objective only when his brand has a lower goodwill than that of the other manufacturer. The intuition behind this behavior is that the manufacturer with the lowest goodwill stock will be given a lowest share of total shelf-space, thus, he reacts by offering an incentive.

3.2 Manufacturers' advertising decisions

The following proposition gives the advertising strategies and value functions for both manufacturers.

PROPOSITION 12.3 *Assuming interior solutions, manufacturers' advertising strategies and value functions are the following:*

(i) Advertising strategies:

$$A_1(G_1, G_2) = \frac{\alpha_1}{u_1}(K_{11}G_1 + K_{13}G_2 + K_{14}), \tag{12.11}$$

$$A_2(G_1, G_2) = \frac{\alpha_2}{u_2}(K_{23}G_1 + K_{22}G_2 + K_{25}), \tag{12.12}$$

where K_{ij}, $i = 1, 2, j = 1, \ldots, 6$ *are parameters given in the Appendix. Furthermore,* $K_{11}, K_{22} \geq 0$ *and*

- $K_{i3} \geq 0$ *if* $K_{ii}^{'}$ *is chosen and* $\Gamma > 0$*;*
- $K_{i3} \leq 0$ *if* $K_{ii}^{'}$ *is chosen and* $\Gamma < 0$ *or* $K_{ii}^{''}$ *is chosen,*

where $i, j = 1, 2, i \neq j$, $K_{ii}^{'}, K_{ii}^{''}$ *are given in the Appendix and*

$$\Gamma = \delta(\delta+\rho)u_{M_i}(b_i\pi_{Ri}+b_j\pi_{Rj})^2 - \alpha_i^2\pi_{Mi}\pi_{Ri}a_i^2(b_i\pi_{Ri}+2b_j\pi_{Rj}).$$

(ii) Manufacturers' value functions are the following:

$$V_{M_i}(G_1, G_2) = \frac{1}{2}K_{i1}G_1^2 + \frac{1}{2}K_{i2}G_2^2 + K_{i3}G_1G_2 + K_{i4}G_1 + K_{i5}G_2 + K_{i6},$$

$i = 1, 2.$

Proof. See Appendix. □

Item (i) indicates that the Markovian advertising strategies in oligopolies are linear in the goodwill levels of both brands in the market. As in situations of monopoly they satisfy the classical rule equating marginal revenues to marginal costs. However, for competitive situations, the results state that each manufacturer reacts to his own goodwill increase by rising his advertising investments, while its reaction to the increase of his competitor's brand image could be an increase or a decrease of his advertising effort. Such reaction differs from that of monopolist manufacturer who must decrease his advertising when his goodwill increases. This reaction is mainly driven by the model's parameters. According to Roberts and Samuelson (1988), it indicates whether manufacturers' advertising effect is informative or predatory. When the advertising effect is informative, the advertising investment by one manufacturer leads to higher sales for both brands and a higher total market size. In case of predatory advertising, each manufacturer increases his advertising effort in order to increase his goodwill stock. This increase leads the competitor to decrease his advertising effort, which leads to a decrease of his goodwill stock[6].

Item (*ii*) in the proposition indicates that the values functions of both manufacturers are quadratic in the goodwill levels of both brands.

3.3 Shelf-space allocation

The shelf-space allocated to brand 1 can be computed once the optimal values for the incentive coefficients in (12.7) have been substituted. The following proposition characterizes the shelf-space allocation decision and value function of the retailer.

PROPOSITION 12.4 *If* $S_1 > 0$, *retailer's shelf-space allocation decision at the equilibrium and value function are the following:*

(*i*) *Shelf-space allocation at the equilibrium:*

$$S_1(G_1, G_2) = \frac{a_1(\pi_{M_1}+\pi_{R_1})G_1 - a_2(\pi_{M_2}+\pi_{R_2})G_2}{b_1(\pi_{M_1}+3\pi_{R_1})+b_2(\pi_{M_2}+3\pi_{R_2})} + \frac{b_2(\pi_{M_2}+2\pi_{R_2})+b_1\pi_{R_1}}{b_1(\pi_{M_1}+3\pi_{R_1})+b_2(\pi_{M_2}+3\pi_{R_2})}. \tag{12.13}$$

(*ii*) *Retailer's value function:*

$$V_R(G_1, G_2) = \frac{1}{2}L_1 G_1^2 + \frac{1}{2}L_2 G_2^2 + L_3 G_1 G_2 + L_4 G_1 + L_5 G_2 + L_6,$$

where constants $L_k, k = 1, \ldots, 6$ *are given in the Appendix.*

[6]Items (*i*) and (*ii*) in the proposition above are similar, qualitatively speaking, to previous results obtained in Martín-Herrán and Taboubi (2004). For more details about the advertising strategies of manufacturers and proofs about the issue of bounded goodwill stocks and stability of the optimal time paths, see Martín-Herrán and Taboubi (2004).

Proof. Substituting into the retailer's reaction function (12.6) the incentive coefficient functions at the equilibrium in (12.7), the expression (12.13) is obtained.

The proof of item (*ii*) follows the same steps than that of Proposition 12.3 and for that reason is omitted here. In the Appendix we state the expressions of the coefficients of the retailer's value function. □

From item (*i*) in the proposition, the shelf-space allocated to brand 1 is always increasing with its goodwill stock and decreasing with the goodwill of the competitor's brand, 2.

Item (*ii*) of the proposition states that the retailer's value function is quadratic in both goodwill stocks. Let us note that the expression of the retailer's value function is not needed to determine her optimal policy, but only necessary to compute the retailer's profit at the equilibrium.

COROLLARY 12.3 *The shelf-space allocated to brand i is greater than that of the competitor if and only if the following condition is fulfilled:*

$$(2a_iG_i - b_i)(\pi_{M_i} + \pi_{R_i}) - (2a_jG_j - b_j)(\pi_{M_j} + \pi_{R_j}) > 0, \; i, j = 1, 2, \; i \neq j. \tag{12.14}$$

Under the symmetry assumption, (12.14) reduces to

$$G_i - G_j > 0, \quad i, j = 1, 2, \; i \neq j. \tag{12.15}$$

Proof. Expression (12.13) can be rewritten as:

$$S_1 = \frac{1}{2} + \frac{(2a_1G_1 - b_1)(\pi_{M1} + \pi_{R1}) - (2a_2G_2 - b_2)(\pi_{M2} + \pi_{R2})}{2(b_1(\pi_{M1} + \pi_{R1}) + b_2(\pi_{M2} + \pi_{R2}))}.$$

Therefore, the retailer allocates a greater shelf-space to brand 1 than to brand 2 if and only if the following condition is satisfied:

$$(2a_1G_1 - b_1)(\pi_{M1} + \pi_{R1}) - (2a_2G_2 - b_2)(\pi_{M2} + \pi_{R2}) > 0.$$

□

In the symmetric case, the condition in (12.15) means that the retailer gives a highest share of her available shelf-space to the brand with the highest goodwill stock.

4. A numerical example

In order to illustrate the behavior of the retailer's and manufacturers' equilibrium strategies, we present a numerical example. The values of the model parameters are shown in Table 12.1, except $\alpha_1 = 2.7, \alpha_2 = 2.74$. The subscript k in Table 12.1 indicates that the same value of the parameter has been fixed for brands 1 and 2.

Table 12.1. Values of model parameters.

Parameters	π_{M_k}	π_{R_k}	a_k	b_k	u_k	α_k	δ	ρ
Fixed values	1	1.8	0.5	1.62	1	2.7	0.5	0.35

We assume that both players choose $K_{ii}^{''}$, implying that K_{i3} is negative. The steady-state equilibrium for the goodwill variables $(G_1^{\infty}, G_2^{\infty}) = (7.9072, 7.7000)$ is a saddle point[7].

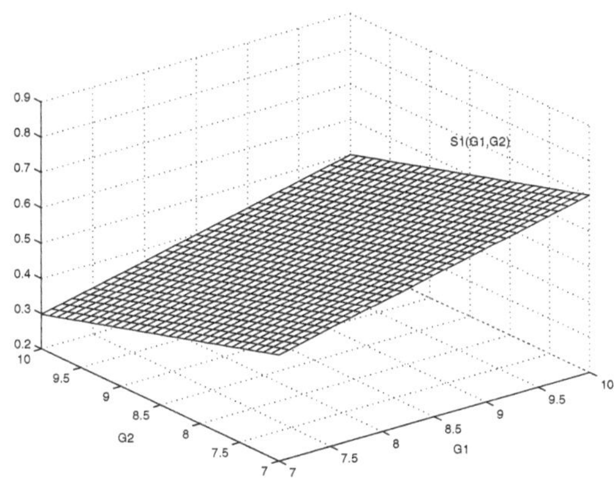

Figure 12.1. Shelf-space feedback strategy

Figure 12.1 shows the retailer's feedback equilibrium strategy and displays how the shelf-space for brand 1 varies according to G_1 and G_2. The slope of the plane shows that the shelf-space for each brand increases with his own goodwill and decreases with that of the competitor. For high values of G_2 and low values of G_1 the highest shelf-space is allocated to brand 2.

Figure 12.2 shows the incentive strategies of both manufacturers. The slopes of the two planes illustrate that the state-dependent coefficient in the incentive strategy of each manufacturer depends negatively on his own goodwill and positively on the goodwill of his competitor. It is easy to verify that both manufacturers choose the same coefficients if and only if G_1 equals G_2. Therefore, as Figure 12.2 depicts for values of G_1 greater than those of G_2, manufacturer 1 chooses an incentive coefficient lower than that of his competitor. The result is reversed when the goodwill stock of the first brand is lower than that of the second brand.

[7]The expressions of the steady-state equilibrium values are shown in the Appendix.

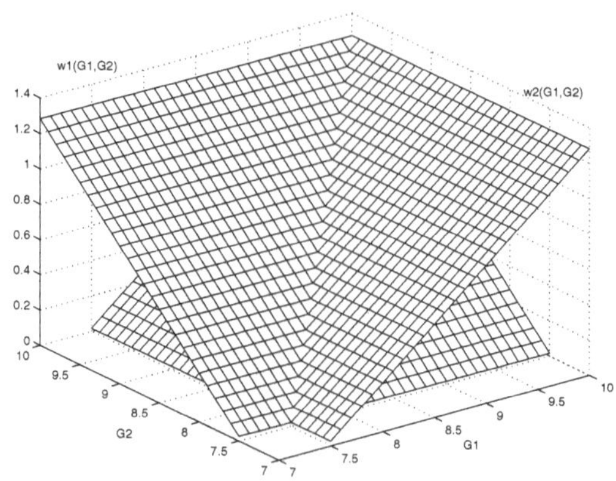

Figure 12.2. Incentive coefficient feedback strategies

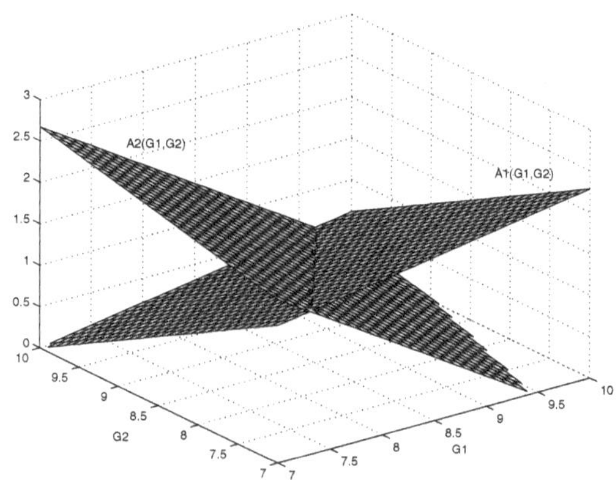

Figure 12.3. Advertising feedback strategies

Figure 12.3 shows the advertising feedback strategies of the manufacturers. The slopes of the two planes illustrate how the advertising strategy of each manufacturer is positively affected by his own goodwill and negatively by the goodwill of his competitor. Note that this behavior is just the opposite of the one presented above for the incentive coefficient feedback strategies. We also notice that for high values of G_1 and low values of G_2, manufacturer 1 invests more in advertising than his competitor, who acts in the opposite way. Both manufacturers invest equally in advertising when their goodwill stocks satisfy the following equality: $G_2 = 1.2744G_1 - 2.3043$. As Figure 12.3 illustrates, manufacturer 1 advertises higher than his competitor if and only if $G_2 < 1.2744G_1 - 2.3043$.

An easy computation allows us to establish that if the goodwill stock for the first brand, G_1 exceeds 8.3976, then whenever $G_2 < G_1$, manufacturer 1 invests more in advertising but chooses a lower incentive coefficient than that of his competitor. When G_1 belongs to the interval $(1.8081, 8.3976)$, the same behavior than before can be guaranteed if the goodwill stocks of the both brands satisfy the following inequality: $G_2 < 1.2744G_1 - 2.3043$.

5. Sensitivity analysis

In order to understand the behavior of the strategies and the outcomes at the steady-state, we use specific values of the parameters. We examine the sensitivity of the strategies and outcomes to these parameters by fixing all except one. The effects of the parameters are identified by comparing the results to a "base" case. The parameter values for the "base" case are given in Table 12.1.

We examined the sensitivity of the strategies and outcomes to the the effect of advertising on goodwill (α_k), the retail margins (π_{Rk}), and the long-term effect of shelf-space on sales (b_k) under symmetric and asymmetric conditions. In Tables 12.2-12.4 we report the results at the steady-states[8]. Moreover, for all the numerical simulations reported below one of the eigenvalues of the Jacobian matrix associated to the dynamical system is negative, leading to asymptotically stable steady-states for the goodwill stocks. All the results we present correspond to values of the parameters for which the incentive coefficient of a manufacturer is a decreasing function of his own goodwill stock [9].

Each table shows the steady-state values for the goodwill stocks, advertising investments, incentive coefficients, shelf-space for the first brand, demand for each brand, and channel members' individual profits.

5.1 Sensitivity to the advertising effect on goodwill

We begin by analyzing the sensitivity of channel members' strategies and outcomes to the variation in the advertising effect on goodwill, under symmetric and asymmetric conditions. The values in the first two columns of Table 12.2 are obtained under the hypotheses of symmetry

[8]For all the numerical simulations it has been verified that the increasing and positiveness conditions on the demand functions leading to conditions (12.2) are satisfied.

[9]The results of the different sensitivity analysis remain qualitatively the same when the values of the parameters lead to incentive coefficient functions which increase with an increment of his own goodwill stock.

in all the parameters, while the values of the last column give the results of a scenario where all the parameters are set equal, except for the parameter that we vary. This case corresponds to a situation of non-symmetry.

Table 12.2. Summary of sensitivity results to the advertising effect on goodwill.

Sensitivity to	$\alpha_k = 2.7$	$\alpha_k = 2.74$	$\alpha_2 = 2.74$
G_1	7.6920	7.9216	7.9072
G_2	7.6920	7.9216	7.7000
A_1	1.4244	1.4456	1.4643
A_2	1.4244	1.4456	1.4051
w_1	0.1200	0.2348	0.1234
w_2	0.1200	0.2348	0.2282
S_1	0.5000	0.5000	0.5140
D_1	1.7205	1.7779	1.8181
D_2	1.7205	1.7779	1.6798
J_{M_1}	1.8456	1.7591	1.9503
J_{M_2}	1.8456	1.7591	1.6621
J_R	18.0395	18.9579	18.4875

The results in the first column of Table 12.2 are intuitive. They indicate that under full symmetry, manufacturers use both pull and push strategies at the steady-state: they invest equally in advertising and give the same incentive to the retailer, who allocates the shelf-space equally to both brands. Interestingly, we can notice that even though the shelf-space is equally shared by both brands, the manufacturers still offer an incentive to the retailer. This behavior can be explained by the fact that both manufacturers act simultaneously without sharing any information about their advertising and incentive decisions. Each manufacturer gives an incentive with the aim of getting a higher share of the shelf-space compared to his competitor.

The second column indicates that a symmetric increase of the advertising effect on the goodwill stock for both brands leads to an increase of advertising and goodwill stocks. The shelf-space is equally shared by both brands, and the manufacturers decide to offer the same incentive coefficient to the retailer, but this coefficient is increased.

Now we remove the hypothesis that the effect of advertising on goodwill is the same for both brands, and suppose that these effects are $\alpha_1 = 2.7$ for brand 1, and $\alpha_2 = 2.74$ for brand 2. The results are reported in the last column of Table 12.2 and state that, when the advertising efficiency of manufacturer 2 is increased[10], compared to that of manufacturer 1, manufacturer 2 allocates his marketing efforts differently (compared to the symmetric case): he invests more in push than in pull strategies. Indeed manufacturer 2 lowers his advertising investment while his competitor invests more in advertising. The resulting goodwill stock for brand 1 becomes higher than that of brand 2. The retailer then gives a highest share of the available shelf-space to brand 1 and the manufacturer of brand 2 has to fix a highest incentive coefficient in order to influence the retailer's shelf-space allocation decision.

5.2 Sensitivity to retailer's profit margins

The results in the first and second columns of Table 12.3 indicate the effects of an increase of retailer's profit margins on channel members' strategies and outcomes under symmetric conditions. We notice that an increase of π_{R_k} leads both manufacturers to allocate more efforts to the pull than the push strategies: both of them increase their advertising investments and decrease their display allowances.

Table 12.3. Summary of sensitivity results to retailer's margin.

Sensitivity to	$\pi_{R_k} = 1.8$	$\pi_{R_k} = 1.85$	$\pi_{R_2} = 1.85$
G_1	7.6920	7.8367	7.9294
G_2	7.6920	7.8367	7.5951
A_1	1.4244	1.4512	1.4684
A_2	1.4244	1.4512	1.4273
w_1	0.1200	0.1113	0.0838
w_2	0.1200	0.1113	0.1455
S_1	0.5000	0.5000	0.5152
D_1	1.7205	1.7567	1.8276
D_2	1.7205	1.7567	1.6507
J_{M_1}	1.8456	1.8513	2.0180
J_{M_2}	1.8456	1.8513	1.6887
J_R	18.0395	18.8886	18.4491

Under non-symmetric conditions, the manufacturer of the brand with the lowest retail margin (the first one in this example) increases his advertising effort, which becomes higher than that of his competitor. This leads to a higher goodwill stock for his brand and a higher share of the

[10]We can imagine a situation where manufacturer 2 chooses a more efficient advertising media or message.

shelf-space. The manufacturer of the other brand reacts by increasing the incentive that he offers to the retailer (compared to the symmetric case). The results indicate that, even though the display allowance of manufacturer 2 is higher than that of manufacturer 1, the retailer's shelf-space allocation decision is driven by the goodwill differential of brands. Sales and profits are higher for the brand with the lowest retail margin, since sales are affected by the shelf-space and the goodwill levels of the brands, which are higher for brand 1.

5.3 Sensitivity to the effect of shelf-space on sales

The results in the first two columns of Table 12.4 indicate that the parameter capturing the long-term effect of shelf-space on sales (b_k) has no effect on the steady-state values of the advertising strategies of the manufacturers. However, it has a positive impact on their incentive strategies which are increased under a decrease of this parameter. Hence, when the long-term effect of shelf-space on sales is decreased, manufacturers are better off when they choose to allocate more marketing efforts in push strategies in order to get an immediate effect on the shelf-space allocation.

Table 12.4. Summary of sensitivity results to the long-term effect of shelf-space on sales.

Sensitivity to	$b_k = 1.62$	$b_k = 1.58$	$b_2 = 1.58$
G_1	7.6920	7.6920	7.7195
G_2	7.6920	7.6920	7.6645
A_1	1.4244	1.4244	1.4295
A_2	1.4244	1.4244	1.4194
w_1	0.1200	0.2120	0.1622
w_2	0.1200	0.2120	0.1698
S_1	0.5000	0.5000	0.5010
D_1	1.7205	1.7255	1.7305
D_2	1.7205	1.7255	1.7155
J_{M_1}	1.8456	1.7285	1.7927
J_{M_2}	1.8456	1.7285	1.6676
J_R	18.0395	18.3538	18.3045

Finally, under non-symmetric conditions, the results in the last column indicate that the manufacturer of the brand that has the lowest long-term effect of shelf-space on sales (in our numerical example, the second one) increases his incentive, but lowers his advertising investment. Thus, his goodwill stock decreases while the goodwill stock of his competitor increases (since the competitor reacts by increasing his advertising investment). The retailer gives a higher share of her shelf-space to the brand with the highest goodwill level and manufacturer 2 reacts by increasing his incentive coefficient and setting it higher than that of his competitor.

6. Concluding remarks

- Manufacturers can affect retailer's shelf-space allocation decisions through the use of incentive strategies (push) and/or advertising investments (pull).

- The numerical results indicate that a manufacturer who wants to influence the retailer's shelf-space allocation decisions can choose between using incentive strategies and/or advertising, this choice depends on the model's parameters.

- In further research, we should remove the hypothesis of myopia and relax the assumption of constant margins by introducing retail and wholesale prices as control variables for the retailer and both manufacturers.

Acknowledgments. Research completed when the first author was visiting professor at GERAD, HEC Montréal. The first author's research was partially supported by MCYT under project BEC2002-02361 and by JCYL under project VA051/03, confinanced by FEDER funds. The second author's research was supported by NSERC, Canada. The authors thank Georges Zaccour and Steffen Jørgensen for helpful comments.

Appendix: Proof of Proposition 12.3

We apply the sufficient condition for a stationary feedback Nash equilibrium and wish to find bounded and continuously functions $V_{M_i}(G_1, G_2), i = 1, 2$, which satisfy, for all $G_i(t) \geq 0, i = 1, 2$, the following HJB equations:

$$\rho V_{M_i}(G_1, G_2) = \max_{A_i} \left\{ \pi_{M_i} D_i - \frac{1}{2} u_i A_i^2 - \omega_i S_i + \frac{\partial V_{M_i}}{\partial G_1}(G_1, G_2)(\alpha_i A_i - \delta G_i) \right\},$$

$i = 1, 2$, where D_i, ω_i and S_i are given in (12.1), (12.7) and (12.6), respectively.

The maximization of the right-hand side of the above equation with respect to A_i leads to

$$A_i(G_1, G_2) = \frac{\alpha_i}{u_{M_i}} \frac{\partial V_{M_i}}{\partial G_i}(G_1, G_2).$$

Substitution of $A_i, i = 1,2$ by their values into the HJB equations leads to conjecture the following quadratic functions for the manufacturers:

$$V_{M_i}(G_i, G_j) = \frac{1}{2}K_{i1}G_1^2 + \frac{1}{2}K_{i2}G_2^2 + K_{i3}G_1G_2 + K_{i4}G_1 + K_{i5}G_2 + K_{i6}.$$

Inserting these expressions as well as their partial derivatives in the HJB equations and identifying coefficients, we obtain the following:

$$K_{11} = \frac{(2\delta+\rho)u_1 \pm \sqrt{[(2\delta+\rho)u_1]^2 - 4\alpha_1^2 u_1 Z}}{2\alpha_1^2},$$

$$Y = (b_1(\pi_{M_1} + 3\pi_{R_1}) + b_2(\pi_{M_2} + 3\pi_{R_2}))^2,$$

$$Z_1 = \frac{a_1^2(\pi_{M_1} + \pi_{R_1})^2(b_1(\pi_{M_1} + 2\pi_{R_1}) + 2b_2\pi_{R_2})}{Y},$$

$$K_{13} = \frac{a_1a_2u_1(\pi_{M_1} + \pi_{R_1})(\pi_{M_2} + \pi_{R_2})(b_1(\pi_{M_1} + 2\pi_{R_1}) + 2b_2\pi_{R_2})}{Y(K_{11}\alpha_1^2 - u_1(\delta+\rho))},$$

$$K_{12} = \frac{(K_{13}\alpha_1)^2Y + a_2^2u_1(\pi_{M_2} + \pi_{R_2})^2(b_1(\pi_{M_1} + 3\pi_{R_1}) + 2b_2\pi_{R_2})}{u_1\rho Y},$$

$$K_{14} = -\frac{a_1u_1(\pi_{M_1} + \pi_{R_1})(b_1(\pi_{M_1} + 2\pi_{R_1}) + 2b_2\pi_{R_2})(b_1\pi_{R_1} + b_2(\pi_{M_2} + 2\pi_{R_2}))}{Y(K_{11}\alpha_1^2 - u_1(\delta+\rho))},$$

$$\begin{aligned} K_{15} = {} & \frac{1}{u_1\rho Y}\left\{K_{13}K_{14}\alpha_1^2Y - a_2u_1(\pi_{M_2} + \pi_{R_2})[b_1^2\pi_{R_1}(\pi_{M_1} + 2\pi_{R_1})\right. \\ & \left. + 2b_2^2\pi_{R_2}(\pi_{M_2} + 2\pi_{R_2}) + b_1b_2(\pi_{M_1}(\pi_{M_2} + 2\pi_{R_2}) + 2\pi_{R_1}(\pi_{M_2} + 3\pi_{R_2}))]\right\}, \end{aligned}$$

$$\begin{aligned} K_{16} = {} & \frac{1}{2u_1\rho Y}\left\{u_1\{b_1^2\pi_{R_1}[b_1\pi_{R_1}(\pi_{M_1} + 2\pi_{R_1}) + 2b_2(\pi_{M_2}(\pi_{M_1} + 2\pi_{R_1})\right. \\ & + \pi_{R_2}(2\pi_{M_1} + 5\pi_{R_1}))] + b_2^2(\pi_{M_2} + \pi_{R_2})[2b_2\pi_{R_2}(\pi_{M_2} + 2\pi_{R_2}) \\ & \left. + b_1(\pi_{M_1}(\pi_{M_2} + 2\pi_{R_2}) + 2\pi_{R_1}(\pi_{M_2} + 4\pi_{R_2}))] + (K_{14}\alpha_1)^2Y\}\right\}. \end{aligned}$$

The coefficients of the value function for the manufacturer 2, as in the standard case, can be obtained following next rule:

$$K_{12} \hookrightarrow K_{21},\ K_{11} \hookrightarrow K_{22},\ K_{13} \hookrightarrow K_{23},\ K_{15} \hookrightarrow K_{24},\ K_{14} \hookrightarrow K_{25},\ K_{16} \hookrightarrow K_{26},$$

where the arrow indicates that in each parameter the subscripts 1 and 2 have been interchanged.

Appendix: Parameters of retailer's value function

Parameters of retailer's value function are the following:

$$N_i = u_i(2\delta+\rho) - 2K_{ii}\alpha_i^2, \quad R_i = \pi_{M_i} + \pi_{R_i},$$

$$P = b_1(\pi_{M_1} + 3\pi_{R_1}) + b_2(\pi_{M_2} + 3\pi_{R_2}), \quad Q = b_1\pi_{R_1} + b_2\pi_{R_2},$$

$$T_i = (2\pi_{M_i} + 3\pi_{R_i})(\pi_{M_i} + 3\pi_{R_i}) + \pi_{R_i}(\pi_{M_i} + 2\pi_{R_i}),$$

$$X_i = \pi_{M_i}^2 + 5\pi_{R_i}\pi_{M_i} + 5\pi_{R_i}^2, \quad Y_i = \pi_{M_j}R_i + \pi_{R_j}(5\pi_{M_i} + 9\pi_{R_i}),$$

$$Z_i = 12\pi_{M_j}\pi_{R_j}(\pi_{M_i} + 2\pi_{R_i}) + \pi_{M_j}^2(2\pi_{M_i} + 3\pi_{R_i}) + 2\pi_{R_j}^2(7\pi_{M_i} + 17\pi_{R_i}),$$
$$i, j = 1, 2, i \neq j,$$

$$L_1 = \frac{u_1(2K_{23}L_3\alpha_2^2P^2 + a_1^2u_2{R_1}^2Q)}{u_2P^2N_1},$$

$$L_2 = \frac{u_2(2K_{13}L_3\alpha_1^2P^2 + a_2^2u_1{R_2}^2Q)}{u_1P^2N_2},$$

$$L_3 = \frac{u_1u_2Q(a_2^2K_{23}\alpha_2^2R_2^2N_1 + a_1^2K_{13}\alpha_1^2R_1^2N_2 - a_1a_2R_1R_2N_1N_2)}{P^2(K_{11}u_2\alpha_1^2 + u_1(K_{22}\alpha_2^2 - u_2(2\delta+\rho)))(4K_{13}K_{23}\alpha_1^2\alpha_2^2 - N_1N_2)},$$

$$L_4 = \frac{u_1[2K_{14}K_{23}L_3\alpha_1^2\alpha_2^2P^2 + \alpha_2^2(K_{25}L_3 + K_{23}L_5)P^2N_1 + a_1^2K_{14}u_2\alpha_1^2R_1^2Q]}{u_2P^2N_1(\alpha_1^2K_{11} - u_1(\delta+\rho))}$$
$$+ \frac{u_1a_1R_1N_1(b_1^2\pi_{R_1}(\pi_{M_1} + 4\pi_{R_1}) + b_2^2X_2 + b_1b_2Y_1}{P^2N_1(\alpha_1^2K_{11} - u_1(\delta+\rho))},$$

$$L_5 = \frac{u_2[N_2P^2\alpha_1^2(L_4K_{13} + L_3K_{14}) + K_{25}\alpha_2^2(Qa_2^2u_1R_2^2 + 2L_3\alpha_1^2P^2)]}{u_1P^2N_2(u_2(\delta+\rho) - K_{22}\alpha_2^2)}$$
$$+ \frac{a_2u_1u_2R_2[b_1^2X_1 + b_2^2(\pi_{M_2} + 4\pi_{R_2}) + b_1b_2Y_2]}{u_1P^2(u_2(\delta+\rho) - K_{22}\alpha_2^2)},$$

$$L_6 = \frac{2P^2(u_1K_{25}L_5\alpha_2^2 + u_2K_{14}L_4\alpha_1^2) - u_1u_2[b_1^3\pi_{R1}T_1 + b_2^3\pi_{R2}T_2 + b_1b_2(b_1Z_1 - b_2Z_2)]}{2\rho u_1u_2P^2}.$$

Appendix: Steady-state equilibrium values for the goodwill stocks

The steady-state values are given by:

$$G_1^\infty = -\frac{\alpha_1^2(K_{13}K_{25}\alpha_2^2 + (u_2\delta - K_{22}\alpha_2^2)K_{14})}{K_{13}K_{23}\alpha_1^2\alpha_2^2 - (u_1\delta - K_{11}\alpha_1^2)(u_2\delta - K_{22}\alpha_2^2)},$$

$$G_2^\infty = -\frac{\alpha_2^2(K_{23}K_{14}\alpha_1^2 + (u_1\delta - K_{11}\alpha_1^2)K_{25})}{K_{13}K_{23}\alpha_1^2\alpha_2^2 - (u_1\delta - K_{11}\alpha_1^2)(u_2\delta - K_{22}\alpha_2^2)}.$$

References

Bergen, M. and John, G. (1997). Understanding cooperative advertising participation rates in conventional channels. *Journal of Marketing Research*, 34:357–369.

Bultez, A. and Naert, P. (1988). SHARP: Shelf-allocation for retailers' profit. *Marketing Science*, 7:211–231.

Chintagunta, P.K. and Jain, D. (1992). A dynamic model of channel member strategies for marketing expenditures. *Marketing Science*, 11:168–188.

Corstjens, M. and Doyle, P. (1981). A model for optimizing retail space allocation. *Management Science*, 27:822–833.

Corstjens, M. and Doyle, P. (1983). A dynamic model for strategically allocating retail space. *Journal of the Operational Research Society*, 34:943–951.

Curhan, R.C. (1973). Shelf space allocation and profit maximization in mass retailing. *Journal of Marketing*, 37:54–60.

Desmet, P. and Renaudin, V. (1998). Estimation of product category sales responsiveness to allocated shelf space. *International Journal of Research in Marketing*, 15:443–457.

Drèze, X., Hoch, S.J., and Purk, M.E. (1994). Shelf management and space elasticity. *Journal of Retailing*, 70:301–326.

Food Marketing Institute Report. (1999). Slotting allowances in the supermarket industry. FMI News Center. htpp://www.fmi.org

Jeuland, A.P. and Shugan, S.M. (1983). Coordination in marketing channels. In: D.A. Gautshi (ed.), *Productivity and Efficiency in Distribution Systems*, Elsevier Publishing Co.

Jørgensen, S., Sigué, S.P., and Zaccour, G. (2000). Dynamic cooperative advertising in a channel. *Journal of Retailing*, 76:71–92.

Jørgensen, S., Sigué, S.P., and Zaccour, G. (2001). Stackelberg leadership in a marketing channel. *International Game Theory Review*, 3:13–26.

Jørgensen, S., Taboubi, S., and Zaccour, G. (2001). Cooperative advertising in a marketing channel. *Journal of Optimization Theory and Applications*, 110:145–158.

Jørgensen, S., Taboubi, S., and Zaccour, G. (2003). Retail promotions with negative brand image effects: Is cooperation possible? *European Journal of Operational Research*, 150:395–405.

Jørgensen, S. and Zaccour, G. (1999). Equilibrium pricing and advertising strategies in a marketing channel. *Journal of Optimization Theory and Applications*, 102:111–125.

Jørgensen, S. and Zaccour, G. (2003). Channel coordination over time: Incentive equilibria and credibility. *Journal of Economic Dynamics and Control*, 27:801–822.

Jørgensen, S. and Zaccour, G. (2004). *Differential Games in Marketing*. Kluwer Academic Publishers.

Martín-Herrán, G. and Taboubi, S. (2004). Shelf-space allocation and advertising decisions in the marketing channel: A differential game approach. To appear in *International Game Theory Review*.

Martín-Herrán, G., Taboubi, S., and Zaccour, G. (2004). *A Time-Consistent Open-Loop Stackelberg Equilibrium of Shelf-space Allocation*. GERAD Discussion Paper G-2004-16.

Nerlove, M. and Arrow, K.J. (1962). Optimal advertising policy under dynamic conditions. *Economica*, 39:129–142.

Roberts, M.J. and Samuelson, L. (1988). An empirical analysis of dynamic nonprice competition in an oligopolistic industry. *RAND Journal of Economics*, 19:200–220.

Taboubi, S. and Zaccour, G. (2002). Impact of retailer's myopia on channel's strategies. In: G. Zaccour (ed.), *Optimal Control and Differential Games, Essays in Honor of Steffen Jørgensen*, pages 179–192, Advances in Computational Management Science, Kluwer Academic Publishers.

Wang, Y. and Gerchack, Y. (2001). Supply chain coordination when demand is shelf-space dependent. *Manufacturing & Service Operations Management*, 3:82–87.

Yang, M.H. (2001). An efficient algorithm to allocate shelf-space. *European Journal of Operational Research*, 131:107–118.

Zufreyden, F.S. (1986). A dynamic programming approach for product selection and supermarket shelf-space allocation. *Journal of the Operational Research Society*, 37:413–422.

Chapter 13

SUBGAME CONSISTENT DORMANT-FIRM CARTELS

David W.K. Yeung

Abstract Subgame consistency is a fundamental element in the solution of cooperative stochastic differential games. In particular, it ensures that the extension of the solution policy to a later starting time and any possible state brought about by prior optimal behavior of the players would remain optimal. Hence no players will have incentive to deviate from the initial plan. Recently a general mechanism for the derivation of payoff distribution procedures of subgame consistent solutions in stochastic cooperative differential games has been found. In this paper, we consider a duopoly in which the firms agree to form a cartel. In particular, one firm has absolute and marginal cost advantage over the other forcing one of the firms to become a dormant firm. A subgame consistent solution based on the Nash bargaining axioms is derived.

1. Introduction

Formulation of optimal behaviors for players is a fundamental element in the theory of cooperative games. The players' behaviors satisfying some specific optimality principles constitute a solution of the game. In other words, the solution of a cooperative game is generated by a set of optimality principles (for instance, the Nash bargaining solution (1953) and the Shapley values (1953)). For dynamic games, an additional stringent condition on their solutions is required: the specific optimality principle must remain optimal at any instant of time throughout the game along the optimal state trajectory chosen at the outset. This condition is known as *dynamic stability or time consistency.* In particular, the dynamic stability of a solution of a cooperative differential game is the property that, when the game proceeds along an "optimal" trajectory, at each instant of time the players are guided by the same optimality principles, and hence do not have any ground for deviation from the previously adopted "optimal" behavior throughout the game.

The question of dynamic stability in differential games has been explored rigorously in the past three decades. Haurie (1976) discussed the problem of instability in extending the Nash bargaining solution to differential games. Petrosyan (1977) formalized mathematically the notion of dynamic stability in solutions of differential games. Petrosyan and Danilov (1979 and 1985) introduced the notion of "imputation distribution procedure" for cooperative solution. Tolwinski et al. (1986) considered cooperative equilibria in differential games in which memory-dependent strategies and threats are introduced to maintain the agreed-upon control path. Petrosyan and Zenkevich (1996) provided a detailed analysis of dynamic stability in cooperative differential games. In particular, the method of regularization was introduced to construct time consistent solutions. Yeung and Petrosyan (2001) designed a time consistent solution in differential games and characterized the conditions that the allocation distribution procedure must satisfy. Petrosyan (2003) used regularization method to construct time consistent bargaining procedures.

A cooperative solution is subgame consistent if an extension of the solution policy to a situation with a later starting time and any feasible state would remain optimal. Subgame consistency is a stronger notion of time-consistency. Petrosyan (1997) examined agreeable solutions in differential games. In the presence of stochastic elements, subgame consistency is required in addition to dynamic stability for a credible cooperative solution. In the field of cooperative stochastic differential games, little research has been published to date due to the inherent difficulties in deriving tractable subgame consistent solutions. Haurie et al. (1994) derived cooperative equilibria of a stochastic differential game of fishery with the use of monitoring and memory strategies. As pointed out by Jørgensen and Zaccour (2001), conditions ensuring time consistency of cooperative solutions could be quite stringent and analytically intractable. The recent work of Yeung and Petrosyan (2004) developed a generalized theorem for the derivation of analytically tractable "payoff distribution procedure" of subgame consistent solution. Being capable of deriving analytical tractable solutions, the work is not only theoretically interesting but would enable the hitherto intractable problems in cooperative stochastic differential games to be fruitfully explored.

In this paper, we consider a duopoly game in which one of the firms enjoys absolute cost advantage over the other. A subgame consistent solution is developed for a cartel in which one firm becomes a dormant partner. The paper is organized as follows. Section 2 presents the formulation of a dynamic duopoly game. In Section 3, Pareto optimal trajectories under cooperation are derived. Section 4 examines the notion of subgame consistency and the subgame consistent payoff distribution. Section 5 presents a subgame consistent cartel based on the Nash bargaining axioms. An illustration is provided in Section 6. Concluding remarks are given in Section 7.

2. A generalized dynamic duopoly game

Consider a duopoly in which two firms are allowed to extract a renewable resource within the duration $[t_0, T]$. The dynamics of the resource is characterized by the stochastic differential equations:

$$dx(s) = f[s, x(s), u_1(s) + u_2(s)]\, ds + \sigma[s, x(s)]\, dz(s),$$
$$x(t_0) = x_0 \in X, \tag{13.1}$$

where $u_i \in U_i$ is the (nonnegative) amount of resource extracted by firm i, for $i \in [1, 2]$, $\sigma[s, x(s)]$ is a scaling function and $z(s)$ is a Wiener process.

The extraction cost for firm $i \in N$ depends on the quantity of resource extracted $u^i(s)$ and the resource stock size $x(s)$. In particular, firm i's extraction cost can be specified as $c^i[u_i(s), x(s)]$. This formulation of unit cost follows from two assumptions: (i) the cost of extraction is proportional to extraction effort, and (ii) the amount of resource extracted, seen as the output of a production function of two inputs (effort and stock level), is increasing in both inputs (See Clark (1976)). In particular, firm 1 has absolute and marginal cost advantage so that $c^1(u_1, x) < c^2(u_2, x)$ and $\partial c^1(u_1, x)/\partial u_1 < \partial c^2(u_2, x)/\partial u_2$.

The market price of the resource depends on the total amount extracted and supplied to the market. The price-output relationship at time s is given by the following downward sloping inverse demand curve $P(s) = g[Q(s)]$, where $Q(s) = u_1(s) + u_2(s)$ is the total amount of resource extracted and marketed at time s. At time T, firm i will receive a termination bonus $q_i(x(T))$. There exists a discount rate r, and profits received at time t has to be discounted by the factor $\exp[-r(t - t_0)]$.

At time t_0, the expected profit of firm $i \in [1, 2]$ is:

$$E_{t_0}\left\{ \int_{t_0}^{T} \left[g[u_1(s) + u_2(s)]\, u_i(s) - c^i[u_i(s), x(s)] \right] \exp[-r(s - t_0)]\, ds \right.$$
$$\left. + \exp[-r(T - t_0)]\, q_i[x(T)] \mid x(t_0) = x_0 \right\}, \tag{13.2}$$

where E_{t_0} denotes the expectation operator performed at time t_0.

We use $\Gamma(x_0, T-t_0)$ to denote the game (13.1)–(13.2) and $\Gamma(x_\tau, T-\tau)$ to denote an alternative game with state dynamics (13.1) and payoff structure (13.2), which starts at time $\tau \in [t_0, T]$ with initial state $x_\tau \in X$. A non-cooperative Nash equilibrium solution of the game $\Gamma(x_\tau, T-\tau)$ can be characterized with the techniques introduced by Fleming (1969), Isaacs (1965) and Bellman (1957) as:

DEFINITION 13.1 *A set of feedback strategies* $\{u_i^{(\tau)*}(t) = \phi_i^{(\tau)*}(t,x),$ *for* $i \in [1,2]\}$, *provides a Nash equilibrium solution to the game* $\Gamma(x_\tau, T-\tau)$, *if there exist twice continuously differentiable functions* $V^{(\tau)i}(t,x) : [\tau, T] \times R \to R$, $i \in [1,2]$, *satisfying the following partial differential equations:*

$$-V_t^{(\tau)i}(t,x) - \frac{1}{2}\sigma(t,x)^2 V_{xx}^{(\tau)i}(t,x) =$$
$$\max_{u_i}\left\{\left[g\left[u_i + \phi_j^{(\tau)*}(t,x)\right]u_i - c^i[u_i,x]\right]\exp[-r(t-\tau)]\right.$$
$$\left.+V_x^{(\tau)i}(t,x) f\left[t,x,,u_i + \phi_j^{(\tau)}(t,x)\right]\right\}, \quad \textit{and}$$
$$V^{(\tau)i}(T,x) = q_i(x)\exp[-r(T-\tau)]\,ds,$$
$$\textit{for } i \in [1,2], \quad j \in [1,2] \quad \textit{and } j \neq i.$$

REMARK 13.1 From Definition 13.1, one can readily verify that $V^{(\tau)i}(t,x) = V^{(s)i}(t,x)\exp[-r(\tau-s)]$ and $\phi_i^{(\tau)*}(t,x) = \phi_i^{(s)*}(t,x)$, for $i \in [1,2]$, $t_0 \le \tau \le s \le t \le T$ and $x \in X$.

3. Dynamic cooperation and Pareto optimal trajectory

Assume that the firms agree to form a cartel. Since profits are in monetary terms, these firms are required to solve the following joint profit maximization problem to achieve a Pareto optimum:

$$E_{t_0}\left\{\int_{t_0}^{T}\left[g\left[u_1(s)+u_2(s)\right]\left[u_1(s)+u_2(s)\right] - c^1\left[u_1(s),x(s)\right]\right.\right.$$
$$\left.-c^2\left[u_2(s),x(s)\right]\right]\exp[-r(s-t_0)]\,ds$$
$$\left.+\exp[-r(T-t_0)]\left(q_i[x(T)]+q_i[x(T)]\right] \mid \; x(t_0)=x_0\right\}, \tag{13.3}$$

subject to dynamics (13.1).

An optimal solution of the problem (13.1) and (13.3) can be characterized with the techniques introduced by Fleming's (1969) stochastic control techniques as:

DEFINITION 13.2 *A set of feedback strategies* $\left[\psi_1^{(t_0)*}(s,x), \psi_2^{(t_0)*}(s,x)\right]$, *for* $s \in [t_0, T]$ *provides an optimal control solution to the problem* (13.1) *and* (13.3), *if there exist a twice continuously differentiable function* $W^{(t_0)}(t,x) : [t_0,T] \times R \to R$ *satisfying the following partial differential equations:*

$$-W_t^{(t_0)}(t,x) - \frac{1}{2}\sigma(t,x)^2 W_{xx}^{(t_0)}(t,x) =$$
$$\max_{u_1,u_2}\Big\{ \left[g(u_1+u_2)(u_1+u_2) - c^1(u_1,x) - c^2(u_2,x)\right]\exp\left[-r(t-\tau)\right]$$
$$+W_x^{(t_0)}(t,x) f(t,x,u_1+u_2)\Big\}, \quad \text{and}$$
$$W^{(t_0)}(T,x) = \left[q_1(x) + q_2(x)\right]\exp\left[-r(T-t_0)\right].$$

Performing the indicated maximization in Definition 13.2 yields:

$$g'(u_1+u_2)u_1 + g(u_1+u_2) + W_x^{(t_0)}(t,x) f_{u_1+u_2}(t,x,u_1+u_2)$$
$$-\partial c^1(u_1,x)/\partial u_1 \le 0, \tag{13.4}$$

and

$$g'(u_1+u_2)u_2 + g(u_1+u_2) + W_x^{(t_0)}(t,x) f_{u_1+u_2}(t,x,u_1+u_2)$$
$$-\partial c^2(u_1,x)/\partial u_2 \le 0. \tag{13.5}$$

Since $\partial c^1(u_1,x)/\partial u_1 < \partial c^2(u_2,x)/\partial u_2$, firm 2 has to refrain from extraction.

Upon substituting $\psi_1^{(t_0)*}(t,x)$ and $\psi_2^{(t_0)*}(t,x)$ into (13.1) yields the optimal cooperative state dynamics as:

$$dx(s) = f\left[s, x(s), \psi_1^{(t_0)*}(s,x(s))\right] ds + \sigma\left[s,x(s)\right] dz(s),$$
$$x(t_0) = x_0 \in X. \tag{13.6}$$

The solution to (13.6) yields a Pareto optimal trajectory, which can be expressed as:

$$x^*(t) = x_0 + \int_{t_0}^{t} f\left[s, x(s), \psi_1^{(t_0)*}(s,x(s))\right] ds + \int_{t_0}^{t} \sigma\left[s,x(s)\right] dz(s). \tag{13.7}$$

We denote the set containing realizable values of $x^*(t)$ by $X_t^{\alpha_1(t_0)}$, for $t \in (t_0, T]$.

We use $\Gamma_c(x_0, T-t_0)$ to denote the cooperative game (13.1)–(13.2) and $\Gamma_c(x_\tau, T-\tau)$ to denote an alternative game with state dynamics (13.1) and payoff structure (13.2), which starts at time $\tau \in [t_0, T]$ with initial state $x_\tau \in X_\tau^*$.

REMARK 13.2 One can readily show that:

$$W^{(\tau)}(s,x) = \exp\left[-r(t-\tau)\right] W^{(t)}(s,x), \text{ and}$$
$$\psi_i^{(\tau)*}(s,x(s)) = \psi_i^{(t)*}(s,x(s)),$$
for $s \in [t,T]$ and $i \in [1,2]$ and $t_0 \leq \tau \leq t \leq s \leq T$.

4. Subgame consistency and payoff distribution

Consider the cooperative game $\Gamma_c(x_0, T-t_0)$ in which total cooperative payoff is distributed between the two firms according to an agree-upon optimality principle. At time t_0, with the state being x_0, we use the term $\xi^{(t_0)i}(t_0,x_0)$ to denote the expected share/imputation of total cooperative payoff (received over the time interval $[t_0,T]$) to firm i guided by the agree-upon optimality principle. We use $\Gamma_c(x_\tau, T-\tau)$ to denote the cooperative game which starts at time $\tau \in [t_0,T]$ with initial state $x_\tau \in X_\tau^*$. Once again, total cooperative payoff is distributed between the two firms according to same agree-upon optimality principle as before. Let $\xi^{(\tau)i}(\tau,x_\tau)$ denote the expected share/imputation of total cooperative payoff given to firm i over the time interval $[\tau,T]$.

The vector $\xi^{(\tau)}(\tau,x_\tau) = \left[\xi^{(\tau)1}(\tau,x_\tau), \xi^{(\tau)2}(\tau,x_\tau)\right]$, for $\tau \in (t_0,T]$, yields valid imputations if the following conditions are satisfied.

DEFINITION 13.3 *The vectors $\xi^{(\tau)}(\tau,x_\tau)$ is an imputation of the cooperative game $\Gamma_c(x_\tau, T-\tau)$, for $\tau \in (t_0,T]$, if it satisfies:*

(i) $\sum_{j=1}^{2} \xi^{(\tau)j}(\tau,x_\tau) = W^{(\tau)}(\tau,x_\tau)$, *and*

(ii) $\xi^{(\tau)i}(\tau,x_\tau) \geq V^{(\tau)i}(\tau,x_\tau)$, *for $i \in [1,2]$ and $\tau \in [t_0,T]$.*

In particular, part (i) of Definition 13.3 ensures Pareto optimality, while part (ii) guarantees individual rationality.

A payoff distribution procedure (PDP) of the cooperative game (as proposed in Petrosyan (1997) and Yeung and Petrosyan (2004)) must be now formulated so that the agreed imputations can be realized. Let the vectors $B^\tau(s) = [B_1^\tau(s), B_2^\tau(s)]$ denote the instantaneous payoff of the cooperative game at time $s \in [\tau,T]$ for the cooperative game $\Gamma_c(x_\tau, T-\tau)$. In other words, firm i, for $i \in [1,2]$, is offered a payoff equaling $B_i^\tau(s)$ at time instant s. A terminal payment $q^i(x(T))$ is given to firm i at time T.

In particular, $B_i^\tau(s)$ and $q^i(x(T))$ constitute a PDP for the game $\Gamma_c(x_\tau, T-\tau)$ in the sense that $\xi^{(\tau)i}(\tau,x_\tau)$ equals:

$$E_\tau\left\{\left(\int_\tau^T B_i^\tau(s)\exp\left[-r(s-\tau)\right] ds\right.\right.$$

$$+q^i(x(T))\exp[-r(T-\tau)]\Big)\mid x(\tau)=x_\tau\Big\}, \quad (13.8)$$

for $i\in[1,2]$ and $\tau\in[t_0,T]$.

Moreover, for $i\in[1,2]$ and $t\in[\tau,T]$, we use $\xi^{(\tau)i}(t,x_t)$ which equals

$$E_\tau\Big\{\Big(\int_t^T B_i^\tau(s)\exp[-r(s-\tau)]\,ds$$
$$+q^i(x(T))\exp[-r(T-\tau)]\Big)\mid x(t)=x_t\Big\}, \quad (13.9)$$

to denote the expected present value of firm i's cooperative payoff over the time interval $[t,T]$, given that the state is x_t at time $t\in[\tau,T]$, for the game which starts at time τ with state $x_\tau\in X_\tau^*$.

Definition 13.4 *The imputation vectors* $\xi^{(t)}(t,x_t)=[\xi^{(t)1}(t,x_t),\xi^{(t)2}(t,x_t)]$, *for* $t\in[t_0,T]$, *as defined by* (13.8) *and* (13.9), *are subgame consistent imputations of* $\Gamma_c(x_\tau,T-\tau)$ *if they satisfy Definition 13.3 and the condition that* $\xi^{(t)i}(t,x_t)=\exp[-r(t-\tau)]\,\xi^{(\tau)i}(t,x_t)$, *where* $t_0\le\tau\le t\le T$, $i\in[1,2]$ *and* $x_t\in X_t^{(\tau)*}$.

The conditions in Definition 13.4 guarantee subgame consistency of the solution imputations throughout the game interval in the sense that the extension of the solution policy to a situation with a later starting time and any possible state brought about by prior optimal behavior of the players remains optimal.

For Definition 13.4 to hold, it is required that $B_i^\tau(s)=B_i^t(s)$, for $i\in[1,2]$ and $\tau\in[t_0,T]$ and $t\in[t_0,T]$ and $r\neq t$. Adopting the notation $B_i^\tau(s)=B_i^t(s)=B_i(s)$ and applying Definition 13.4, the PDP of the subgame consistent imputation vectors $\xi^{(\tau)}(\tau,x_\tau)$ has to satisfy the following condition.

Corollary 13.1 *The PDP with* $B(s)$ *and* $q(x(T))$ *corresponding to the subgame consistent imputation vectors* $\xi^{(\tau)}(\tau,x_\tau)$ *must satisfy the following conditions:*

(i) $$\sum_{j=1}^{2}B_i(s)=\Big[g\big[\psi_1^{(\tau)*}(s)+\psi_2^{(\tau)*}(s)\big]\big[\psi_1^{(\tau)*}(s)+\psi_2^{(\tau)*}(s)\big]$$
$$-c^1\big[\psi_1^{(\tau)*}(s),x(s)\big]-c^1\big[\psi_1^{(\tau)*}(s),x(s)\big]\Big],$$
for $s\in[t_0,T]$;

(ii) $$E_\tau\Big\{\Big(\int_\tau^T B_i(s)\exp[-r(s-\tau)]\,ds$$
$$+q^i(x(T))\exp[-r(T-\tau)]\Big)\mid x(\tau)=x_\tau\Big\}\ge V^{(\tau)i}(\tau,x_\tau),$$
for $i\in[1,2]$ *and* $\tau\in[t_0,T]$; *and*

(iii) $\xi^{(\tau)i}(\tau, x_\tau) =$
$$E_\tau\left\{\left(\int_\tau^{\tau+\Delta t} B_i(s)\exp[-r(s-\tau)]\,ds + \exp\left[-\int_\tau^{\tau+\Delta t} r(y)\,dy\right]\right.\right.$$
$$\left.\times\xi^{(\tau+\Delta t)i}(\tau+\Delta t, x_\tau+\Delta x_\tau) \mid x(\tau)=x_\tau\right\},$$
for $\tau \in [t_0, T]$ *and* $i \in [1,2]$;

where
$\Delta x_\tau = f\left[\tau, x_\tau, \psi_1^{(\tau)*}(\tau, x_\tau)\right]\Delta t + \sigma[\tau, x_\tau]\Delta z_\tau + o(\Delta t)$,
$x(\tau) = x_\tau \in X_\tau^*$, $\Delta z_\tau = z(\tau+\Delta t) - z(\tau)$, *and* $E_\tau[o(\Delta t)]/\Delta t \to 0$ *as* $\Delta t \to 0$.

Consider the following condition concerning subgame consistent imputations $\xi^{(\tau)}(\tau, x_\tau)$, for $\tau \in [t_0, T]$:

CONDITION 13.1 *For* $i \in [1,2]$ *and* $t \geq \tau$ *and* $\tau \in [t_0, T]$, *the terms* $\xi^{(\tau)i}(t, x_t)$ *are functions that are continuously twice differentiable in* t *and* x_t.

If the subgame consistent imputations $\xi^{(\tau)}(\tau, x_\tau)$, *for* $\tau \in [t_0, T]$, *satisfy Condition 13.1, a PDP with* $B(s)$ *and* $q(x(T))$ *will yield the relationship:*

$$E_\tau\left\{\int_\tau^{\tau+\Delta t} B_i(s)\exp\left[-\int_\tau^s r(y)\,dy\right]ds \mid x(\tau)=x_\tau\right\}$$
$$= E_\tau\left\{\xi^{(\tau)i}(\tau, x_\tau) - \exp\left[-\int_\tau^{\tau+\Delta t} r(y)\,dy\right]\xi^{(\tau+\Delta t)i}(\tau+\Delta t, x_\tau+\Delta x_\tau)\right\}$$
$$= E_\tau\left\{\xi^{(\tau)i}(\tau, x_\tau) - \xi^{(\tau)i}(\tau+\Delta t, x_\tau+\Delta x_\tau)\right\},$$
$$\text{for all } \tau \in [t_0, T] \text{ and } i \in [1,2]. \tag{13.10}$$

With $\Delta t \to 0$, condition (13.10) can be expressed as:

$$E_\tau\{B_i(\tau)\Delta t + o(\Delta t)\}$$
$$= E_\tau\left\{-\left[\xi_t^{(\tau)i}(t, x_t)\,|_{t=\tau}\right]\Delta t\right.$$
$$-\left[\xi_{x_t}^{(\tau)i}(t, x_t)\,|_{t=\tau}\right] f\left[\tau, x_\tau, \psi_1^{(\tau)*}(\tau, x_\tau)\right]\Delta t$$
$$-\frac{1}{2}\sigma[\tau, x_\tau]^2\left[\xi_{x_t^h x_t^\zeta}^{(\tau)i}(t, x_t)\,|_{t=\tau}\right]\Delta t$$
$$\left.-\left[\xi_{x_t}^{(\tau)i}(t, x_t)\,|_{t=\tau}\right]\sigma[\tau, x_\tau]\Delta z_\tau - o(\Delta t)\right\}. \tag{13.11}$$

Taking expectation and dividing (13.11) throughout by Δt, with $\Delta t \to 0$, yield

$$B_i(\tau) = -\left[\xi_t^{(\tau)i}(t, x_t)\,|_{t=\tau}\right]$$

$$- \left[\xi_{x_t}^{(\tau)i} (t, x_t) \,|_{t=\tau} \right] f \left[\tau, x_\tau, \psi_1^{(\tau)*} (\tau, x_\tau) \right]$$
$$- \frac{1}{2} \sigma \left[\tau, x_\tau \right]^2 \left[\xi_{x_t^h x_t^\varsigma}^{(\tau)i} (t, x_t) \,|_{t=\tau} \right]. \tag{13.12}$$

Therefore, one can establish the following theorem.

THEOREM 13.1 *(Yeung-Petrosyan Equation (2004)). If the solution imputations $\xi^{(\tau)i}(\tau, x_\tau)$, for $i \in [1,2]$ and $\tau \in [t_0, T]$, satisfy Definition 13.4 and Condition 13.1, a PDP with a terminal payment $q^i(x(T))$ at time T and an instantaneous imputation rate at time $\tau \in [t_0, T]$:*

$$B_i(\tau) = - \left[\xi_t^{(\tau)i} (t, x_t) \,|_{t=\tau} \right]$$
$$- \left[\xi_{x_t}^{(\tau)i} (t, x_t) \,|_{t=\tau} \right] f \left[\tau, x_\tau, \psi_1^{(\tau)*} (\tau, x_\tau) \right]$$
$$- \frac{1}{2} \sigma \left[\tau, x_\tau \right]^2 \left[\xi_{x_t^h x_t^\varsigma}^{(\tau)i} (t, x_t) \,|_{t=\tau} \right], \qquad \textit{for } i \in [1,2],$$

yielda a subgame consistent solution to the cooperative game $\Gamma_c(x_0, T - t_0)$.

5. A Subgame Consistent Cartel

In this section, we present a subgame consistent solution in which the firms agree to maximize the sum of their expected profits and divide the total cooperative profit satisfying the Nash bargaining outcome – that is, they maximize the product of expected profits in excess of the noncooperative profits. The Nash bargaining solution is perhaps the most popular cooperative solution concept which possesses properties not dominated by those of any other solution concepts. Assume that the agents agree to act and share the total cooperative profit according to an optimality principle satisfying the Nash bargaining axioms: (i) Pareto optimality, (ii) symmetry, (iii) invariant to affine transformation, and (iv) independence from irrelevant alternatives. In economic cooperation where profits are measured in monetary terms, Nash bargaining implies that agents agree to maximize the sum of their profits and then divide the total cooperative profit satisfying the Nash bargaining outcome – that is, they maximize the product of the agents' gains in excess of the noncooperative profits. In the two player case with transferable payoffs, the Nash bargaining outcome also coincides with the Shapley value.

Let S^i denote the aggregate cooperative gain imputed to agent i, the Nash product can be expressed as

$$[S^i]\left[W^{(t_0)}(t_0,x_0)-\sum_{j=1}^{2}V^{(t_0)j}(t_0,x_0)-S^i\right].$$

Maximization of the Nash product yields

$$S^i=\frac{1}{2}\left[W^{(t_0)}\left(t_0,x_0\right)-\sum_{j=1}^{2}V^{(t_0)j}\left(t_0,x_0\right)\right],\qquad \text{for } i\in[1,2].$$

The sharing scheme satisfies the so-called Nash formula (see Dixit and Skeath (1999)) for splitting a total value $W^{(t_0)}\left(t_0,x_0\right)$ symmetrically.

To extend the scheme to a dynamic setting, we first propose that the optimality principle guided by Nash bargaining outcome be maintained not only at the outset of the game but at every instant within the game interval. Dynamic Nash bargaining can therefore be characterized as: The firms agree to maximize the sum of their expected profits and distribute the total cooperative profit among themselves so that the Nash bargaining outcome is maintained at every instant of time $\tau\in[t_0,T]$.

According the optimality principle generated by dynamic Nash bargaining as stated in the above proposition, the imputation vectors must satisfy:

Proposition 13.1 *In the cooperative game $\Gamma\left(x_\tau,T-\tau\right)$, for $\tau\in[t_0,T]$, under dynamic Nash bargaining,*
$\xi^{(\tau)i}\left(\tau,x_\tau\right)=V^{(\tau)i}\left(\tau,x_\tau\right)+\frac{1}{2}\left[W^{(\tau)}\left(\tau,x_\tau\right)-\sum_{j=1}^{2}V^{(\tau)j}\left(\tau,x_\tau\right)\right],$
for $i\in[1,2]$.

Note that each firm will receive an expected profit equaling its expected noncooperative profit plus half of the expected gains in excess of expected noncooperative profits over the period $[\tau,T]$, for $\tau\in[t_0,T]$.

The imputations in Proposition 13.1 satisfy Condition 13.1 and Definition 13.4. Note that:

$$\xi^{(t)i}\left(t,x_t\right)=\exp\left[\int_\tau^t r\left(y\right)dy\right]\xi^{(\tau)i}\left(t,x_t\right)\equiv$$

$$\exp\left[\int_\tau^t r\left(y\right)dy\right]\left\{V^{(\tau)i}\left(t,x_t\right)+\frac{1}{2}\left[W^{(\tau)}\left(t,x_t\right)-\sum_{j=1}^{2}V^{(\tau)j}\left(t,x_t\right)\right]\right\},$$

for $t_0\leq\tau\leq t$, (13.13)

and hence the imputations satisfy Definition 13.4. Therefore Proposition 13.1 gives the imputations of a subgame consistent solution to the cooperative game $\Gamma_c(x_0, T - t_0)$.

Using Theorem 13.1 we obtain a PDP with a terminal payment $q^i(x(T))$ at time T and an instantaneous imputation rate at time $\tau \in [t_0, T]$:

$$\begin{aligned}
B_i(\tau) &= \frac{-1}{2}\Bigg[\left[V_t^{(\tau)i}(t, x_t)\mid_{t=\tau}\right] \\
&+ \left[V_{x_t}^{(\tau)i}(t, x_t)\mid_{t=\tau}\right] f\left[\tau, x_\tau, \psi_1^{(\tau)*}(\tau, x_\tau)\right] \\
&+ \frac{1}{2}\sigma[\tau, x_\tau]^2\left[V_{x_t^h x_t^\varsigma}^{(\tau)i}(t, x_t)\mid_{t=\tau}\right]\Bigg] \\
&- \frac{1}{2}\Bigg[\left[W_t^{(\tau)}(t, x_t)\mid_{t=\tau}\right] \\
&+ \left[W_{x_t}^{(\tau)}(t, x_t)\mid_{t=\tau}\right] f\left[\tau, x_\tau, \psi_1^{(\tau)*}(\tau, x_\tau)\right] \\
&+ \frac{1}{2}\sigma[\tau, x_\tau]^2\left[W_{x_t^h x_t^\varsigma}^{(\tau)}(t, x_t)\mid_{t=\tau}\right]\Bigg] \\
&+ \frac{1}{2}\Bigg[\left[V_t^{(\tau)j}(t, x_t)\mid_{t=\tau}\right] \\
&+ \left[V_{x_t}^{(\tau)j}(t, x_t)\mid_{t=\tau}\right] f\left[\tau, x_\tau, \psi_1^{(\tau)*}(\tau, x_\tau)\right] \\
&+ \frac{1}{2}\sigma[\tau, x_\tau]^2\left[V_{x_t^h x_t^\varsigma}^{(\tau)j}(t, x_t)\mid_{t=\tau}\right]\Bigg], \quad \text{for } i \in [1, 2]. \qquad (13.14)
\end{aligned}$$

6. An Illustration

Consider a duopoly in which two firms are allowed to extract a renewable resource within the duration $[t_0, T]$. The dynamics of the resource is characterized by the stochastic differential equations:

$$\begin{aligned}
&dx(s) = \left[ax(s)^{1/2} - bx(s) - u_1(s) - u_2(s)\right] ds + \sigma x(s)\, dz(s), \\
&x(t_0) = x_0 \in X, \qquad (13.15)
\end{aligned}$$

where $u_i \in U_i$ is the (nonnegative) amount of resource extracted by firm i, for $i \in [1, 2]$, a, b and σ are positive constants, and $z(s)$ is a Wiener process. Similar stock dynamics of a biomass of renewable resource had been used in Jørgensen and Yeung (1996 and 1999), Yeung (2001 and 2003).

The extraction cost for firm $i \in N$ depends on the quantity of resource extracted $u^i(s)$, the resource stock size $x(s)$, and a parameter c_i. In particular, firm i's extraction cost can be specified as $c_i u^i(s)\, x(s)^{-1/2}$.

This formulation of unit cost follows from two assumptions: (i) the cost of extraction is proportional to extraction effort, and (ii) the amount of resource extracted, seen as the output of a production function of two inputs (effort and stock level), is increasing in both inputs (See Clark (1976)). In particular, firm 1 has absolute cost advantage and $c_1 < c_2$.

The market price of the resource depends on the total amount extracted and supplied to the market. The price-output relationship at time s is given by the following downward sloping inverse demand curve $P(s) = Q(s)^{-1/2}$, where $Q(s) = u_1(s) + u_2(s)$ is the total amount of resource extracted and marketed at time s. At time T, firm i will receive a termination bonus with satisfaction $q_i x(T)^{1/2}$, where q_i is nonnegative. There exists a discount rate r, and profits received at time t has to be discounted by the factor $\exp[-r(t-t_0)]$.

At time t_0, the expected profit of firm $i \in [1,2]$ is:

$$E_{t_0}\left\{\int_{t_0}^{T}\left[\frac{u_i(s)}{[u_1(s)+u_2(s)]^{1/2}} - \frac{c_i}{x(s)^{1/2}}u_i(s)\right]\exp[-r(s-t_0)]\,ds \right.$$
$$\left. + \exp[-r(T-t_0)]\,q_i x(T)^{\frac{1}{2}} \mid x(t_0) = x_0\right\}, \tag{13.16}$$

where E_{t_0} denotes the expectation operator performed at time t_0.

A set of feedback strategies $\{u_i^{(\tau)*}(t) = \phi_i^{(\tau)*}(t,x)$, for $i \in [1,2]\}$ provides a Nash equilibrium solution to the game $\Gamma(x_\tau, T-\tau)$, if there exist twice continuously differentiable functions $V^{(\tau)i}(t,x) : [\tau,T]\times R \to R$, $i \in [1,2]$, satisfying the following partial differential equations:

$$-V_t^{(\tau)i}(t,x) - \frac{1}{2}\sigma^2 x^2 V_{xx}^{(\tau)i}(t,x) =$$
$$\max_{u_i}\left\{\left[\frac{u_i}{(u_i+\phi_j(t,x))^{1/2}} - \frac{c_i}{x^{1/2}}u_i\right]\exp[-r(t-\tau)]\right.$$
$$\left. + V_x^{(\tau)i}(t,x)\left[ax^{1/2} - bx - u_i - \phi_j(t,x)\right]\right\}, \text{ and}$$
$$V^{(\tau)i}(T,x) = q_i x^{1/2}\exp[-r(T-\tau)]\,ds,$$
$$\text{for } i \in [1,2],\ j \in [1,2] \text{ and } j \neq i. \tag{13.17}$$

Proposition 13.2 *The value function of firm i in the game $\Gamma(x_\tau, T-\tau)$ is:*

$$V^{(\tau)i}(t,x) = \exp[-r(t-\tau)]\left[A_i(t)\,x^{1/2} + B_i(t)\right],$$
$$\textit{for } i \in [1,2] \textit{ and } t \in [\tau,T], \tag{13.18}$$

where $A_i(t)$, $B_i(t)$, $A_j(t)$ *and* $B_j(t)$, *for* $i \in [1,2]$ *and* $j \in [1,2]$ *and* $i \neq j$, *satisfy:*

$$\dot{A}_i(t) = \left[r + \frac{1}{8}\sigma^2 + \frac{b}{2}\right] A_i(t) - \left(\frac{3}{2}\right) \frac{[2c_j - c_i + A_j(t) - A_i(t)/2]}{[c_1 + c_2 + A_1(t)/2 + A_2(t)/2]^2}$$
$$+ \left(\frac{3}{2}\right)^2 \frac{c_i [2c_j - c_i + A_j(t) - A_i(t)/2]}{[c_1 + c_2 + A_1(t)/2 + A_2(t)/2]^3}$$
$$+ \left(\frac{9}{8}\right) \frac{A_i(t)}{[c_1 + c_2 + A_1(t)/2 + A_2(t)/2]^2},$$
$$A_i(T) = q_i;$$
$$\dot{B}_i(t) = rB_i(t) - \frac{a}{2}A_i(t), \quad \textit{and} \quad B_i(t) = 0.$$

Proof. Perform the indicated maximization in (13.17) and then substitute the results back into the set of partial differential equations. Solving this set equations yields Proposition 13.2. □

Assume that the firms agree to form a cartel and seek to solve the following joint profit maximization problem to achieve a Pareto optimum:

$$E_{t_0}\left\{ \int_{t_0}^{T} \left[[u_1(s) + u_2(s)]^{1/2} - \frac{c_1 u_1(s) + c_2 u_2(s)}{x(s)^{1/2}} \right] \exp[-r(s - t_0)]\, ds \right.$$
$$\left. + \exp[-r(T - t_0)]\, [q_1 + q_2]\, x(T)^{1/2} \mid x(t_0) = x_0 \right\}, \tag{13.19}$$

subject to dynamics (13.15).

A set of feedback strategies $\left[\psi_1^{(t_0)*}(s,x), \psi_2^{(t_0)*}(s,x)\right]$, for $s \in [t_0, T]$ provides an optimal control solution to the problem (13.15) and (13.19), if there exist a twice continuously differentiable function $W^{(t_0)}(t,x) : [t_0, T] \times R \to R$ satisfying the following partial differential equations:

$$-W_t^{(t_0)}(t,x) - \frac{1}{2}\sigma^2 x^2 W_{xx}^{(t_0)}(t,x) =$$
$$\max_{u_i, u_2} \left\{ \left[(u_1 + u_2)^{1/2} - (c_1 u_1 + c_2 u_2)\, x^{-1/2} \right] \exp[-r(t - t_0)] \right.$$
$$\left. + W_x^{(t_0)}(t,x) \left[a x^{1/2} - bx - u_1 - u_2 \right] \right\}, \quad \text{and}$$
$$W^{(t_0)}(T,x) = (q_1 + q_2)\, x^{1/2} \exp[-r(T - t_0)].$$

The indicated maximization operation in the above definition requires:

$$\psi_1^{(t_0)*}(t,x) = \frac{x}{4\left[c_1 + W_x \exp\left[r(t-t_0)\right] x^{1/2}\right]^2} \text{ and } \psi_2^{(t_0)*}(t,x) = 0. \tag{13.20}$$

Along the optimal trajectory, firm 2 has to refrain from extraction.

PROPOSITION 13.3 *The value function of the stochastic control problem* (13.15) *and* (13.19) *can be obtained as:*

$$W^{(t_0)}(t,x) = \exp\left[-r(t-t_0)\right]\left[A(t)\,x^{1/2} + B(t)\right], \tag{13.21}$$

where $A(t)$ *and* $B(t)$ *satisfy:*

$$\dot{A}(t) = \left[r + \frac{1}{8}\sigma^2 + \frac{b}{2}\right]A(t) - \frac{1}{4\left[c_1 + A(t)/2\right]},$$
$$A(T) = q_1 + q_2;$$
$$\dot{B}(t) = rB(t) - \frac{a}{2}A(t), \quad \text{and } B(T) = 0.$$

Proof. Substitute the results from (13.20) into the partial differential equations in (13.19). Solving this equation yields Proposition 13.3. □

Upon substituting $\psi_1^{(t_0)*}(t,x)$ and $\psi_2^{(t_0)*}(t,x)$ into (13.15) yields the optimal cooperative state dynamics as:

$$dx(s) = \left[ax(s)^{1/2} - bx(s) - \frac{x(s)}{4\left[c_1 + A(s)/2\right]^2}\right]ds + \sigma x(s)\,dz(s),$$
$$x(t_0) = x_0 \in X. \tag{13.22}$$

The solution to (13.22) yields a Pareto optimal trajectory, which can be expressed as:

$$x^*(t) = \left\{\Phi(t,t_0)\left[x_0^{1/2} + \int_{t_0}^{t} \Phi^{-1}(s,t_0)\frac{a}{2}ds\right]\right\}^2, \tag{13.23}$$

where

$$\Phi(t,t_0) = \exp\left[\int_{t_0}^{t}\left(\frac{-b}{2} - \frac{1}{8\left[c_1 + A(s)/2\right]^2} - \frac{3\sigma^2}{8}\right)ds + \int_{t_0}^{t}\frac{\sigma}{2}dz(s)\right].$$

We denote the set containing realizable values of $x^*(t)$ by $X_t^{\alpha_1(t_0)}$, for $t \in (t_0, T]$.

Using Theorem 13.1 we obtain a PDP with a terminal payment $q^i(x(T))$ at time T and an instantaneous imputation rate at time $\tau \in [t_0, T]$:

$$\begin{aligned}
B_i(\tau) = {} & \frac{-1}{2}\Bigg[\left[V_t^{(\tau)i}(t,x_t)\,|_{t=\tau}\right] \\
& + \left[V_{x_t}^{(\tau)i}(t,x_t)\,|_{t=\tau}\right]\left[a x_\tau^{1/2} - b x_\tau - \frac{x_\tau}{4\left[c_1 + A(\tau)/2\right]^2}\right] \\
& + \frac{\sigma^2 x^2}{2}\left[V_{x_t^h x_t^\xi}^{(\tau)i}(t,x_t)\,|_{t=\tau}\right]\Bigg] \\
& - \frac{1}{2}\Bigg[\left[W_t^{(\tau)}(t,x_t)\,|_{t=\tau}\right] \\
& + \left[W_{x_t}^{(\tau)}(t,x_t)\,|_{t=\tau}\right]\left[a x_\tau^{1/2} - b x_\tau - \frac{x_\tau}{4\left[c_1 + A(\tau)/2\right]^2}\right] \\
& + \frac{\sigma^2 x^2}{2}\left[W_{x_t^h x_t^\xi}^{(\tau)}(t,x_t)\,|_{t=\tau}\right]\Bigg] \\
& + \frac{1}{2}\Bigg[\left[V_t^{(\tau)j}(t,x_t)\,|_{t=\tau}\right] \\
& + \left[V_{x_t}^{(\tau)j}(t,x_t)\,|_{t=\tau}\right]\left[a x_\tau^{1/2} - b x_\tau - \frac{x_\tau}{4\left[c_1 + A(\tau)/2\right]^2}\right] \\
& + \frac{\sigma^2 x^2}{2}\left[V_{x_t^h x_t^\xi}^{(\tau)j}(t,x_t)\,|_{t=\tau}\right]\Bigg], \quad \text{for } i \in [1,2]. \qquad (13.24)
\end{aligned}$$

yields a subgame consistent solution to the cooperative game $\Gamma_c(x_0, T - t_0)$, in which the firms agree to divide their cooperative gains according to Proposition 13.1.

Using (13.19), we obtain:

$$\left[V_{x_t}^{(\tau)i}(t,x_t)\,|_{t=\tau}\right] = \frac{1}{2}A_i(\tau)\,x_\tau^{-1/2},$$

$$\left[V_{x_t^h x_t^\xi}^{(\tau)i}(t,x_t)\,|_{t=\tau}\right] = \frac{-1}{4}A_i(\tau)\,x_\tau^{-3/2},$$

and

$$\left[V_t^{(\tau)i}(t,x_t)\,|_{t=\tau}\right] = -r\left[A_i(\tau)\,x_\tau^{1/2} + B_i(\tau)\right] + \left[\dot{A}_i(\tau)\,x_\tau^{1/2} + \dot{B}_i(\tau)\right],$$

$$\text{for } i \in [1,2], \qquad (13.25)$$

where $\dot{A}_i(\tau)$ and $\dot{B}_i(\tau)$ are given in Proposition 13.2.

Using (13.21), we obtain:

$$\left[W_{x_t}^{(\tau)}(t,x_t)\,|_{t=\tau}\right] = \frac{1}{2}A(\tau)\,x_\tau^{-1/2},$$

$$\left[W_{x_t^h x_t^\xi}^{(\tau)}(t,x_t)\,|_{t=\tau}\right] = \frac{-1}{4}A(\tau)\,x_\tau^{-3/2},$$

and

$$\left[W_t^{(\tau)}(t, x_t)\mid_{t=\tau}\right] = -r\left[A(\tau)\, x_\tau^{1/2} + B(\tau)\right] + \left[\dot{A}(\tau)\, x_\tau^{1/2} + \dot{B}(\tau)\right]$$

where $\dot{A}(\tau)$ and $\dot{B}(\tau)$ are given in Proposition 13.3.

$B_i(\tau)$ in (13.25) yields the instantaneous imputation that will be offered to firm i given that the state is x_τ at time τ.

7. Concluding Remarks

The complexity of stochastic differential games generally leads to great difficulties in the derivation of game solutions. The stringent requirement of subgame consistency imposes additional hurdles to the derivation of solutions for cooperative stochastic differential games. In this paper, we consider a duopoly in which the firms agree to form a cartel. In particular, one firm has absolute cost advantage over the other forcing one of the firms to become a dormant firm. A subgame consistent solution based on the Nash bargaining axioms is derived. The analysis can be readily applied to the deterministic version of the duopoly game by setting σ equal zero.

Acknowledgments. This research was supported by the Research Grant Council of Hong Kong (Grant # HKBU2056/99H) and Hong Kong Baptist University FRG grants Grant # FRG/02-03/II16).

References

Bellman, R. (1957). *Dynamic Programming.* Princeton, Princeton University Press, NJ.

Clark, C.W. (1976). *Mathematical Bioeconomics: The Optimal Management of Renewable Resources.* John Wiley, New York.

Dixit, A. and Skeath, S. (1999). *Games of Strategy.* W.W. Norton & Company, New York.

Fleming, W.H. (1969). Optimal continuous-parameter stochastic control. *SIAM Review*, 11:470–509.

Haurie, A. (1976). A Note on nonzero-sum differential games with bargaining solution. *Journal of Optimization Theory and Applications*, 18:31–39.

Haurie, A., Krawczyk, J.B., and Roche, M. (1994). Monitoring cooperative equilibria in a stochastic differential game. *Journal of Optimization Theory and Applications*, 81:79–95.

Isaacs, R. (1965). *Differential Games.* Wiley, New York.

Jørgensen, S. and Yeung, D.W.K. (1996). Stochastic differential game model of a common property fishery. *Journal of Optimization Theory and Applications*, 90:391–403.

Jørgensen, S. and Yeung, D.W.K. (1999). Inter- and intragenerational renewable resource extraction. *Annals of Operations Research*, 88:275–289.

Jørgensen, S. and Zaccour, G. (2001). Time consistency in cooperative differential games. In: G. Zaccour (ed.), *Decision and Control in Management Sciences: Essays in Honor of Alain Haurie*, pages 349–366, Kluwer Academic Publishers, London.

Nash, J.F. (1953). Two-person cooperative games. *Econometrica*, 21: 128–140.

Petrosyan, L.A. (1977). Stable solutions of differential games with many participants. *Viestnik of Leningrad University*, 19:46–52.

Petrosyan, L.A. (1997). Agreeable solutions in differential games. *International Journal of Mathematics, Game Theory and Algebra*, 7:165–177.

Petrosyan, L.A. (2003). Bargaining in dynamic games. In: L. Petrosyan and D. Yeung (eds.), *ICM Millennium Lectures on Games*, pages 139-1-43, Springer-Verlag, Berlin, Germany.

Petrosyan, L. and Danilov, N. (1979). Stability of solutions in non-zero sum differential games with transferable payoffs. *Viestnik of Leningrad Universtiy*, 1:52–59.

Petrosyan, L.A. and Zenkevich, N.A. (1996). *Game Theory*. World Scientific, Republic of Singapore.

Shapley, L.S. (1953). A value for N-person games. In: H.W. Kuhn and A.W. Tucker (eds.), *Contributions to the Theory of Games*, vol. 2, pages 307–317, Princeton University Press, Princeton, New Jersey.

Tolwinski, B., Haurie, A., and Leitmann, G. (1986). Cooperative equilibria in differential games. *Journal of Mathematical Analysis and Applications*, 119:182–202.

Yeung, D.W.K. (2001). Infinite horizon stochastic differential games with branching payoffs. *Journal of Optimization Theory and Applications*, 111:445–460.

Yeung, D.W.K. (2003). Randomly-furcating stochastic differential games. In: L. Petrosyan and D. Yeung (eds.), *ICM Millennium Lectures on Games*, pages 107–126, Springer-Verlag, Berlin, Germany.

Yeung, D.W.K. and Petrosyan, L.A. (2001). Proportional time-consistent solutions in differential games. *International Conference on Logic, Game Theory and Social Choice*, pages 254–256, St. Petersburg, Russia.

Yeung, D.W.K. and Petrosyan, L.A. (2004). Subgame consistent cooperative solutions in stochastic differential games. *Journal of Optimization Theory and Applications*, 120:651–666.

FSC
www.fsc.org
MIX
Papier | Fördert gute Waldnutzung
FSC® C083411

Kolibri 360 GmbH
Ferdinand-Jühlke-Straße 7
99095 Erfurt, Deutschland
produktsicherheit@kolibri360.de